Study Guide for
Stewart/Redlin/Watson's

Precalculus
Mathematics for Calculus

Fourth Edition

John A. Banks
Evergreen Valley College,
San Jose City College

BROOKS/COLE
TM
THOMSON LEARNING

Australia • Canada • Mexico • Singapore • Spain • United Kingdom • United States

Assistant Editor: *Carol Ann Benedict*
Marketing Manager: *Karin Sandberg*
Marketing Communications: *Samantha Cabaluna*
Marketing Assistant: *Darcie Pool*
Editorial Assistant: *Molly Nance*

Production Coordinator: *Dorothy Bell*
Cover Design: *Roy R. Neuhaus*
Cover Illustration: *Bill Ralph*
Print Buyer: *Christopher Burnham*
Printing and Binding: *Webcom Limited*

For more information about this or any other Brooks/Cole product, contact:
BROOKS/COLE
511 Forest Lodge Road
Pacific Grove, CA 93950 USA
www.brookscole.com
1-800-423-0563 (Thomson Learning Academic Resource Center)

Printed in Canada

10 9 8 7 6 5 4 3 2 1

ISBN 0-534-38545-1

Contents

Chapter 11 Counting and Probability

Chapter 12 Limits: A Preview of Calculus

Preface to Fourth Edition

To master algebra and trigonometry and move on to further studies, you need to practice solving problems. The more problems you do, the better your problem solving skills will become. You also need exposure to many different types of problems so you can develop strategies for solving each type.

This study guide provides additional problems for each section of the main text, *Precalculus: Math for Calculus, Fourth Edition*. Each problem includes a detailed solution located adjacent to it. Where appropriate, the solution includes commentary on what steps are taken and why those particular steps were taken. Some problems are solved using more than one approach, allowing you to compare the different methods.

Developing a strategy on how to solve a type of problem is essential to doing well in mathematics. Therefore, it is extremely important that you work out each problem before looking at the solution.

If you immediately look at the solution before attempting to work it out, you short change yourself. You skip the important steps of planning the strategy and deciding on the method to use to carry out this plan. These problem solving skills cannot be developed by looking at the answer first.

Another method you can use to help study is to use index cards. Put a problem that you find challenging on one side of the card and the solution and its location in the text on the other side of the card. Add new problems to your collection whenever you do your homework. Once a week test yourself with these cards. Be sure to shuffle the cards each time.

Finally, I would like to acknowledge the many useful comments and suggestions provided by Anna Fox, Lothar Redlin, and Saleem Watson.

I hope you find that this study guide helps in your further understanding the concepts of algebra and trigonometry.

John A. Banks
Evergreen Valley College
San Jose City College

Chapter 1
Fundamentals

Section 1.1   Real Numbers

Key Ideas

A. Real numbers and converting a repeating decimal to a fraction.
B. Properties of real numbers.
C. The real line.
D. Set notations and interval notation.
E. Absolute value.

A. There are many different types of numbers that make up the **real number** system. Some of these special sets are shown below.

Symbol	Name	Set
$\mathbb{N}$	Natural (counting)	$\{1, 2, 3, 4, \ldots\}$
$\mathbb{Z}$	Integers	$\{\ldots, -2, -1, 0, 1, 2, \ldots\}$
$\mathbb{Q}$	Rational	$\left\{ r = \dfrac{p}{q} \,\middle|\, p, q \text{ are integers, } q \neq 0 \right\}$
$\mathbb{R}$	Reals	Numbers that can be represented by a point on a line.

Every natural number is an integer, and every integer is a rational number. For example:
$3 = \dfrac{3}{1} = \dfrac{6}{2} = \cdots$. But not every rational number is an integer and not every integer is a natural number. Real numbers that cannot be expressed as a ratio of integers are called **irrational**. π and $\sqrt{2}$ are examples of irrational numbers. Every real number has a decimal representation. When the decimal representation of a number has a sequence of digits that repeats forever it is a rational number. When the repeating sequence is different from '0' you can convert a rational number from its decimal representation to a fraction representation by following these steps:

1.	Set $x =$ the repeating decimal.
2.	Multiply x and the decimal representation by enough powers of 10 to bring one repeating sequence to the left of the decimal point.
3.	Multiply x and the decimal representation by enough powers of 10 so that the first repeating sequence starts immediately after the decimal point.
4.	Subtract the results of Step 2 from the results of Step 1. This creates an equation of the form *integer* $\times\, x = $ *an integer*.
5.	Divide both sides by the coefficient and reduce.

1. Convert each repeating decimal to its fractional representation.

    (a) $0.\overline{4578}$

Let $x = 0.\overline{4578}$.
The repeating sequence has 4 digits in it and no digits before the repeating sequence. Start by multiplying both sides by 10^4.

$$
\begin{aligned}
10000x &= 4578.\overline{4578} \\
1x &= 0.\overline{4578} \qquad \textit{Subtract.} \\
\hline
9999x &= 4578
\end{aligned}
$$

$$
x = \frac{4578}{9999} = \frac{1526}{3333}. \qquad \textit{Divide.}
$$

So the fraction representation of $0.\overline{4578}$ is $\dfrac{1526}{3333}$.

(b) $3.12\overline{3}$

Let $x = 3.12\overline{3}$
The repeating sequence has 1 digit in it and there are 2 digits before the repeating sequence starts, so we multiply both sides by 10^3: $1000x = 3123.\overline{3}$.
Next we need to multiple both sides by 10^2 to bring the 2 non-repeating digits to the other side of the decimal point: $100x = 312.\overline{3}$.
Thus we get:

$$\begin{array}{rl} 1000x = & 3123.\overline{3} \\ 100x = & 312.\overline{3} \qquad \textit{Subtract.} \\ \hline 900x = & 2811 \end{array}$$

$$x = \frac{2811}{900} = \frac{937}{300}. \quad \textit{Divide.}$$

So the fraction representation of $3.12\overline{3}$ is $\dfrac{937}{300}$.

B. The basic properties used in combining real numbers are:

Commutative Laws	$a + b = b + a$	$ab = ba$
Associative Laws	$(a + b) + c = a + (b + c)$	$(ab)c = a(bc)$
Distributive Laws	$a(b + c) = ab + ac$	$(b + c)a = ab + ac$

The number 0, called the **additive identity**, is special for addition because $a + 0 = a$ for any real number a. Every real number a has a negative, $-a$, that satisfies $a + (-a) = 0$. **Subtraction** is the operation that undoes addition and we define $a - b = a + (-b)$. We use the following properties to handle negatives:

$(-1)a = -a$	$-(-a) = a$	$(-a)b = a(-b) = -(ab)$
$(-a)(-b) = ab$	$-(a + b) = -a - b$	$-(a - b) = b - a$

The number 1, called the **multiplicative identity**, is special for multiplication because $a \cdot 1 = a$ for any real number a. Every nonzero real number a has an inverse, $1/a$, that satisfies $a \cdot (1/a) = 1$. **Division** is the operation that undoes multiplication and we define $a \div b = a \cdot \dfrac{1}{b}$. We use the following properties to deal with quotients:

$\dfrac{a}{b} \cdot \dfrac{c}{d} = \dfrac{ac}{bd}$	$\dfrac{a}{b} \div \dfrac{c}{d} = \dfrac{a}{b} \cdot \dfrac{d}{c}$	$\dfrac{a}{c} + \dfrac{b}{c} = \dfrac{a + b}{c}$
$\dfrac{a}{b} + \dfrac{c}{d} = \dfrac{ad + bc}{bd}$	$\dfrac{ac}{bc} = \dfrac{a}{b}$	$\dfrac{-a}{b} = \dfrac{a}{-b} = -\dfrac{a}{b}$
If $\dfrac{a}{b} = \dfrac{c}{d}$, then $ad = bc$.		

2. Use the properties of real numbers to write the given expression without parentheses.

(a) $4(3 + m) - 2(4 + 3m)$

$$4(3 + m) - 2(4 + 3m) = 12 + 4m - 8 - 6m$$
$$= 4 - 2m$$

Remember to properly distribute the -2 over $(4 + 3m)$.

(b) $(2a + 3)(5a - 2b + c)$

$$(2a + 3)(5a - 2b + c)$$
$$= (2a + 3)5a + (2a + 3)(-2b) + (2a + 3)c$$
$$= 10a^2 + 15a - 4ab - 6b + 2ac + 3c$$

C. The real numbers can be represented by points on a line, with the positive direction towards the right. We choose an arbitrary point for the real number 0 and call it the **origin**. Each positive number x is represented by the point on the line that is a distance of x units to the right of the origin, and each negative number $-x$ is represented by the point x units to the left of the origin. This line is called the **real number line** or the **coordinate line**. An important property of real numbers is **order**. Order is used to compare two real numbers and determine their relative position.

Order	Symbol	Geometrically	Algebraically
a is less than b	$a < b$	a lies to the left of b	$b - a$ is positive
a is greater than b	$a > b$	a lies to the right of b	$a - b$ is positive
a is less than or equal b	$a \leq b$	a lies to the left of b or on b	$b - a$ is nonnegative
a is greater than or equal b	$a \geq b$	a lies to the right of b or on b	$a - b$ is nonnegative

3. Classify each real number as a *natural, integer, rational,* or *irrational* number.

(a) 17.312

This decimal number terminates. So the repeating sequence of digits is '0' and thus this number is rational.

(b) $-9.1\overline{27}$

Since 27 repeats, this number is rational.

(c) $-18.101001000100001\ldots$

Since there is not a pattern where a portion is repeated, this number is irrational.
Here the pattern is $0 \cdots 01$ where the number of zeros grows, so no one sequence is repeated.

4

(d) 1.5765780...

Since there is not a pattern in which a portion is repeated, this number is irrational.

4. State whether the given inequality is true or false.
   (a) $-3.1 > -3$

False. $-3.1 - (-3) = -0.1$.

   (b) $2 \leq 2$

True. $2 - 2 = 0$.

   (c) $15.3 \geq -16.3$

True. $15.3 - (-16.3) = 31.6$.

5. Write the statement in terms of an inequality.
   (a) w is negative.

$w < 0$. w is negative is the same as saying that w is less than 0. You can also express this as $0 > w$.

   (b) m is greater than -3.

$m > -3$ or $-3 < m$.

   (c) k is at least 6.

$k \geq 6$ or $6 \leq k$.

   (d) x is at most 7 and greater than -2.

$-2 < x \leq 7$.

D. A **set** is a collection of objects, or **elements**. A capital letter is usually used to denote sets and lower case letters to represent the elements of the set. There are two main ways to write a set. We can **list** all the elements of the set enclosed in { }, *brackets*, or list a few elements and then use ... to represent that the set continues in the same pattern. Or we can use **set builder notation**, "$\{x \mid x$ has property $P\}$." This is read as "*the set of x such that x has the property P.*" The two key binary operations for sets are called **union** and **intersection**. The *union* of two sets is the set that consists of the elements that are in *either* set. The *intersection* of two sets is the set that consists of the elements in *both* sets. The **empty set**, $\emptyset$, is a set that contains no elements. **Intervals** are sets of real numbers that correspond geometrically to line segments. Study the table on page 15 of the main text and note the differences between the symbols (and) as well as the symbols [and].

5

6. Let $A = \{2, 4, 6, 8, 10, 12\}$, $B = \{3, 6, 9, 12\}$, and $C = \{1, 3, 5, 7, 9, 11\}$.
   (a) Find $A \cup B$, $A \cup C$, and $B \cup C$.

$A \cup B = \{2, 3, 4, 6, 8, 9, 10, 12\}$.

$A \cup C = \{1, 2, 3, 4, 5, 6, 7, 8, 9, 10, 11, 12\}$.

$B \cup C = \{1, 3, 5, 6, 7, 9, 11, 12\}$.

   (b) Find $A \cap B$, $A \cap C$, and $B \cap C$.

$A \cap B = \{6, 12\}$.

$A \cap C = \emptyset$.

$B \cap C = \{3, 9\}$.

7. Let $A = \{x \mid x < 8\}$, $B = \{x \mid 3 \le x < 7\}$, and $C = \{x \mid 5 < x\}$.
   (a) Find $A \cup B$, $A \cup C$, and $B \cup C$.

$A \cup B = \{x \mid x < 8\}$.

$A \cup C = \{x \mid -\infty < x < \infty\} = \mathbb{R}$.

$B \cup C = \{x \mid 3 \le x\}$.

   (b) Find $A \cap B$, $A \cap C$, and $B \cap C$.

$A \cap B = \{x \mid 3 \le x < 7\}$.

$A \cap C = \{x \mid 5 < x < 8\}$.

$B \cap C = \{x \mid 5 < x < 7\}$.

8. Let $T = (3, 6)$, $V = [2, 4)$, and $W = (4, 6]$.
   (a) Find $T \cup V$, $T \cup W$, and $V \cup W$. Graph your results.

For unions, the solution consists of the numbers that are in either set. So drawing each set on a number line, we have:

$T \cup V = (3, 6) \cup [2, 4) = [2, 6)$.

$T \cup W = (3, 6) \cup (4, 6] = (3, 6]$.

Note: The only difference between $T \cup W$ and T is that $T \cup W$ contains the point 6.

$V \cup W = [2, 4) \cup (4, 6]$.

Note: This is as far as we can go. The sets V and W do not have any elements in common. See $V \cap W$ on the next part.

(b) Find $T \cap V$, $T \cap W$, and $V \cap W$. Graph your results.

For intersections, the solution consists of the numbers that are in both intervals.
$T \cap V = (3, 6) \cap [2, 4) = (3, 4)$.

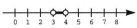

$T \cap W = (3, 6) \cap (4, 6] = (4, 6)$.

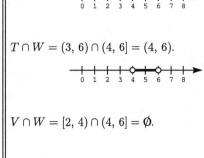

$V \cap W = [2, 4) \cap (4, 6] = \emptyset$.

E. The **absolute value** of a number a, denoted by $|a|$, is the distance from the number a to 0 on the real number line. Remember absolute value is always positive or zero. It is also defined as

$|a| = \begin{cases} a, & \text{if } a \geq 0 \\ -a, & \text{if } a < 0 \end{cases}$. If a and b are real numbers, then the distance between the points a and b on the

real line is $d(a, b) = |b - a|$. By Property 6 of negatives $|b - a| = |a - b|$

9. Find each absolute value.

(a) $|-7|$

$|-7| = -(-7) = 7$.

(b) $|8|$

$|8| = 8$.

(c) $|\pi - 5|$

$|\pi - 5| = -(\pi - 5) = 5 - \pi$, since $\pi - 5 < 0$.

(d) $\left|\sqrt{10} - 3\right|$

$\left|\sqrt{10} - 3\right| = \sqrt{10} - 3$.
Since $\sqrt{10} > \sqrt{9} = 3$, so $\sqrt{10} - 3 > 0$.

10. Find the distance between each pair of numbers.

(a) 4 and 10

$|4 - 10| = |-6| = 6$.
Remember to do the work on the inside of the absolute value bars *first*!

(b) 5 and -3.

$|5 - (-3)| = |5 + 3| = |8| = 8$.

(c) -2 and 9

$|-2 - 9| = |-2 - 9| = |-11| = 11$.

Section 1.2   Exponents and Radicals

Key Ideas

A. Key exponent definitions and rules.
B. Scientific notation.
C. Square roots and nth roots.
D. Rational exponents.
E. Rationalizing the denominator.

A. The key exponent definitions are:

$$a^n = \underbrace{a \cdot a \cdot a \cdot \cdots \cdot a}_{n \text{ factors of } a} \qquad\qquad a^0 = 1,\ a \neq 0 \qquad\qquad a^{-n} = \frac{1}{a^n},\ a \neq 0$$

In addition to the exponent definitions, the following key exponent laws should be mastered.

$a^m a^n = a^{m+n}$	To multiply two powers of the same number, add the exponents.
$\dfrac{a^m}{a^n} = a^{m-n},\ a \neq 0$	To divide two power of the same number, subtract the exponents.
$(a^m)^n = a^{mn}$	To raise a power to a new power, multiply the exponents.
$(ab)^n = a^n b^n$	To raise a product to a power, raise each factor to the power.
$\left(\dfrac{a}{b}\right)^n = \dfrac{a^n}{b^n},\ b \neq 0$	To raise a quotient to a power, raise both numerator and denominator to the power.
$\left(\dfrac{a}{b}\right)^{-n} = \left(\dfrac{b}{a}\right)^n$	To raise a fraction to a negative power, invert the fraction and change the sign of the exponent.
$\dfrac{a^{-n}}{b^{-m}} = \dfrac{b^m}{a^n}$	To move a number raised to a power from numerator to denominator or from the denominator to numerator, change the sign of the exponent.

1. Simplify.

(a) $x^3 x^5$

$\qquad\qquad\qquad\qquad x^3 x^5 = x^{3+5} = x^8$

(b) $w^{12} w^{-8}$

$\qquad\qquad\qquad\qquad w^{12} w^{-8} = w^{12 + (-8)} = w^4$

(c) $\dfrac{y^8}{y^{15}}$

$\qquad\qquad\qquad\qquad \dfrac{y^8}{y^{15}} = y^{8-15} = y^{-7} = \dfrac{1}{y^7}$ or

$\qquad\qquad\qquad\qquad \dfrac{y^8}{y^{15}} = \dfrac{1}{y^{15-8}} = \dfrac{1}{y^7}$

(d) $(5x)^3$

$\qquad\qquad\qquad\qquad (5x)^3 = 5^3 x^3 = 125 x^3$

(e) $(m^3)^7$

$$(m^3)^7 = m^{3 \cdot 7} = m^{21}$$

(f) $\dfrac{2^9}{2^7}$

$$\dfrac{2^9}{2^7} = 2^{9-7} = 2^2 = 4 \text{ or}$$

$$\dfrac{2^9}{2^7} = \dfrac{1}{2^{7-9}} = \dfrac{1}{2^{-2}} = 2^2 = 4$$

A. Shortcut: In problems like 1c and 1f above, compare the exponents in the numerator and denominator. If the exponent in the numerator is larger, we simplify the expression by bringing the factor up into the numerator. If the exponent in the denominator is larger, we simplify the expression by bringing the factor down into the denominator.

2. Simplify $\dfrac{27^3}{9^5}$.

Since both 27 and 9 are powers of 3, first express each number as a power of 3.

$$\dfrac{27^3}{9^5} = \dfrac{(3^3)^3}{(3^2)^5} = \dfrac{3^{3 \cdot 3}}{3^{2 \cdot 5}} = \dfrac{3^9}{3^{10}} = \dfrac{1}{3^{10-9}} = \dfrac{1}{3}$$

3. Simplify $(2x^7y^5)(3x^2y^3)^4$.

$$(2x^7y^5)(3x^2y^3)^4 = (2x^7y^5)[(3)^4(x^2)^4(y^3)^4]$$
$$= (2x^7y^5)(3^4x^8y^{12})$$
$$= (2)(81)(x^7x^8)(y^5y^{12})$$
$$= 162x^{15}y^{17}$$

4. Simplify $\left(\dfrac{x^4y}{z^6}\right)^3\left(\dfrac{xz^3}{y^4}\right)^5$.

$$\left(\dfrac{x^4y}{z^6}\right)^3\left(\dfrac{xz^3}{y^4}\right)^5 = \dfrac{(x^4)^3y^3}{(z^6)^3}\dfrac{x^5(z^3)^5}{(y^4)^5}$$
$$= \dfrac{x^{12}y^3}{z^{18}}\dfrac{x^5z^{15}}{y^{20}}$$
$$= (x^{12}x^5)\left(\dfrac{y^3}{y^{20}}\right)\left(\dfrac{z^{15}}{z^{18}}\right)$$
$$= \dfrac{x^{17}}{y^{17}z^3}$$

5. Eliminate negative exponents and simplify.

(a) $\dfrac{3a^3b^{-4}}{2a^{-2}b^{-1}}$

$$\dfrac{3a^3b^{-4}}{2a^{-2}b^{-1}} = \dfrac{3}{2}a^{3-(-2)}b^{-4-(-1)}$$

$$= \dfrac{3}{2}a^5b^{-3} = \dfrac{3a^5}{2b^3}$$

(b) $\left(\dfrac{4w^3v^{-4}}{2w^2v^5}\right)^{-3}$

Method 1. First distribute the exponent -3, then simplify.

$$\left(\dfrac{4w^3v^{-4}}{2w^2v^5}\right)^{-3} = \dfrac{4^{-3}w^{3(-3)}v^{-4(-3)}}{2^{-3}w^{2(-3)}v^{5(-3)}}$$

$$= \dfrac{2^{-6}w^{-9}v^{12}}{2^{-3}w^{-6}v^{-15}}$$

$$= 2^{-6-(-3)}w^{-9-(-6)}v^{12-(-15)}$$

$$= 2^{-3}w^{-3}v^{27} = \dfrac{v^{27}}{2^3w^3} = \dfrac{v^{27}}{8w^3}$$

Method 2. Simplify inside the parentheses, then simplify.

$$\left(\dfrac{4w^3v^{-4}}{2w^2v^5}\right)^{-3} = \left(2w^{3-2}v^{-4-5}\right)^{-3}$$

$$= \left(2wv^{-9}\right)^{-3}$$

$$= 2^{-3}w^{-3}v^{-9(-3)} = 2^{-3}w^{-3}v^{27}$$

$$= \dfrac{v^{27}}{2^3w^3} = \dfrac{v^{27}}{8w^3}$$

There are additional strategies that could have been used to solve this exercise.

(c) $\dfrac{(5^3m^7n^2)^3}{(3^4m^{-2}n^3)^4}$

$$\dfrac{(5^3m^7n^2)^3}{(3^4m^{-2}n^3)^4} = \dfrac{5^{3(3)}m^{7(3)}n^{2(3)}}{3^{4(4)}m^{-2(4)}n^{3(4)}} = \dfrac{5^9m^{21}n^6}{3^{16}m^{-8}n^{12}}$$

$$= \dfrac{5^9m^{21-(-8)}}{3^{16}n^{12-6}}$$

$$= \dfrac{5^9m^{29}}{3^{16}n^6}$$

B. Scientific notation is used to express very large numbers or very small numbers in a more compact way. The goal is to express the positive number x in the form $a \times 10^n$, where $1 \le a < 10$ and n is an integer.

6. Write the following numbers in scientific notation.

   (a) $23,500,000,000$

$$2.35 \times 10^{10}$$

   (b) 0.000000000067

$$6.7 \times 10^{-11}$$

7. If $a \approx 4.1 \times 10^{-7}$, $b \approx 1.97 \times 10^9$, and $c \approx 3.24 \times 10^{-3}$ find the quotient $\dfrac{a}{bc}$.

$$\frac{a}{bc} \approx \frac{4.1 \times 10^{-7}}{(1.97 \times 10^9)(3.24 \times 10^{-3})}$$

$$\approx \frac{4.1}{(1.97)(3.24)} \times 10^{-7-9-(-3)}$$

$$\approx 0.64 \times 10^{-13}$$

$$\approx 6.4 \times 10^{-14}$$

Notice that the answer is stated to two significant digits because the least accurate of the numbers has two significant digits.

C. The **radical** symbol, $\sqrt{}$, means "the positive square root of." So $\sqrt{a} = b$ means $b^2 = a$ and $b \geq 0$. Similarly, the nth root of a is the number that, when raised to the nth power, gives a. So if n is a positive integer then $\sqrt[n]{a} = b$ means $b^n = a$ and b is called the principal nth root of a. When n is even, $a \geq 0$ and $b \geq 0$. The key rules of exponents to master are:

$$\sqrt{a^2} = |a|$$

If $a \geq 0$ and $b \geq 0$, then $\sqrt{a}\sqrt{b} = \sqrt{ab}$.   If $a \geq 0$ and $b > 0$, then $\sqrt{\dfrac{a}{b}} = \dfrac{\sqrt{a}}{\sqrt{b}}$.

$$\sqrt[n]{ab} = \sqrt[n]{a}\sqrt[n]{b}$$

$$\sqrt[n]{\frac{a}{b}} = \frac{\sqrt[n]{a}}{\sqrt[n]{b}}$$

$$\sqrt[m]{\sqrt[n]{a}} = \sqrt[mn]{a}$$

$$\sqrt[n]{a^n} = \begin{cases} a & \text{if } n \text{ is odd} \\ |a| & \text{if } n \text{ is even.} \end{cases}$$

8. Simplify the following expressions.

   (a) $\sqrt{28}$

$$\sqrt{28} = \sqrt{4}\sqrt{7} = 2\sqrt{7}$$

   (b) $\sqrt{50}\sqrt{8}$

$$\sqrt{50}\sqrt{8} = \sqrt{50 \cdot 8} = \sqrt{400} = 20 \text{ or}$$

$$\sqrt{50}\sqrt{8} = 5\sqrt{2} \cdot 2\sqrt{2} = 10\sqrt{4} = 20$$

   (c) $\dfrac{\sqrt{54}}{\sqrt{6}}$

$$\frac{\sqrt{54}}{\sqrt{6}} = \sqrt{\frac{54}{6}} = \sqrt{9} = 3$$

9. Simplify.

(a) $\sqrt[4]{81}$

$$\sqrt[4]{81} = \sqrt[4]{3^4} = 3$$

(b) $\sqrt[5]{-32}$

$$\sqrt[5]{-32} = \sqrt[5]{(-2)^5} = -2$$

Odd roots of negative numbers are negative.

(c) $\sqrt[3]{32}$

$$\sqrt[3]{32} = \sqrt[3]{2^5} = \sqrt[3]{2^3}\sqrt[3]{2^2} = 2\sqrt[3]{2^2} = 2\sqrt[3]{4}$$

D. The definitions of rational exponents are: $a^{1/n} = \sqrt[n]{a}$ and $a^{m/n} = \left(\sqrt[n]{a}\right)^m = \sqrt[n]{a^m}$.

10. Simplify.

(a) $8^{2/3}$

$$8^{2/3} = \left(\sqrt[3]{8}\right)^2 = 2^2 = 4 \text{ or}$$

$$8^{2/3} = \sqrt[3]{8^2} = \sqrt[3]{64} = 4$$

Both yield the same solution. However, most students find the first way easier to do.

(b) $9^{3/2}$

$$9^{3/2} = \left(\sqrt{9}\right)^3 = 3^3 = 27$$

(c) $(8x^3y^9)^{5/3}$

$$(8x^3y^9)^{5/3} = 8^{5/3}x^{3(5/3)}y^{9(5/3)}$$
$$= 2^{3(5/3)}x^{3(5/3)}y^{9(5/3)} = 2^5x^5y^{15}$$
$$= 32x^5y^{15}$$

E. Radicals are eliminated from the denominators by multiplying *both* the numerator and denominator by an appropriate expression. This procedure is called **rationalizing the denominator**.

11. Rationalize the denominator in each of the following expressions.

(a) $\dfrac{1}{\sqrt{2}}$

$$\frac{1}{\sqrt{2}} = \frac{1}{\sqrt{2}} \cdot \frac{\sqrt{2}}{\sqrt{2}} = \frac{\sqrt{2}}{2}$$

(b) $\sqrt{\dfrac{7}{3}}$

$$\sqrt{\frac{7}{3}} = \frac{\sqrt{7}}{\sqrt{3}} = \frac{\sqrt{7}}{\sqrt{3}} \cdot \frac{\sqrt{3}}{\sqrt{3}} = \frac{\sqrt{21}}{3}$$

Section 1.3   Algebraic Expressions

Key Ideas

A. Polynomial.
B. Operations of algebraic expressions.
C. Special product formulas.
D. Factoring.
E. Special factoring formulas.
F. Factoring by grouping.

A. A **variable** is a letter that can represent any number in a given set of numbers (called the **domain**). A **constant** represents a *fixed* number. Algebraic expressions using only addition, subtraction, and multiplication are called **polynomials**. The *general form of a polynomial of degree n in the variable x* is:

$$a_n x^n + a_{n-1} x^{n-1} + \cdots + a_2 x^2 + a_1 x + a_0$$

where $a_0, a_1, a_2, \ldots, a_{n-1}, a_n$ are constants with $a_n \neq 0$. A polynomial is the sum of **terms** of the form $a_k x^k$ (called **monomials**) where a_k is a constant and k is a nonnegative integer.

1. Determine the degree of each polynomial.

   (a) $5x^3 + 3x^2 - 6$ | Degree 3

   (b) $9y - 6y^7 + 200y^5$ | Degree 7

   (c) -78 | Degree 0

   (d) $\sqrt{5}w^4 + 3w^3 - \frac{3}{4}w^2$ | Degree 4

B. Terms with the *same variables* raised to the *same powers* are called **like terms**. Polynomials are added and subtracted using a process called *combining like terms* which utilizes the Distributive Law.

2. Find the sum:
   $(5x^3 + 3x^2 - 2x + 1) + (x^3 - 6x^2 + 5x - 1)$.

   First group like terms, then combine like terms.
   $(5x^3 + 3x^2 - 2x + 1) + (x^3 - 6x^2 + 5x - 1)$
   $\quad = 5x^3 + x^3 + 3x^2 - 6x^2 + -2x + 5x + 1 - 1$
   $\quad = (5+1)x^3 + (3-6)x^2 + (-2+5)x$
   $\quad = 6x^3 - 3x^2 + 3x$

3. Find the difference:
   $(7x^3 - x^2 + 5) - (2x^3 - 6x^2 + 9x - 6)$.

   First, group like terms, then combine like terms.
   $(7x^3 - x^2 + 5) - (2x^3 - 6x^2 + 9x - 6)$
   $\quad = 7x^3 - 2x^3 - x^2 - (-6x^2) - 9x + 5 - (-6)$
   $\quad = 5x^3 + 5x^2 - 9x + 11$

13

B. The product of two polynomials is found by repeated use of the Distributive Law.

4. Find the product: $(x^2 - 4xy + 2y^2)(x + 4y)$.

Start by treating the first expression as a single number and distribute each term over $(x + 4y)$.

$(x^2 - 4xy + 2y^2)(x + 4y)$
$$= (x^2 - 4xy + 2y^2)x + (x^2 - 4xy + 2y^2)4y$$

Now distribute x over $(x^2 - 4xy + 2y^2)$ and distribute $4y$ over $(x^2 - 4xy + 2y^2)$.

$$= x^3 - 4x^2y + 2xy^2 + 4x^2y - 16xy^2 + 8y^3$$
$$= x^3 - 14xy^2 + 8y^3$$

5. Find the product: $\left(\sqrt{x} - \dfrac{4}{\sqrt{x}}\right)(x + \sqrt{x})$.

Use the same methods used for multiplying polynomials.

$$\left(\sqrt{x} - \frac{4}{\sqrt{x}}\right)(x + \sqrt{x})$$

$$= \left(\sqrt{x} - \frac{4}{\sqrt{x}}\right)x + \left(\sqrt{x} - \frac{4}{\sqrt{x}}\right)\sqrt{x}$$

$$= x\sqrt{x} - \frac{4x}{\sqrt{x}} + \sqrt{x}\sqrt{x} - \frac{4\sqrt{x}}{\sqrt{x}}$$

$$= x\sqrt{x} - \frac{4x}{\sqrt{x}} + x - 4$$

$$= x\sqrt{x} - 4\sqrt{x} + x - 4$$

C. The following products frequently occur. It is advisable to recognize each by name.

Formula	Name
$(a - b)(a + b) = a^2 - b^2$	Difference of Squares
$(a + b)^2 = a^2 + 2ab + b^2$	Perfect Square
$(a - b)^2 = a^2 - 2ab + b^2$	Perfect Square
$(a + b)^3 = a^3 + 3a^2b + 3ab^2 + b^3$	Perfect Cube
$(a - b)^3 = a^3 - 3a^2b + 3ab^2 - b^3$	Perfect Cube

The operation of multiplying algebraic expressions is referred to as *expanding*.

6. Use the special product formulas to find the following products.

(a) $(3x - 2y^3)^2$

Perfect square:
$$(3x - 2y^3)^2 = (3x)^2 - 2(3x)(2y^3) + (2y^3)^2$$
$$= 9x^2 - 12xy^3 + 4y^6$$

(b) $(w^2 + 2w)^3$

Perfect cube:
$$(w^2 + 2w)^3$$
$$= (w^2)^3 + 3(w^2)^2(2w) + 3(w^2)(2w)^2 + (2w)^3$$
$$= w^6 + 6w^5 + 12w^4 + 8w^3$$

(c) $\left(\sqrt{x} + \dfrac{1}{\sqrt{x}}\right)\left(\sqrt{x} - \dfrac{1}{\sqrt{x}}\right)$

Difference of squares:
$$\left(\sqrt{x} + \dfrac{1}{\sqrt{x}}\right)\left(\sqrt{x} - \dfrac{1}{\sqrt{x}}\right)$$
$$= \left(\sqrt{x}\right)^2 - \left(\dfrac{1}{\sqrt{x}}\right)^2 = x - \dfrac{1}{x}$$

D. Factoring is the process of writing an algebraic expression (a sum) as a product of simpler ones. *Factoring* is one of the most important tools in algebra. *Factoring* is the opposite operation of *expanding*. The simplest factoring occurs when each term has a *common factor*. To factor a **quadratic** polynomial, a second-degree polynomial of the form $x^2 + bx + c$, we first observe that $(x + r)(x + s) = x^2 + (r + s)x + rs$, so that $r + s = b$ and $rs = c$. If c is negative, then r and s have different signs and we look for a *difference* of b. If c is positive, then r and s have the same sign as b, either both negative or both positive, and we look for r and s to *sum* to b.

7. Factor each polynomial.

(a) $6x^4 - 2x^7 + 8x^8$

Since the greatest common factor (gcf) of 6, -2, and 8 is 2 and the gcf of x^4, x^7, and x^8 is x^4, we have:
$$6x^4 - 2x^7 + 8x^8 = 2x^4(3 - x^3 + 4x^4)$$

(b) $4x^2w^3 - 16x^5w^4 + 28x^6w^6$

The gcf of 4, -16, and 28 is 4; the gcf of x^2, x^5, and x^6 is x^2, and the gcf of w^3, w^4, and w^6 is w^3.
$$4x^2w^3 - 16x^5w^4 + 28x^6w^6$$
$$= 4x^2w^3(1 - 4x^3w + 7x^4w^3)$$

15

(c) $x^2 - 12x + 20$

Since c is positive and b is negative, r and s are both negative factors of 20 and their sum is 12.

factors	$1 \cdot 20$	$2 \cdot 10$	$4 \cdot 5$
corresponding sum	21	12	9

Therefore, taking $r = -2$ and $s = -10$, we get the factorization:
$$x^2 - 12x + 20 = (x - 2)(x - 10).$$

(d) $x^2 + 6x - 16$

Since c is negative, r and s have different signs and we look for a difference of 6.

factors	$1 \cdot 16$	$2 \cdot 8$	$4 \cdot 4$
corresponding sum	15	6	0

Since b is positive, the larger factor is positive, so taking $r = 8$ and $s = -2$, we get the factorization:
$$x^2 + 6x - 16 = (x + 8)(x - 2).$$

D. To factor a *general quadratic* of the form $ax^2 + bx + c$, we look for factors of the form $px + r$ and $qx + s$, so $ax^2 + bx + c = (px + r)(qx + s) = pqx^2 + (ps + qr)x + rs$. Therefore, we must find numbers p, q, r, and s such that $pq = a$; $ps + qr = b$; and $rs = c$. If a is positive, then p and q are positive and the signs of r and s are determined as before by the sign of b and c. If a is negative, first factor out -1.

8. Factor $3x^2 - 7x + 2$

Since c is positive and b is negative, r and s are both negative.

factors of a	$1 \cdot 3$	$1 \cdot 3$
factors of c	$1 \cdot 2$	$2 \cdot 1$
corresponding sum	$1 + 6 = 7$	$2 + 3 = 5$

Since, $1 \cdot 1 + 3 \cdot 2 = 7$ the 1's must go into different factors and the 3 and 2 must go into different factors. And we get the factorization:
$$3x^2 - 7x + 2 = (x - 2)(3x - 1).$$

16

E. Formulas 1-3 are used to factor special quadratics, while formulas 4 and 5 are used on the sum or difference of cubes (third degree polynomials).

	Formula	Name
1.	$a^2 - b^2 = (a - b)(a + b)$	Difference of Squares
2.	$a^2 + 2ab + b^2 = (a + b)^2$	Prefect Square
3.	$a^2 - 2ab + b^2 = (a - b)^2$	Prefect Square
4.	$a^3 - b^3 = (a - b)(a^2 + ab + b^2)$	Difference of Cubes
5.	$a^3 + b^3 = (a + b)(a^2 - ab + b^2)$	Sum of Cubes

Notice that formulas 1, 4, and 5 have only *two terms* while formulas 2 and 3 have *three terms* each. When factoring a quadratic with three terms and the first and third terms are perfect squares, check the middle term to see if it is $2ab$, if so, then it is a perfect square. Also notice this is almost the same table that appeared earlier in the section. In this version of the table the sums are listed first followed by the product equivalent. In many cases, factoring techniques are combined with the special factoring formulas.

9. Factor.

(a) $x^2 - 8x + 16$

Since x^2 and 16 are *perfect squares*, we check the middle term and find that it fits the form: $\pm 2ab$, that is, $-8x = -2(x)(4)$. It does, so: $x^2 - 8x + 16 = (x - 4)^2$.

(b) $25x^2 - 64$

Since this quadratic is the difference of *two terms*, which are both *perfect squares*, we have: $25x^2 - 64 = (5x - 8)(5x + 8)$.

(c) $9x^2 + 12x + 4$

Again, since $9x^2$ and 4 are *perfect squares*, we check the middle term and find: $12x = 2(3x)(2)$. Factoring, we get: $9x^2 + 12x + 4 = (3x + 2)^2$.

(d) $8x^3 - 64$

Here we have the difference of two terms, which are both *perfect cubes*. Using the difference of cubes, we get:
$$8x^3 - 64 = (2x)^3 - (4)^3$$
$$= (2x - 4)[(2x)^2 + (2x)(4) + (4)^2]$$
$$= (2x - 4)(4x^2 + 8x + 16).$$
Note: $4x^2 + 8x + 16$ *does not factor any further.*

10. Factor.
    (a) $8x^4 - 24x^3 + 18x^2$

First factor out the common factor of $2x^2$:
$8x^4 - 24x^3 + 18x^2 = 2x^2(4x^2 - 12x + 9)$
Then factor the second factor (which is a perfect square). So:
$8x^4 - 24x^3 + 18x^2 = 2x^2(2x - 3)^2$.

    (b) $3x^2 - 27$

First factor out the common factor 3, then notice that the result is the difference of two squares:
$3x^2 - 27 = 3(x^2 - 9) = 3(x - 3)(x + 3)$.

F. Polynomials with four terms can sometimes be factored by grouping the terms into groups of 1 and 3 terms, or 2 and 2 terms, or 3 and 1 terms.

11. Factor $x^6 + 2x^4 + 2x^2 + 4$.

First group into groups of two, then factor out the common factor.
$$x^6 + 2x^4 + 2x^2 + 4 = (x^6 + 2x^4) + (2x^2 + 4)$$
$$= x^4(x^2 + 2) + 2(x^2 + 2)$$
$$= (x^4 + 2)(x^2 + 2)$$

Remember, the sum of squares *does not* factor.

12. Factor $x^4 - x^2 + 6x - 9$.

Grouping into groups of two terms each yields no common factor. However, if we group $-x^2 + 6x - 9$ together and factor out -1 we get:
$$x^4 - x^2 + 6x - 9 = x^4 + (-x^2 + 6x - 9)$$
$$= x^4 - (x^2 - 6x + 9)$$
$$= x^4 - (x - 3)^2$$
$$= [x^2 - (x - 3)][x^2 + (x - 3)]$$
$$= (x^2 - x + 3)(x^2 + x - 3).$$

Section 1.4   Fractional Expressions

Key Ideas

A. Rational expressions, simplifying fractional expressions.
B. Multiplying fractional expressions.
C. Dividing fractional expressions.
D. Adding and subtracting fractional expressions.
E. Handling compound fractional expressions.
F. Rationalizing the denominator or the numerator.

A. A quotient of two algebraic expressions is called a **fractional expression**. We deal only with values of the variables that do *not* make the denominator zero. **Rational expressions** are fractional expressions where the numerator and denominator are both polynomials. We simplify fractional expressions in the same way we simplify common fractions: First factor and then cancel common factors.

1. Simplify $\dfrac{x^2 + 2x - 3}{x^2 - 1}$.

$$\frac{x^2 + 2x - 3}{x^2 - 1} = \frac{(x+3)(x-1)}{(x+1)(x-1)} \qquad \text{Factor}$$

$$= \frac{x+3}{x+1} \qquad \text{Cancel}$$

B. The rule $\dfrac{a}{b} \cdot \dfrac{c}{d} = \dfrac{a \cdot c}{b \cdot d}$ is used to multiply two fractional expressions. However, before multiplying the numerator and denominator, first factor each expression, simplify by canceling common factors, and then multiply the remaining factors.

2. Multiply $\dfrac{x^2 - 2x + 1}{x^2 - 4} \cdot \dfrac{x^2 + 4x + 4}{x^2 + x - 2}$.

Factor, cancel common factors.

$$\frac{x^2 - 2x + 1}{x^2 - 4} \cdot \frac{x^2 + 4x + 4}{x^2 + x - 2}$$

$$= \frac{(x-1)(x-1)}{(x+2)(x-2)} \cdot \frac{(x+2)(x+2)}{(x+2)(x-1)} \qquad \text{Factor}$$

$$= \frac{x-1}{x-2} \qquad \text{Cancel}$$

C. The rule $\dfrac{a}{b} \div \dfrac{c}{d} = \dfrac{\dfrac{a}{b}}{\dfrac{c}{d}} = \dfrac{a}{b} \cdot \dfrac{d}{c} = \dfrac{a \cdot d}{b \cdot c}$ is used to divide two fractional expressions. This rule is commonly referred to as *"invert and multiply."*

3. Divide $\dfrac{x^2 - x - 6}{x + 4} \div \dfrac{x^2 - 9}{x^2 + 3x - 4}$.

$$\dfrac{x^2 - x - 6}{x + 4} \div \dfrac{x^2 - 9}{x^2 + 3x - 4}$$

$$= \dfrac{x^2 - x - 6}{x + 4} \cdot \dfrac{x^2 + 3x - 4}{x^2 - 9} \qquad \text{Invert \& multiply}$$

$$= \dfrac{(x - 3)(x + 2)}{(x + 4)} \cdot \dfrac{(x + 4)(x - 1)}{(x - 3)(x + 3)} \qquad \text{Factor}$$

$$= \dfrac{(x + 2)(x - 1)}{x + 3} \qquad \text{Cancel}$$

$$= \dfrac{x^2 + x - 2}{x + 3} \qquad \text{Simplify}$$

4. Divide $\dfrac{x^3 - 7x^2 - 8x}{x^2 - 2x - 15} \div \dfrac{x^2 - 9x + 8}{x^3 - 9x^2 + 20x}$.

$$\dfrac{x^3 - 7x^2 - 8x}{x^2 - 2x - 15} \div \dfrac{x^2 - 9x + 8}{x^3 - 9x^2 + 20x}$$

$$= \dfrac{x^3 - 7x^2 - 8x}{x^2 - 2x - 15} \cdot \dfrac{x^3 - 9x^2 + 20x}{x^2 - 9x + 8} \qquad \text{Invert \& multiply}$$

$$= \dfrac{x(x + 1)(x - 8)}{(x - 5)(x + 3)} \cdot \dfrac{x(x - 4)(x - 5)}{(x - 1)(x - 8)} \qquad \text{Factor}$$

$$= \dfrac{x^2(x + 1)(x - 4)}{(x + 3)(x - 1)} \qquad \text{Cancel}$$

$$= \dfrac{x^4 - 3x^3 - 4x^2}{x^2 + 2x - 3} \qquad \text{Simplify}$$

D. Just as in the addition of two rational numbers, two fractional expressions must first have a common denominator before they can be added. Then use $\dfrac{a}{b} + \dfrac{c}{b} = \dfrac{a + c}{b}$.

5. Combine and simplify.

(a) $\dfrac{2}{x - 4} + \dfrac{3x}{4 - x}$

Since $4 - x = -(x - 4)$ and $\dfrac{a}{-b} = -\dfrac{a}{b}$,

$$\dfrac{2}{x - 4} + \dfrac{3x}{4 - x} = \dfrac{2}{x - 4} + \dfrac{3x}{-(x - 4)}$$

$$= \dfrac{2}{x - 4} - \dfrac{3x}{x - 4}$$

$$= \dfrac{2 - 3x}{x - 4}$$

(b) $\dfrac{2}{x+3} + \dfrac{x}{x+5}$

The LCD (least common denominator) of $x+3$ and $x+5$ is $(x+3)(x+5)$.

$$\dfrac{2}{x+3} + \dfrac{x}{x+5}$$

$$= \dfrac{2}{(x+3)}\dfrac{(x+5)}{(x+5)} + \dfrac{x}{(x+5)}\dfrac{(x+3)}{(x+3)}$$

$$= \dfrac{2x+10}{(x+3)(x+5)} + \dfrac{x^2+3x}{(x+3)(x+5)}$$

$$= \dfrac{x^2+5x+10}{(x+3)(x+5)}$$

Notice that the numerator is multiplied and simplified while the denominator is left in factored form.

(c) $\dfrac{2}{(x-3)(x+2)} + \dfrac{3}{(x+3)(x+2)}$

The LCD of $(x-3)(x+2)$ and $(x+3)(x+2)$ is $(x-3)(x+3)(x+2)$.

$$\dfrac{2}{(x-3)(x+2)} + \dfrac{3}{(x+3)(x+2)}$$

$$= \dfrac{2}{(x-3)(x+2)}\dfrac{(x+3)}{(x+3)}$$

$$+ \dfrac{3}{(x+3)(x+2)}\dfrac{(x-3)}{(x-3)}$$

$$= \dfrac{2x+6}{(x-3)(x+3)(x+2)}$$

$$+ \dfrac{3x-9}{(x-3)(x+3)(x+2)}$$

$$= \dfrac{5x-3}{(x-3)(x+3)(x+2)}$$

(d) $\dfrac{-2}{x^2+4x+4} + \dfrac{1}{x^2+5x+6}$

Start by factoring each denominator, then find the LCD.

$x^2+4x+4 = (x+2)(x+2)$
$x^2+5x+6 = (x+2)(x+3)$
So the LCD is $(x+2)(x+2)(x+3)$.

$$\dfrac{-2}{x^2+4x+4} + \dfrac{1}{x^2+5x+6}$$

$$= \dfrac{-2}{(x+2)(x+2)} + \dfrac{1}{(x+2)(x+3)}$$

$$= \dfrac{-2}{(x+2)(x+2)}\dfrac{(x+3)}{(x+3)}$$

$$+ \dfrac{1}{(x+2)(x+3)}\dfrac{(x+2)}{(x+2)}$$

$$= \dfrac{-2x-6}{(x+2)(x+2)(x+3)}$$

$$+ \dfrac{x+2}{(x+2)(x+2)(x+3)}$$

$$= \dfrac{-x-4}{(x+2)(x+2)(x+3)}$$

E. To simplify a compound fractional expression, first get a single term in the numerator and a single term denominator, by using a common denominator.

6. Simplify $\dfrac{\dfrac{y}{x}+1}{\dfrac{x}{y}-\dfrac{y}{x}}$.

$$\dfrac{\dfrac{y}{x}+1}{\dfrac{x}{y}-\dfrac{y}{x}}=\dfrac{\dfrac{y}{x}+1\left(\dfrac{x}{x}\right)}{\left(\dfrac{x}{y}\right)\left(\dfrac{x}{x}\right)-\left(\dfrac{y}{x}\right)\left(\dfrac{y}{y}\right)}$$

$$=\dfrac{\dfrac{y}{x}+\dfrac{x}{x}}{\dfrac{x^2}{xy}-\dfrac{y^2}{xy}}=\dfrac{\dfrac{y+x}{x}}{\dfrac{x^2-y^2}{xy}}$$

$$=\dfrac{y+x}{x}\div\dfrac{x^2-y^2}{xy}$$

$$=\dfrac{y+x}{x}\cdot\dfrac{xy}{x^2-y^2}$$

$$=\dfrac{(y+x)}{x}\cdot\dfrac{xy}{(x-y)(x+y)}$$

$$=\dfrac{y}{x-y}$$

7. Simplify $\dfrac{\dfrac{1}{x}+\dfrac{x}{x+1}}{x-\dfrac{2}{x-1}}$.

$$\dfrac{\dfrac{1}{x}+\dfrac{x}{x+1}}{x-\dfrac{2}{x-1}}=\dfrac{\dfrac{1}{x}\dfrac{(x+1)}{(x+1)}+\dfrac{x}{(x+1)}\dfrac{x}{x}}{x\cdot\dfrac{(x-1)}{(x-1)}-\dfrac{2}{x-1}}$$

$$=\dfrac{\dfrac{x+1}{x(x+1)}+\dfrac{x^2}{x(x+1)}}{\dfrac{x^2-x}{x-1}-\dfrac{2}{x-1}}$$

$$=\dfrac{\dfrac{x^2+x+1}{x(x+1)}}{\dfrac{x^2-x-2}{x-1}}$$

$$=\dfrac{x^2+x+1}{x(x+1)}\div\dfrac{x^2-x-2}{x-1}$$

$$=\dfrac{x^2+x+1}{x(x+1)}\cdot\dfrac{(x-1)}{(x^2-x-2)}$$

$$=\dfrac{(x^2+x+1)}{x(x+1)}\cdot\dfrac{(x-1)}{(x+1)(x-2)}$$

$$=\dfrac{x^3-1}{x(x+1)^2(x-2)}$$

F. Radicals are eliminated from the denominators by multiplying *both* the numerator and denominator by an appropriate expression. This procedure is called **rationalizing the denominator**. If a fraction has a denominator of the form $A + B\sqrt{C}$, we may rationalize the denominator by multiplying numerator and denominator by the **conjugate radical**, $A - B\sqrt{C}$. Then by the *difference of squares*, we have

$$\left(A + B\sqrt{C}\right)\left(A - B\sqrt{C}\right) = A^2 - B^2 C.$$

8. Rationalize the denominator in each of the following expressions.

(a) $\dfrac{3}{1 - \sqrt{2}}$

We take advantage of the difference of squares formula to rationalize the denominator.

$$\frac{3}{1 - \sqrt{2}} = \frac{3}{\left(1 - \sqrt{2}\right)} \cdot \frac{\left(1 + \sqrt{2}\right)}{\left(1 + \sqrt{2}\right)}$$

$$= \frac{3 + 3\sqrt{2}}{1 - 2} = \frac{3 + 3\sqrt{2}}{-1}$$

$$= -3 - 3\sqrt{2}$$

(b) $\dfrac{\sqrt{3}}{\sqrt{5} + \sqrt{3}}$

$$\frac{\sqrt{3}}{\sqrt{5} + \sqrt{3}} = \frac{\sqrt{3}}{\left(\sqrt{5} + \sqrt{3}\right)} \cdot \frac{\left(\sqrt{5} - \sqrt{3}\right)}{\left(\sqrt{5} - \sqrt{3}\right)}$$

$$= \frac{\sqrt{15} - 3}{5 - 3}$$

$$= \frac{\sqrt{15} - 3}{2}$$

9. Rationalize the numerator.

$$\frac{\sqrt{3} - \sqrt{x}}{7}$$

$$\frac{\sqrt{3} - \sqrt{x}}{7} = \frac{\sqrt{3} - \sqrt{x}}{7} \cdot \frac{\sqrt{3} + \sqrt{x}}{\sqrt{3} + \sqrt{x}}$$

$$= \frac{3 - x}{7\left(\sqrt{3} + \sqrt{x}\right)}$$

Section 1.5   Equations

Key Ideas

A. Solving linear equations.
B. Solving for one variable in terms of others.
C. Quadratic equations and the zero factor property.
D. Completing the square.
E. Quadratic formula.
F. Other equations.

A. A **linear equation** is a first degree equation. To solve an equation means to find all solutions (roots) of the equation. Two equations are **equivalent** if they have exactly the same solutions. You go between equivalent equations by using the two Properties of Equality:

1.	You can add or subtract the same quantity to *both sides* of the equation.
2.	You can multiply or divide *both sides* of the equation by the same nonzero quantity.

Remember to always check each solution in the original equation. Rational equations can be transformed to an equivalent linear equation by multiplying both sides by the LCD (provided that the LCD is not zero). This process can introduce extraneous solutions, so it is extremely important to check all solutions in the original equation.

1. Solve.

(a) $3x - 9 = 0$

$$3x - 9 = 0$$
$$3x = 9 \qquad\qquad \textit{Add 9}$$
$$x = 3 \qquad\qquad \textit{Multiply by 1/3}$$
Check: $3(3) - 9 \overset{?}{=} 0$
$$9 - 9 = 0 \checkmark$$
$$x = 3$$

(b) $5x + 7 = x - 1$

$$5x + 7 = x - 1$$
$$4x = -8 \qquad\qquad \textit{Subtract x and 7}$$
$$x = -2 \qquad\qquad \textit{Multiply by 1/4}$$
Check: $5(-2) + 7 \overset{?}{=} (-2) - 1$
$$-10 + 7 \overset{?}{=} -3$$
$$-3 = -3 \checkmark$$

(c) $x - \dfrac{2}{3} = \dfrac{x}{6}$

$$\left(x - \frac{2}{3}\right)6 = \left(\frac{x}{6}\right)6 \qquad\qquad \textit{Multiply by LCD}$$
$$(x)6 - \left(\frac{2}{3}\right)6 = x \qquad\qquad \textit{Distributive property}$$
$$6x - 4 = x$$
$$-4 = -5x \qquad\qquad \textit{Subtract 6x}$$
$$\tfrac{4}{5} = x$$
Check: $\left(\dfrac{4}{5}\right) - \dfrac{2}{3} \overset{?}{=} \dfrac{\left(\frac{4}{5}\right)}{6}$
$$\tfrac{12}{15} - \tfrac{10}{15} \overset{?}{=} \tfrac{4}{30}$$
$$\tfrac{2}{15} = \tfrac{2}{15} \checkmark$$

2. Solve.

(a) $\dfrac{2x+6}{3x} + \dfrac{1}{2} = \dfrac{1}{6}$

$\left(\dfrac{2x+6}{3x} + \dfrac{1}{2}\right)6x = \left(\dfrac{1}{6}\right)6x$ *Multiply by LCD*

$\left(\dfrac{2x+6}{3x}\right)6x + \left(\dfrac{1}{2}\right)6x = x$ *Simplify*

$4x + 12 + 3x = x$

$7x + 12 = x$

$6x = -12$

$x = -2$ *Potential solution*

Check: $\dfrac{2(-2)+6}{3(-2)} + \dfrac{1}{2} \overset{?}{=} \dfrac{1}{6}$

$\dfrac{2}{-6} + \dfrac{1}{2} \overset{?}{=} \dfrac{1}{6}$

$\dfrac{1}{6} = \dfrac{1}{6}$ ✓

(b) $\dfrac{x+2}{x-3} + 2 = \dfrac{10}{2x-6}$

Since $2x - 6 = 2(x-3)$, the LCD is $2x - 6$.

$\dfrac{x+2}{x-3} + 2 = \dfrac{10}{2x-6}$

$\left(\dfrac{x+2}{x-3} + 2\right)(2x-6) = \left(\dfrac{10}{2x-6}\right)(2x-6)$

$\left(\dfrac{x+2}{x-3}\right)2(x-3) + 2(2x-6) = \dfrac{10}{2x-6}(2x-6)$

$2x + 4 + 4x - 12 = 10$

$6x - 8 = 10$

$6x = 18$

$x = 3$ *Potential solution*

Check: $\dfrac{(3)+2}{(3)-3} + 2 \overset{?}{=} \dfrac{10}{2(3)-6}$

$\dfrac{6}{0} + 2 \overset{?}{=} \dfrac{10}{0}$

Since the fractions in the last equation are undefined, there is NO solution.

B. In many formulas, it is often necessary to express one of the variables in terms of the others. As in other equations in this section, we first isolate the terms involving this variable on one side of the equation and then use one of the techniques discussed so far.

3. Solve for b_1 in terms of the other variables.

$A = \frac{1}{2}(b_1 + b_2)h$

$A = \frac{1}{2}(b_1 + b_2)h$ *Isolated the term with b_1*

$\dfrac{2A}{h} = b_1 + b_2$

$b_1 + b_2 = \dfrac{2A}{h}$

$b_1 = \dfrac{2A}{h} - b_2$

4. Solve for h in terms of the other variables.

$A = 2\pi rh + 2\pi r^2$

$A = 2\pi rh + 2\pi r^2$ *Isolated the term*

$A - 2\pi r^2 = 2\pi rh$ *with h*

$2\pi rh = A - 2\pi r^2$

$h = \dfrac{A - 2\pi r^2}{2\pi r}$

C. Quadratic equations are second-degree equations. Quadratic equations can be solved by first getting one side equal to zero and factoring the quadratic equation. Then use the **zero factor property**: $ab = 0$ if and only if $a = 0$ or $b = 0$. Refer back to Section 1.3 for a review of factoring. Some equations can also be solved by using the form $x^2 = c$, where c is a constant and $c \geq 0$, and then taking the square root of both sides. So $x = \pm\sqrt{c}$.

5. Solve the equation $x^2 + 3x + 2 = 0$.

$x^2 + 3x + 2 = 0$ *Factor*

$(x + 1)(x + 2) = 0$ *Set each factor $= 0$*

$x + 1 = 0$ $x + 2 = 0$ *Solve*

$x = -1$ $x = -2$

Check:

$(-1)^2 + 3(-1) + 2 = 1 - 3 + 2 = 0$ ✓

$(-2)^2 + 3(-2) + 2 = 4 - 6 + 2 = 0$ ✓

6. Solve the equation $x^2 + 12 = 7x$.

$x^2 + 12 = 7x$ *Move all terms to LHS*

$x^2 - 7x + 12 = 0$ *Factor*

$(x - 3)(x - 4) = 0$ *Set each factor $= 0$*

$x - 3 = 0$ $x - 4 = 0$ *Solve*

$x = 3$ $x = 4$

Check:

$(3)^2 + 12 \overset{?}{=} 7(3)$ $(4)^2 + 12 \overset{?}{=} 7(4)$

$9 + 12 = 21$ ✓ $16 + 12 = 28$ ✓

7. Solve the equation $x^2 = 2x + 3$.

$$x^2 = 2x + 3 \qquad \textit{Move all terms to LHS}$$
$$x^2 - 2x - 3 = 0 \qquad \textit{Factor}$$
$$(x-3)(x+1) = 0 \qquad \textit{Set each factor} = 0$$
$$x - 3 = 0 \qquad x + 1 = 0 \qquad \textit{Solve}$$
$$x = 3 \qquad\qquad x = -1$$

Check:
$$(3)^2 \overset{?}{=} 2(3) + 3 \qquad\qquad (-1)^2 \overset{?}{=} 2(-1) + 3$$
$$9 = 6 + 3 \ \checkmark \qquad\qquad 1 = -2 + 3 \ \checkmark$$

8. Solve the equation $x^2 = 9$.

$$x^2 = 9$$
$$x = \pm\sqrt{9}$$
$$x = \pm 3, \text{ that is, the solutions are } x = 3 \text{ and } x = -3.$$

Check each solution.

9. Solve the equation $x^2 - 8 = 0$.

$$x^2 - 8 = 0 \qquad\qquad \textit{Isolate } x$$
$$x^2 = 8$$
$$x = \pm\sqrt{8} = \pm 2\sqrt{2}$$

Check each solution.

10. Solve the equation $4x^2 = 24$.

$$4x^2 = 24$$
$$x^2 = 6 \qquad \textit{Divide by } 4$$
$$x = \pm\sqrt{6}$$

Another way to solve this problem is:
$$4x^2 = 24$$
$$(2x)^2 = 24$$
$$2x = \pm\sqrt{24} = \pm 2\sqrt{6}$$
$$x = \pm\sqrt{6}$$

Again, check each solution.

D. The goal in finding the solution to a quadratic equation by **completing the square** is to add the appropriate constant to make a perfect square, $(x + m)^2$ or $(x - m)^2$. Step one is to isolate the terms involving the variables, x^2 and bx, on one side of the equation. Step two is to determine the constant needed to add to both sides to complete the square. Since $(x \pm m)^2 = x^2 \pm 2mx + m^2$, the coefficient of the middle term, b, must equal $2m$, that is, $m = \frac{1}{2}b$; so add $\left(\frac{b}{2}\right)^2$ to both sides.

11. Solve by completing the square: $x^2 - 6x + 4 = 0$.

$x^2 - 6x + 4 = 0$ *Isolate the terms with x*

$x^2 - 6x \quad = -4$ *on one side.*

$x^2 - 6x + \mathbf{9} = -4 + \mathbf{9}$ *Add $\left(\frac{6}{2}\right)^2 = 9$ to both sides.*

$(x - 3)^2 = 5$ *Simplify.*

$x - 3 = \pm\sqrt{5}$ *Take sq. root of both sides.*

$x = 3 \pm \sqrt{5}$

Check each solution.

12. Solve by completing the square: $4x^2 = 12x + 1$.

$$4x^2 = 12x + 1$$

$4x^2 - 12x \quad = 1$ *Isolate the terms with x on one side.*

$x^2 - 3x \quad = \frac{1}{4}$ *Divide both sides by 4.*

$x^2 - 3x + \frac{9}{4} = \frac{1}{4} + \frac{9}{4}$ *Add $\left(\frac{3}{2}\right)^2 = \frac{9}{4}$ to both sides.*

$\left(x - \frac{3}{2}\right)^2 = \frac{10}{4}$ *Simplify.*

$x - \frac{3}{2} = \frac{\pm\sqrt{10}}{2}$ *Take sq. root of both sides.*

$x = \frac{3}{2} \pm \frac{\sqrt{10}}{2}$

$x = \frac{3 \pm \sqrt{10}}{2}$

Check each solution.

E. The **quadratic formula**, $x = \dfrac{-b \pm \sqrt{b^2 - 4ac}}{2a}$, gives the roots (solutions) to the quadratic equation $ax^2 + bx + c = 0$. The radical, $\sqrt{b^2 - 4ac}$, is called the **discriminant**, D. When $D > 0$, the quadratic equation has two real solutions. When $D = 0$, there is exactly one solution. And when $D < 0$, the quadratic equation has no real solution.

13. Solve each equation by using the quadratic formula.

(a) $4x^2 - 4x - 3 = 0$

Setting $a = 4$, $b = -4$, $c = -3$ and substituting into the quadratic formula:

$$x = \frac{-b \pm \sqrt{b^2 - 4ac}}{2a}$$

$$= \frac{-(-4) \pm \sqrt{(-4)^2 - 4(4)(-3)}}{2(4)}$$

$$= \frac{4 \pm \sqrt{16 + 48}}{8} = \frac{4 \pm \sqrt{64}}{8} = \frac{4 \pm 8}{8}.$$

So the solutions are $x = \frac{4+8}{8} = \frac{3}{2}$ and $x = \frac{4-8}{8} = -\frac{1}{2}$.

Note: This equation factors: $(2x - 3)(2x + 1) = 0$.

(b) $3x^2 - 7x + 3 = 0$

Setting $a = 3$, $b = -7$, $c = 3$ and substituting into the quadratic formula:

$$x = \frac{-b \pm \sqrt{b^2 - 4ac}}{2a}$$

$$= \frac{-(-7) \pm \sqrt{(-7)^2 - 4(3)(3)}}{2(3)}$$

$$= \frac{7 \pm \sqrt{49 - 36}}{6} = \frac{7 \pm \sqrt{13}}{6}.$$

So the solutions are $x = \frac{7 + \sqrt{13}}{6}$ and $x = \frac{7 - \sqrt{13}}{6}$.

(c) $4x^2 + 20x + 25 = 0$

Setting $a = 4$, $b = 20$, $c = 25$ and substituting into the quadratic formula:

$$x = \frac{-b \pm \sqrt{b^2 - 4ac}}{2a}$$

$$= \frac{-(20) \pm \sqrt{(20)^2 - 4(4)(25)}}{2(4)}$$

$$= \frac{-20 \pm \sqrt{400 - 400}}{8} = \frac{-20 \pm \sqrt{0}}{8} = \frac{-20 \pm 0}{8}$$

So the only solution is $x = \frac{-20}{8} = -\frac{5}{2}$.
Since $4x^2 + 20x + 25 = (2x + 5)^2$, the solution checks.

(d) $\dfrac{3}{x - 2} - \dfrac{2}{x + 4} = 1$

$$\frac{3}{x - 2} - \frac{2}{x + 4} = 1 \qquad \textit{Multiply both sides by LCD}$$

$$(x - 2)(x + 4)\left(\frac{3}{x - 2} - \frac{2}{x + 4}\right) = (x - 2)(x + 4)$$

$$3(x + 4) - 2(x - 2) = x^2 + 2x - 8$$

$$3x + 12 - 2x + 4 = x^2 + 2x - 8$$

$$x + 16 = x^2 + 2x - 8$$

$$0 = x^2 + x - 24$$

Setting $a = 1$, $b = 1$, $c = -24$ and substituting into the quadratic formula:

$$x = \frac{-b \pm \sqrt{b^2 - 4ac}}{2a}$$

$$= \frac{-(1) \pm \sqrt{(1)^2 - 4(1)(-24)}}{2(1)}$$

$$= \frac{-1 \pm \sqrt{1 + 96}}{2} = \frac{-1 \pm \sqrt{97}}{2}$$

So the solutions are $x = \frac{-1 + \sqrt{97}}{2}$ and $x = \frac{-1 - \sqrt{97}}{2}$.

Since neither of these solutions make a denominator zero, both solutions are valid.
You should check each solution.

F. Higher degree equations are solved by getting one side equal to zero and factoring.

14. Find all solutions of the equation $x^3 - 4x^2 = 5x$.

$$x^3 - 4x^2 = 5x \qquad \textit{Move all terms to LHS}$$
$$x^3 - 4x^2 - 5x = 0 \qquad \textit{Factor}$$
$$x(x^2 - 4x - 5) = 0$$
$$x(x - 5)(x + 1) = 0 \qquad \textit{Set each factor} = 0$$

$$x = 0 \qquad x - 5 = 0 \qquad x + 1 = 0$$
$$x = 5 \qquad x = -1$$

The three solutions are $x = 0$, $x = 5$, and $x = -1$.

NOTE: Many students forget the lone x factor and will lose the solution $x = 0$. Another common mistake is to divide both sides by x at the first step. This also loses the solution $x = 0$.

15. Find all solutions of the equation $x^4 - 16 = 0$.

$$x^4 - 16 = 0 \qquad \textit{Factor}$$
$$(x^2 - 4)(x^2 + 4) = 0 \qquad \textit{Factor}$$
$$(x - 2)(x + 2)(x^2 + 4) = 0 \qquad \textit{Set each factor} = 0$$

$$x - 2 = 0 \qquad x + 2 = 0 \qquad x^2 + 4 = 0$$
$$x = 2 \qquad x = -2$$

There is no real solution to $x^2 + 4 = 0$, so only two real solutions are $x = 2$ and $x = -2$.

F. An equation of the form $aw^2 + bw + c = 0$, where w is an algebraic expression, is an equation of the **quadratic type**. We solve equation of the quadratic type by substituting for the algebraic expression.

16. Find all solutions of the equation $x^4 - 3x^2 + 2 = 0$.

Since all variables are raised to even powers, we use the substitution $w = x^2$. The equation then becomes
$$x^4 - 3x^2 + 2 = (x^2)^2 - 3(x^2) + 2$$
$$= w^2 - 3w + 2$$
$$w^2 - 3w + 2 = 0 \qquad \textit{Factor.}$$
$$(w - 2)(w - 1) = 0 \qquad \textit{Set each factor} = 0.$$

$$w - 2 = 0 \qquad w - 1 = 0$$
$$x^2 - 2 = 0 \qquad x^2 - 1 = 0$$
$$x^2 = 2 \qquad x^2 = 1$$
$$x = \pm\sqrt{2} \qquad x = \pm 1$$

The four real solutions are $x = 1$, $x = -1$, $x = \sqrt{2}$, and $x = -\sqrt{2}$.

17. Find all solutions of the equation

$$\left(x - \frac{1}{x}\right)^2 + 4\left(x + \frac{1}{x}\right) - 5 = 0$$

We substitute $w = x + \frac{1}{x}$, to get

$$\left(x - \frac{1}{x}\right)^2 + 4\left(x + \frac{1}{x}\right) - 5 = w^2 + 4w - 5.$$

$$w^2 + 4w - 5 = 0$$
$$(w - 1)(w + 5) = 0$$

$$w - 1 = 0 \qquad\qquad w + 5 = 0$$
$$x + \frac{1}{x} - 1 = 0 \qquad x + \frac{1}{x} + 5 = 0 \quad \textit{Multiply by } x.$$
$$x^2 + 1 - x = 0 \qquad x^2 + 1 + 5x = 0$$
$$x^2 - x + 1 = 0 \qquad x^2 + 5x + 1 = 0$$

Applying the quadratic formula to $x^2 - x + 1 = 0$

$$x = \frac{-(-1) \pm \sqrt{(-1)^2 - 4(1)(1)}}{2(1)}$$

$$= \frac{1 \pm \sqrt{1 - 4}}{2} = \frac{1 \pm \sqrt{-3}}{2}$$

Thus there are no real solutions to $x^2 - x + 1 = 0$.

Applying the quadratic formula to $x^2 + 5x + 1 = 0$

$$x = \frac{-(5) \pm \sqrt{(5)^2 - 4(1)(1)}}{2(1)}$$

$$= \frac{-5 \pm \sqrt{25 - 4}}{2} = \frac{-5 \pm \sqrt{21}}{2}$$

$$= -\frac{5}{2} \pm \frac{\sqrt{21}}{2}.$$

There are two real solutions

$$x = -\frac{5}{2} + \frac{\sqrt{21}}{2} \text{ and } x = -\frac{5}{2} - \frac{\sqrt{21}}{2}.$$

F. We solve basic equations of the form $X^n = a$ by taking radicals of both sides. The rules applying to radicals are:

n even	$a \geq 0$: two nth roots are $\sqrt[n]{a}$ and $-\sqrt[n]{a}$.	$a < 0$: nth root does not exists.
n odd	$\sqrt[n]{a}$ exists and is unique.	

18. Solve the equation.
    (a) $x^5 = -32$

$$x^5 = -32$$

$$x = \sqrt[5]{-32} = -2$$

Check the answer.

31

(b) $x^3 + 8 = 0$

$x^3 + 8 = 0$

$\qquad x^3 = -8$

$\qquad x = \sqrt[3]{-8} = -2$

Check each answer.

(c) $x^4 - 16 = 0$

$x^4 - 16 = 0$

$\qquad x^4 = 16$

$\qquad x = \pm\sqrt[4]{16} = \pm 2$

Check each answer.

19. Solve each equation.
   (a) $(x + 3)^2 - 5 = 0$

$(x + 3)^2 - 5 = 0$

$\qquad (x + 3)^2 = 5$

$\qquad x + 3 = \pm\sqrt{5}$

$\qquad x = -3 \pm \sqrt{5}$

Check each answer.

   (b) $(2x - 5)^3 + 64 = 0$

$(2x - 5)^3 + 64 = 0$

$\qquad (2x - 5)^3 = -64$

$\qquad 2x - 5 = \sqrt[3]{-64} = -4$

$\qquad 2x = 1$

$\qquad x = \frac{1}{2}$

Check each answer.

Section 1.6   Modeling with Equations

Key Ideas
A. Guidelines for modeling with equations.
B. Modeling with quadratic equations
C. Mixture and concentrations models.

A. The steps involved in solving a word problem are:

1.	Read the entire question. Identify the variable.
2.	Express all unknown quantities in terms of the variable.
3.	Set up an equation or model. Relate the quantities.
4.	Solve the equation and check your solutions.

1. A train leaves station A at 7 am and travels north to Chicago at 42 miles per hour. Two hours later, an express train leaves station A and travels north to Chicago at 70 miles per hour. When will the express overtake the slower train?

After reading through the problem, we find that we are looking for the number of hours the first train travels.

Let $t =$ the number of hours the first train travels. Then $t - 2 =$ the number of hours the second train travels.

The trains will have traveled the same distance when the express train overtakes the other train, so compare the distance traveled by each train using the formula $D = RT$.

1st train distance $= 42t$
2nd train distance $= 70(t - 2)$

$$42t = 70(t - 2)$$
$$42t = 70t - 140$$
$$-28t = -140$$
$$t = 5 \text{ hours}$$

So the express train overtakes the slower train at 12 noon.

Check:
1st train travels $42(5) = 210$ miles
2nd train travels $70(5 - 2) = 70(3) = 210$ miles $\checkmark$

2. Use the distance formula $D = RT$.
A bicyclist leaves home and pedals to school at 20 miles per hour. Fifteen minutes later, her roommate leaves home and drives her car to school at 35 miles per hour. When will the car overtake the bicyclist?

In this problem we seek the time when the distance traveled by the bicyclist is equal to the distance traveled by the car.

Let $t =$ time the car drives, in hours.

Then $t + \frac{1}{4} =$ time bicyclist rides, since the bicyclist leaves 15 minutes before the car.

Using the formula $D = RT$,

$$20\left(t + \tfrac{1}{4}\right) = 35t$$
$$20t + 5 = 35t$$
$$5 = 15t$$
$$t = \tfrac{5}{15} = \tfrac{1}{3} \text{ hours or 20 minutes.}$$

Check:
Bicyclist: $20\left(\tfrac{1}{3} + \tfrac{1}{4}\right) = 20\left(\tfrac{7}{12}\right) = \tfrac{35}{3}$ miles
Car: $35\left(\tfrac{1}{3}\right) = \tfrac{35}{3}$ miles $\checkmark$

3. Wanda has 75 coins. Some of the coins are quarters and the rest are dimes. If the total amount is \$12.90. How many of each type of coin does she have?

Here we are after the number of quarters and the number of dimes. We let the variable equal one of the quantities.

Let $x =$ number of quarters

$75 - x =$ number of dimes

$\overline{\qquad 75 \text{ coins total}}$

Now translate the fact that the value of the coins adds up to \$12.90.

$$\begin{pmatrix} \text{value of} \\ \text{the quarters} \end{pmatrix} + \begin{pmatrix} \text{value of} \\ \text{the dimes} \end{pmatrix} = 12.90$$

$$0.25x + 0.10(75 - x) = 12.90$$

Solve for x:

$$0.25x + 7.5 - 0.10x = 12.90$$

$$0.15x + 7.5 = 12.90$$

$$0.15x = 5.40$$

$$x = \frac{5.40}{0.15} = 36.$$

So Wanda has 36 quarters and $75 - 36 = 39$ dimes.

Check:

$0.25(36) + 0.10(39) = 9.00 + 3.90 = 12.90$ $\checkmark$

B. Applied problems often lead to models that involve quadratic equations.

4. Linda takes 2 hours more than Greg does to service a photo copier. Together they can service the copier in 5 hours. How long does each take to service the copier?

Let $t =$ the time it takes Greg to service the copier; $t + 2 =$ the time it takes Linda.

Each hour Greg does $\dfrac{1}{t}$ and Linda does $\dfrac{1}{t+2}$ of the job, and together they can do $\frac{1}{5}$ of the job, that is,

$$\frac{1}{t} + \frac{1}{t+2} = \frac{1}{5} \qquad \begin{array}{l} \textit{Multiply by the} \\ \textit{LCD } 5t(t+2) \end{array}$$

$$5t(t+2)\left(\frac{1}{t} + \frac{1}{t+2}\right) = \left(\frac{1}{5}\right)5t(t+2)$$

$$5(t+2) + 5t = t(t+2)$$

$$5t + 10 + 5t = t^2 + 2t$$

$$10t + 10 = t^2 + 2t$$

$$0 = t^2 - 8t - 10$$

Setting $a = 1$, $b = -8$, $c = -10$ and substituting into the quadratic formula:

$$t = \frac{-b \pm \sqrt{b^2 - 4ac}}{2a}$$

$$= \frac{-(-8) \pm \sqrt{(-8)^2 - 4(1)(-10)}}{2(1)}$$

$$= \frac{8 \pm \sqrt{64+40}}{2} = \frac{8 \pm \sqrt{104}}{2} = \frac{8 \pm 2\sqrt{26}}{2} = 4 \pm \sqrt{26}.$$

Since $4 - \sqrt{26} < 0$, the only solution is $4 + \sqrt{26}$.

So it takes Greg $4 + \sqrt{26}$ hours ≈ 9 hours and 6 minutes.

It takes Linda $4 + \sqrt{26} + 2 = 6 + \sqrt{26} \approx 11$ hours and 6 minutes

5. A tug tows a barge 24 miles up a river at 10 miles per hour and returns down the river at 12 miles per hour. The entire trip take $5\frac{1}{2}$ hours. What is the rate of the river's current?

Let $r = $ the rate of the river's current.
Then $10 - r$ is the true rate of the barge up river and $12 + r$ is the true rate of the barge down river.
Use the distance formula and $D = RT$ and solve for time: $T = \dfrac{D}{R}$.

The time it takes the barge to go up the river is
$\dfrac{24}{10 - r}$.

The time it takes the barge to go down the river is
$\dfrac{24}{12 + r}$.

$$\frac{24}{10 - r} + \frac{24}{12 + r} = 5\frac{1}{2} = \frac{11}{2}$$

Multiply by the LCD: $2(10 - r)(12 + r)$

$$2(10 - r)(12 + r)\left(\frac{24}{10 - r} + \frac{24}{12 + r}\right)$$

$$= \left(\frac{11}{2}\right)2(10 - r)(12 + r)$$

$$48(12 + r) + 48(10 - r) = 11(10 - r)(12 + r)$$

$$576 + 48r + 480 - 48r = 1320 - 22r - 11r^2$$

$$1056 = 1320 - 22r - 11r^2$$

$$0 = -11r^2 - 22r + 264$$

$$0 = -11(r^2 + 2r - 24)$$

$$0 = -11(r - 4)(r + 6)$$

$r - 4 = 0 \qquad\qquad r + 6 = 0$

$r = 4 \qquad\qquad\qquad r = -6$

$r = -6$ does not make sense, since it requires the river to flow backwards. Thus the rate of the river's current is 4 miles per hour.

Check: $r = 4$
When $r = 4$, the rate up river is $10 - 4 = 6$ miles per hour, so the trip up river takes $\frac{24}{6}$ or 4 hours.
The rate down river is $12 + 4 = 16$ miles per hour, so the trip down river takes $\frac{24}{16}$ hours or $1\frac{1}{2}$ hours.

C. Mixture and concentration problems are solved by relating the values of some quantity put into the mixture with the final value of the quantity removed from the mixture. In these problems, it is helpful to draw pictures and fill in a table like the one shown below.

	Value in	Value out
Type or name		
%		
amount		
value		

6. A butcher at a supermarket has 50 pounds of ground meat that contains 36% fat. How many pounds of ground meat containing 15% fat must be added to obtain a mixture that contains 22% fat?

Let x = the number of pounds of 15% fat ground meat added to make the mixture.

In this case, the amount of fat put into the mixture must equal the amount of fat taken out.

	Value in		Value out
Type	36% fat	15% fat	22% fat
%	36%	15%	22%
amount	50	x	$50 + x$
value	$0.36(50)$	$0.15x$	$0.22(50 + x)$

$$0.36(50) + 0.15x = 0.22(50 + x) \quad \textit{Clear decimals,}$$
$$1800 + 15x = 1100 + 22x \quad \textit{Multiply by } 100$$
$$700 = 7x$$
$$100 = x$$

So $x = 100$ pounds. Check:

The amount of fat put into the mixture is:
$0.36(50) + 0.15(100) = 18 + 15 = 33$ pounds.
The amount of fat in the final product is:
$0.22[50 + (100)] = 0.22(150) = 33$ pounds. $\checkmark$

7. A winery makes a variety wine that, according to the label, contains 95% Cabernet grape juice. Due to a valve error, too much Zinfandel juice is added to the blend. As a result, the winery ends up with 1500 gallons of wine that now contains 93% Cabernet juice. The winery has only 500 gallons of pure Cabernet juice, so they remove a portion of the 93% blend and add 500 gallons of 100% pure Cabernet to reach 95% Cabernet concentration. How many gallons do they need to remove and replace with pure Cabernet juice?

Let x = gallons of 93% Cabernet blend that is removed. Then $1500 - x$ is the gallons of the 93% blend that remains. When 500 gallons are added, the resulting blend will have $2000 - x$ gallons.

	Value in		Value out
Type	old blend	Pure Cabernet	final blend
%	93%	100%	95%
gal.	$1500 - x$	500	$2000 - x$
value	$0.93(1500 - x)$	$1(500)$	$0.95(2000 - x)$

$$0.93(1500 - x) + 500 = 0.95(2000 - x)$$
$$93(1500 - x) + 50,000 = 95(2000 - x)$$
$$139,500 - 93x + 50,000 = 190,000 - 95x$$
$$189,500 - 93x = 190,000 - 95x$$
$$2x = 500$$
$$x = 250 \text{ gallons}$$

Check:
If 250 gallons of the 93% blend are removed, we will have 1250 gallons left, to which 500 gallons of 100% juice are added. The result will be 1750 gallons of wine. The gallons of Cabernet juice put into the blend are:
$$0.93(1250) + 1.00(500) = 1162.5 + 500$$
$$= 1662.5 \text{ gallons}$$
The gallons of Cabernet juice in the final blend are:
$$0.95(1750) = 1662.5 \text{ gallons} \checkmark$$

Section 1.7   Inequalities

Key Ideas
- **A.** Rules for manipulating inequalities and linear inequalities.
- **B.** Simultaneous inequalities.
- **C.** Sign of a product or quotient.
- **D.** Solving nonlinear inequalities.
- **E.** Properties of absolute value inequalities.
- **F.** Modeling with inequalities.

A. **Inequalities** are statements involving the symbols $<$, $>$, $\geq$, and $\leq$. These rules should be mastered to the point that you know what rule to use and how to use it.

Rules	Description
If $a \leq b$, then $a + c \leq b + c$.	Add/subtract same value from each sides.
If $a \leq b$ and $c > 0$, then $ac \leq bc$	Multiplying each side by a positive constant maintains the direction of the inequality sign.
If $a \leq b$ and $c < 0$, then $ac \geq bc$.	Multiplying each side by a negative reverses the direction of the inequality sign.
If $0 < a \leq b$, then $\dfrac{1}{a} \geq \dfrac{1}{b}$.	Taking reciprocal of both sides of an inequality of positive numbers reverses the direction of the inequality sign.
If $a \leq b$ and $c \leq d$, then $a + c \leq b + d$.	Inequalities can be added.

An inequality is **linear** if each term is constant or a multiple of the variable.

1. Solve and graph the solution on the number line below.
   $4x - 5 < 3$

$4x - 5 < 3$	*Add 5 to each side.*
$4x < 8$	*Divide each side by 4.*
$x < 2$	

   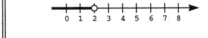

2. Solve and graph the solution on the number line below.
   $-\frac{1}{3}x + 7 \geq 2x$

   $-\frac{1}{3}x + 7 \geq 2x$ *Add $\frac{1}{3}x$ to each side.*

   $7 \geq 2x + \frac{1}{3}x$

   $7 \geq \frac{7}{3}x$ *Multiply both sides by $\frac{3}{7}$.*

   $3 \geq x$

3. Solve and graph the solution on the number line below.
   $-8x < 2x - 15$

   $-8x < 2x - 15$ *Subtract $2x$ from each side.*

   $-10x < -15$ *Divide each side by -10.*

   $x > \frac{3}{2}$ *Reverse the direction of the inequality sign.*

B. Simultaneous inequalities use the same rules as simple inequalities. Whichever inequality rule you apply to one of the inequality expressions, you *must* apply to each of the other expressions. These inequalities are also usually expressed in increasing order.

4. Solve the inequality $3 \leq 2x - 7 < 8$.

$3 \leq 2x - 7 < 8$ *Add 7 to each side.*

$10 \leq 2x < 15$ *Divide by 2.*

$5 \leq x < \frac{15}{2}$

5. Solve the inequality $3 < 8 + 4x \leq 8$.

$3 < 8 + 4x \leq 8$ *Subtract 8 from each side*

$-5 < 4x \leq 0$ *Divide by 4.*

$-\frac{5}{4} < x \leq 0$

6. Solve the inequality $-7 < 2 - 3x < 11$.

$-7 < 2 - 3x < 11$ *Subtract 2 from each side.*

$-9 < -3x < 9$ *Divide by -3, reverse the direction of the inequality sign.*

$3 > x > -3$ *Rewrite in ascending order.*

$-3 < x < 3$

C. The sign of a product or quotient is governed by the following principle:

> If a product or a quotient has an *even* number of *negative* factors, then its value is *positive*.
> If a product or a quotient has an *odd* number of *negative* factors, then its value is *negative*.

Since a factor can only change signs when it is 0, we first set each factor equal to 0 and determine these points. These points determine intervals which we can check. To determine the signs of the factors in each region we use a **test value**, a number we choose from within each interval.

7. Solve the inequality $(x - 2)(x + 1) < 0$.

$(x - 2)(x + 1) < 0$

$x - 2 = 0$ $x + 1 = 0$

$x = 2$ $x = -1$

Interval	$(-\infty, -1)$	$(-1, 2)$	$(2, \infty)$
Test value	-2	0	3
Sign of $x - 2$	$-$	$-$	$+$
Sign of $x + 1$	$-$	$+$	$+$
Sign of $(x - 2)(x + 1)$	$+$	$-$	$+$

Since we want the interval where the product is negative, the solution is $(-1, 2)$.

8. Solve the inequality $\dfrac{x-5}{x+7} > 0$.

$$\dfrac{x-5}{x+7} > 0$$

$$x - 5 = 0 \qquad\qquad x + 7 = 0$$
$$x = 5 \qquad\qquad\quad x = -7$$

Interval	$(-\infty, -7)$	$(-7, 5)$	$(5, \infty)$
Test value	-10	0	10
Sign of $x - 5$	$-$	$-$	$+$
Sign of $x + 7$	$-$	$+$	$+$
Sign of $\dfrac{x-5}{x+7}$	$+$	$-$	$+$

Since we want the interval where the quotient is positive, the solution is $(-\infty, -7) \cup (5, \infty)$.

D. These steps are used to solve inequalities that are not linear.

1. Move all terms to one side of the inequality sign. If the nonzero side of the inequality involves quotients, bring them to a common denominator.
2. Factor the nonzero side of the inequality.
3. List the intervals determined by the factorization.
4. Make a table or diagram of the signs of each factor on each interval. In the last row of the table we determine the sign of the product (or quotient) of these factors.
5. Determine the solution set from the last row of the table. Be sure to check whether the inequality is satisfied by some or all of the endpoints of the intervals (only necessary when $\leq$ and $\geq$ are involved).

9. Solve the inequality $x^2 + 6 > 5x$.

$$x^2 + 6 > 5x \qquad \textit{Move all terms to one side.}$$
$$x^2 - 5x + 6 > 0 \qquad \textit{Factor}$$
$$(x-2)(x-3) > 0$$
$$x - 2 = 0 \qquad\qquad x - 3 = 0$$
$$x = 2 \qquad\qquad\quad x = 3$$

Interval	$(-\infty, 2)$	$(2, 3)$	$(3, \infty)$
Test value	0	2.5	4
Sign of $x - 2$	$-$	$+$	$+$
Sign of $x - 3$	$-$	$-$	$+$
Sign of $(x-2)(x-3)$	$+$	$-$	$+$

Since we want the interval where the product is positive, the solution is $(-\infty, 2) \cup (3, \infty)$.

10. Solve the inequality $\dfrac{2x}{x^2+1} > 0$.

$\dfrac{2x}{x^2+1} > 0$

$2x = 0 \quad \Leftrightarrow \quad x = 0$

Since $x^2 + 1 \neq 0$, this does not factor any further.

Interval	$(-\infty, 0)$	$(0, \infty)$
Test value	-1	1
Sign of $2x$	$-$	$+$
Sign of $x^2 + 1$	$+$	$+$
Sign of $\dfrac{2x}{x^2+1}$	$-$	$+$

Since we want the interval where the quotient is positive, the solution is $(0, \infty)$.

11. Solve the inequality $\dfrac{9}{x} \leq x$.

$\dfrac{9}{x} \leq x \quad \Leftrightarrow \quad \dfrac{9}{x} - x \leq 0$

Bring all terms to a common dominator, then factor.

$\dfrac{9 - x^2}{x} \leq 0 \quad \Leftrightarrow \quad \dfrac{(3-x)(3+x)}{x} \leq 0$

$\begin{aligned} 3 - x &= 0 & 3 + x &= 0 & x &= 0 \\ 3 &= x & x &= -3 \end{aligned}$

Interval	$(-\infty, -3)$	$(-3, 0)$	$(0, 3)$	$(3, \infty)$
Test value	-5	-1	1	5
Sign of $3 - x$	$+$	$+$	$+$	$-$
Sign of $3 + x$	$-$	$+$	$+$	$+$
Sign of x	$-$	$-$	$+$	$+$
Sign of $\dfrac{9 - x^2}{x}$	$+$	$-$	$+$	$-$

Since the inequality included the equality, we first check the endpoints. The endpoint $x = 0$ is excluded from the solution since it makes a denominator 0. Since we want the interval where the quotient is negative, the solution is $[-3, 0) \cup [3, \infty)$.

12. Solve the inequality $\dfrac{4}{x-1} \geq \dfrac{3}{x-4}$.

$\dfrac{4}{x-1} \geq \dfrac{3}{x-4} \quad\Leftrightarrow\quad \dfrac{4}{x-1} - \dfrac{3}{x-4} \geq 0 \quad\Leftrightarrow$

$\dfrac{4(x-4)}{(x-1)(x-4)} - \dfrac{3(x-1)}{(x-4)(x-1)} \geq 0$

$\dfrac{4x-16-3x+3}{(x-1)(x-4)} \geq 0 \quad\Leftrightarrow\quad \dfrac{x-13}{(x-1)(x-4)} \geq 0$

$x - 13 = 0 \qquad x - 1 = 0 \qquad x - 4 = 0$

$x = 13 \qquad x = 1 \qquad x = 4$

Interval	$(-\infty, 1)$	$(1, 4)$	$(4, 13)$	$(13, \infty)$
Test value	0	2	10	15
Sign of $x - 13$	$-$	$-$	$-$	$+$
Sign of $x - 1$	$-$	$+$	$+$	$+$
Sign of $x - 4$	$-$	$-$	$+$	$+$
Sign of $\dfrac{x-13}{(x-1)(x-4)}$	$-$	$+$	$-$	$+$

Since the inequality included the equality, we first check the endpoints. The endpoints $x = 1$ and $x = 4$ are excluded from the solution since each makes a denominator 0. Since we want the interval where the quotient is positive, the solution is $(1, 4) \cup [13, \infty)$.

13. Solve the inequality $\dfrac{x^2 - 11}{x + 1} \leq 1$.

$\dfrac{x^2 - 11}{x+1} \leq 1 \quad\Leftrightarrow\quad \dfrac{x^2-11}{x+1} - 1 \leq 0 \quad\Leftrightarrow$

Move all terms to one side.

$\dfrac{x^2-11}{x+1} - \dfrac{x+1}{x+1} \leq 0 \quad\Leftrightarrow\quad \dfrac{x^2-x-12}{x+1} \leq 0$

Get a common denominator, simplify, and factor.

$\dfrac{(x-4)(x+3)}{x+1} \leq 0$

$x - 4 = 0 \qquad x + 3 = 0 \qquad x + 1 = 0$

$x = 4 \qquad x = -3 \qquad x = -1$

Interval	$(-\infty, -3)$	$(-3, -1)$	$(-1, 4)$	$(4, \infty)$
Test value	-5	-2	0	5
Sign of $x - 4$	$-$	$-$	$-$	$+$
Sign of $x + 3$	$-$	$+$	$+$	$+$
Sign of $x + 1$	$-$	$-$	$+$	$+$
Sign of $\dfrac{x^2-x-12}{x+1}$	$-$	$+$	$-$	$+$

Since the inequality included the equality, we first check the endpoints. The endpoints $x = 1$ is excluded from the solution since it makes a denominator 0. Since we want the interval where the quotient is negative, the solution is $(-\infty, -3] \cup (-1, 4]$.

E. Equations involving absolute values are solved using the property $|x| = c$ is equivalent to $x = \pm c$. To solve inequalities that involve absolute value we use the following properties.

Inequality	Equivalent form	Graph
1. $\quad |x| < c$	$-c < x < c$	
2. $\quad |x| \leq c$	$-c \leq x \leq c$	
3. $\quad |x| > c$	$x < -c$ or $c < x$	
4. $\quad |x| \geq c$	$x \leq -c$ or $c \leq x$	

14. Solve $|x + 6| = 8$.

$$x + 6 = 8 \quad \text{or} \quad x + 6 = -8$$
$$x = 2 \qquad\qquad x = -14$$

15. Solve $|3x + 5| < 7$ and graph the solution.

$$|3x + 5| < 7$$
$$-7 < 3x + 5 < 7$$
$$-12 < 3x < 2$$
$$-4 < x < \tfrac{2}{3}$$

Check by substituting values of x that are inside the interval and outside the interval.

16. Solve $|5 - 2x| \geq 9$ and graph the solution.

$$|5 - 2x| \geq 9$$
$$5 - 2x \geq 9 \quad \text{or} \quad 5 - 2x \leq -9$$
$$-2x \geq 4 \qquad\qquad -2x \leq -14$$
$$x \leq -2 \qquad\qquad x \geq 7$$

17. Solve $\left| \dfrac{3}{2x+1} \right| \geq 2$ and graph the solution.

$\left| \dfrac{3}{2x+1} \right| \geq 2$

$\dfrac{|3|}{|2x+1|} \geq 2$

$3 \geq 2|2x+1|$

$\tfrac{3}{2} \geq |2x+1|$ *Express in the $|\ \ | \leq c$ form.*

$|2x+1| \leq \tfrac{3}{2}$

$-\tfrac{3}{2} \leq 2x+1 \leq \tfrac{3}{2}$

$-\tfrac{5}{2} \leq 2x \leq \tfrac{1}{2}$

$-\tfrac{5}{4} \leq x \leq \tfrac{1}{4}$

However, $\left| \dfrac{3}{2x+1} \right|$ is not defined for $x = -\tfrac{1}{2}$.

Therefore, the solution is $-\tfrac{5}{4} \leq x \leq \tfrac{1}{4}$, $x \neq -\tfrac{1}{2}$.

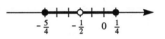

F. Applied problems sometimes lead to nonlinear inequalities.

18. After a night of gambling, a person finds they have only 20 coins left. If they only have quarters and dimes left, what is the most (and least) they have left?

They have between

$20 \times 0.10 \leq x \leq 20 \times 0.25$

$2 \leq x \leq 5$.

So they have between \$2 and \$5.

Section 1.8   Coordinate Geometry

Key Ideas

A. Coordinate plane.

B. Distance formula and Midpoint formula.

C. The graph of an equation.

D. Intercepts.

E. Equation of a circle.

F. Symmetry of graphs.

A. The x-y coordinate system is called the **rectangular coordinate system** or the **Cartesian coordinate system**. The plane used with this coordinate system is called the **coordinate plane** or the **Cartesian plane**. The points are given by a unique **ordered pair** of numbers (x-coordinate, y-coordinate).

B. The **distance** between the points $P_1(x_1, y_1)$ and $P_2(x_2, y_2)$ is given by the formula
$d(P_1, P_2) = \sqrt{(x_2 - x_1)^2 + (y_2 - y_1)^2}$.

The **midpoint** of the line segment joining $P_1(x_1, y_1)$ and $P_2(x_2, y_2)$ is $\left(\dfrac{x_1 + x_2}{2}, \dfrac{y_1 + y_2}{2} \right)$.

1. Find the distance between the following pairs of points.

(a) $(4, 7)$ and $(5, 5)$

$$
\begin{aligned}
d &= \sqrt{(x_2 - x_1)^2 + (y_2 - y_1)^2} \\
&= \sqrt{(5 - 4)^2 + (5 - 7)^2} \\
&= \sqrt{(1)^2 + (-2)^2} = \sqrt{1 + 4} \\
&= \sqrt{5}
\end{aligned}
$$

(b) $(-2, 3)$ and $(1, -4)$

$$
\begin{aligned}
d &= \sqrt{(x_2 - x_1)^2 + (y_2 - y_1)^2} \\
&= \sqrt{(1 - (-2))^2 + (-4 - 3)^2} \\
&= \sqrt{(3)^2 + (-7)^2} = \sqrt{9 + 49} \\
&= \sqrt{58}
\end{aligned}
$$

(c) $(-5, -1)$ and $(7, -6)$

$$
\begin{aligned}
d &= \sqrt{(x_2 - x_1)^2 + (y_2 - y_1)^2} \\
&= \sqrt{(7 - (-5))^2 + ((-6) - (-1))^2} \\
&= \sqrt{(12)^2 + (-5)^2} = \sqrt{144 + 25} \\
&= \sqrt{169} = 13
\end{aligned}
$$

44

2. Which point is closer to the origin?
   (a) $(5, 4)$ or $(-3, -3)$

   Distance from $(5, 4)$ to the origin is
   $$d = \sqrt{(5-0)^2 + (4-0)^2} = \sqrt{25+16} = \sqrt{41}.$$

   Distance from $(-3, -3)$ to the origin is
   $$d = \sqrt{(-3-0)^2 + (-3-0)^2} = \sqrt{9+9} = \sqrt{18}.$$
   $$= 3\sqrt{2}$$

   Since $\sqrt{18} < \sqrt{41}$, $(-3, -3)$ is closer to the origin.

   (b) $(7, 3)$ or $(-6, 5)$

   Distance from $(7, 3)$ to the origin is
   $$d = \sqrt{(7-0)^2 + (3-0)^2} = \sqrt{49+9} = \sqrt{56}.$$

   Distance from $(-6, 5)$ to the origin is
   $$d = \sqrt{(-6-0)^2 + (5-0)^2} = \sqrt{36+25} = \sqrt{61}.$$

   Since $\sqrt{56} < \sqrt{61}$, $(7, 3)$ is closer to the origin.

3. Find the midpoint of the segment joining P_1 and P_2.
   (a) $P_1 = (1, 5)$ and $P_2 = (-3, -3)$

   $$\left(\frac{1 + (-3)}{2}, \frac{5 + (-3)}{2} \right) = \left(\frac{-2}{2}, \frac{2}{2} \right) = (-1, 1)$$

   (b) $P_1 = (7, -2)$ and $P_2 = (5, 8)$

   $$\left(\frac{7 + 5}{2}, \frac{(-2) + 8}{2} \right) = \left(\frac{12}{2}, \frac{6}{2} \right) = (6, 3)$$

   (c) $P_1 = (-5, 4)$ and $P_2 = (-2, 1)$

   $$\left(\frac{(-5) + (-2)}{2}, \frac{4 + 1}{2} \right) = \left(\frac{-7}{2}, \frac{5}{2} \right)$$

C. The graph of an equation in x and y is the set of all points (x, y) that satisfy the equation. The point (x, y) lies on the graph of the equation if and only if its coordinates satisfy the equation.

4. Make a table and sketch the graph of the equation $2x + y = 5$.

Start by solving $2x + y = 5$ for y: $y = -2x + 5$.
Next put in different values for x and find the corresponding y value.
Make a list and plot these points.

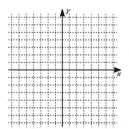

x	y	(x, y)
-2	9	$(-2, 9)$
-1	7	$(-1, 7)$
0	5	$(0, 5)$
1	3	$(1, 3)$
2	1	$(2, 1)$
3	-1	$(3, -1)$

5. Make a table and sketch the graph of the equation $x = 2y^2 - 1$.

Since the equation is already solved for x, place in values for y and find the corresponding x value.
Make a list and plot these points.

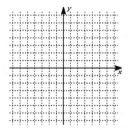

y	x	(x, y)
-2	7	$(7, -2)$
-1	1	$(1, -1)$
0	-1	$(-1, 0)$
1	1	$(1, 1)$
2	7	$(7, 2)$
3	17	$(17, 3)$

D. The x-coordinates of points where the graph of an equation intersects the x-axis is called the **x-intercept**. The y-coordinates of points where the graph of an equation intersects the y-axis is called the **y-intercept**.

	Intercepts	How to find them	Where they are on the graph
x-intercept	The x-coordinates of the points where the graph intersects the x-axis.	Set $y = 0$ and solve for x.	
y-intercept	The y-coordinates of the points where the graph intersects the y-axis.	Set $x = 0$ and solve for y.	

6. Find the x- and y-intercepts of the graph of the given equation.

(a) $y = 2x - 3$

x-intercept: set $y = 0$ and solve for x.
$0 = 2x - 3 \quad \Leftrightarrow \quad 2x = 3 \quad \Leftrightarrow \quad x = \frac{3}{2}$
So the x-intercept is $\frac{3}{2}$.

y-intercept: set $x = 0$ and solve for y.
$y = 2(0) - 3 = -3$
So the y-intercept is -3.

(b) $x^2 - 4x - 2xy = y + 5$

x-intercept: set $y = 0$ and solve for x.
$x^2 - 4x - 2x(0) = (0) + 5 \quad \Leftrightarrow \quad x^2 - 4x = 5$
$x^2 - 4x - 5 = (x - 5)(x + 1) = 0$

$x - 5 = 0 \qquad\qquad x + 1 = 0$
$\quad x = 5 \qquad\qquad\quad x = -1$

So the x-intercepts are 5 and -1.

y-intercept: set $x = 0$ and solve for y.
$(0)^2 - 4(0) - 2(0)y = y + 5 \quad \Leftrightarrow \quad 0 = y + 5 \Leftrightarrow$
$y = -5$
So the y-intercept is -5.

(c) $x = 2y^2 - 1$

x-intercept: set $y = 0$ and solve for x.
$x = 2(0)^2 - 1 \quad \Leftrightarrow \quad x = -1$
So the x-intercept is -1.

y-intercept: set $x = 0$ and solve for y.
$0 = 2y^2 - 1 \quad \Leftrightarrow \quad 2y^2 = 1 \quad \Rightarrow \quad y = \pm\sqrt{\frac{1}{2}}$
So the y-intercepts are $\pm\sqrt{\frac{1}{2}}$.

E. The **general equation of a circle** with radius r and centered at (h, k) is: $\mathbf{(x - h)^2 + (y - k)^2 = r^2}$. When the center is the origin, $(0, 0)$, this equation becomes $x^2 + y^2 = r^2$.

7. Determine the graph of $x^2 + y^2 = 49$.

Write as $x^2 + y^2 = 7^2$.
We see that this is the equation of a circle of radius 7, centered at the origin.

8. Find an equation of the circle with radius 8 units centered at $(2, -4)$.

Placing $h = 2$, $k = -4$, and $5 = 8$, we have:
$$(x - 2)^2 + (y - (-4))^2 = 8^2$$
$$(x - 2)^2 + (y + 4)^2 = 64$$

9. Determine the center and the radius of the circle given by $x^2 + y^2 + 6x - 4y - 68 = 0$.

Group terms by variable; complete the square.
$$x^2 + y^2 + 6x - 4y - 68 = 0$$
$$x^2 + 6x + \underline{} + y^2 - 4y + \underline{} = 68$$
$$x^2 + 6x + \mathbf{9} + y^2 - 4y + \mathbf{4} = 68 + \mathbf{9} + \mathbf{4}$$
$$(x + 3)^2 + (y - 2)^2 = 81 = 9^2$$

Center is at $(-3, 2)$.
Radius is 9 units.

F. Symmetry is an important tool that can be used in graphing equations.

> (1) A curve is **symmetric with respect to the x-axis** if its equation is unchanged when y is replaced by $-y$.
> (2) A curve is **symmetric with respect to the y-axis** if its equation is unchanged when x is replaced by $-x$.
> (3) A curve is **symmetric with respect to the origin** if its equation is unchanged when x is replaced by $-x$ and y is replaced by $-y$.

10. Determine which kind of symmetry the graph of the equation $x - y^2 = 6$ has.

Check each type of symmetry.
"with respect to the x-axis"
Yes, since $x - (-y)^2 = 6 \iff x - y^2 = 6$
is the same equation.

"with respect to the y-axis"
NO. $(-x) - y^2 = -x - y^2 \neq x - y^2$

"with respect to the origin"
NO. $(-x) - (-y)^2 = -x - y^2 \neq x - y^2$

11. Determine which kind of symmetry the graph of the equation $x^2 + y^2 - 6y = 2$ has.

Check each type of symmetry.
"with respect to the x-axis"
NO, since
$x^2 + (-y)^2 - 6(-y) = x^2 + y^2 + 6y$ and
$x^2 + y^2 + 6y \neq x^2 + y^2 - 6y$.

"with respect to the y-axis"
Yes, since $(-x)^2 + y^2 - 6y = x^2 + y^2 - 6y = 2$.

"with respect to the origin"
NO. $(-x)^2 + (-y)^2 - 6(-y) = x^2 + y^2 + 6y$
and $x^2 + y^2 + 6y \neq x^2 + y^2 - 6y$

12. Determine which kind of symmetry the graph of the equation $x^4 + 3y^2 = 6$ has.

Check each type of symmetry.

"with respect to the x-axis"
Yes, since $x^4 + 3(-y)^2 = 6 \quad \Leftrightarrow \quad x^4 + 3y^2 = 6$ is the same equation.

"with respect to the y-axis"
Yes, since $(-x)^4 + 3y^2 = x^4 + 3y^2 = 6$.

"with respect to the origin"
Yes, since $(-x)^4 + 3(-y)^2 = x^4 + 3y^2 = 6$.

Section 1.9   Solving Equations and Inequalities Graphically

Key Ideas
A. Using a graphing calculator.
B. Solving equations graphically.
C. Solving an inequality graphically.

A. Calculators and computers graph by plotting points. Sometimes the points are connected by straight lines, other times just the points are displayed. The viewing area of a calculator or computer is referred to as the **viewing rectangle**. This is an $[a, b]$ by $[c, d]$ portion of the coordinate axis. Choosing the *appropriate* viewing rectangle is one of the most important aspects of using a graphing calculator. Choose a relatively large viewing rectangle to obtain a global view of the graphs. Then to investigate details you must zoom to a smaller viewing rectangle that shows just the details of interest.

1. Use a graphing calculator to draw the graph of the equation $f(x) = \sqrt{x^2 + 9}$ in the following viewing rectangles.

    (a) $[-3, 3]$ by $[-3, 3]$

    (b) $[-5, 5]$ by $[-5, 5]$

    (c) $[-5, 5]$ by $[0, 10]$

The solutions shown below are representations of what you should see on your graphing calculator.

Only the point $(0, 3)$ should appear, and it will be lost in the y-axis.

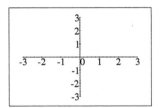

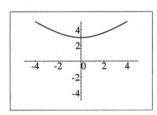

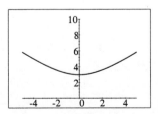

2. Graph the equation below on a graphing calculator.
   $y = |x^2 - 4|$

We graph the equation $y = \text{abs}(x^2 - 4)$ in the viewing rectangle $[-5, 5]$ by $[-5, 5]$.

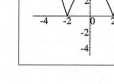

3. Graph the equation $9x^2 + 4y^2 = 36$ using a graphing calculator with each of the viewing rectangles. Then sketch the graph below.
   (a) $[-2, 2]$ by $[-3, 3]$

We solve for y: $9x^2 + 4y^2 = 36$ $\Leftrightarrow$
$4y^2 = 36 - 9x^2$ $\Leftrightarrow$ $y^2 = 9 - \frac{9}{4}x^2$ $\Rightarrow$
$y = \pm\sqrt{9 - \frac{9}{4}x^2}$. Top half: $y = \sqrt{9 - \frac{9}{4}x^2}$;

Bottom half: $y = -\sqrt{9 - \frac{9}{4}x^2}$.

This example shows how the viewing rectangle can give different impressions of the same equation.

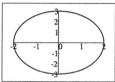

   (b) $[-3, 3]$ by $[-6, 6]$

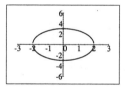

   (c) $[-6, 6]$ by $[-4, 4]$

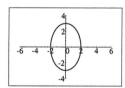

   (d) Now graph the equation.

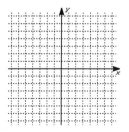

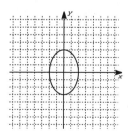

B. In the **algebraic method**, we use the rules of algebra to isolate x on one side of the equation. In the **graphical method**, we move all terms to one side and set the expression equal to y. We then sketch the graph of the equation and find the values of x for which $y = 0$. Another method is to set each side of the equation equal to y, then sketch the graph of the equations and determine where the graphs intersect. Solutions found using a graphing calculator are only approximations. Exact solutions can only be confirmed by checking.

4. Solve both algebraically and graphically.
$$y = 4x^2 - 3$$
$$y = x^2 - 6x + 9$$
Algebraically.

Set the equations equal to each other.
$$4x^2 - 3 = x^2 - 6x + 9$$
$$3x^2 + 6x - 12 = 0 \qquad \textit{Set equal to zero.}$$
$$x^2 + 2x - 4 = 0 \qquad \textit{Use the quadratic equation.}$$
$$x = \frac{-2 \pm \sqrt{2^2 - (4)(1)(4)}}{2(1)} = \frac{-2 \pm \sqrt{20}}{2}$$
$$= -1 \pm \sqrt{5}$$
Thus $x = -1 + \sqrt{5}$ or $x = -1 - \sqrt{5}$.

Graphically.

We start by graphing the two equations in the viewing rectangle
$[-10, 10]$ by $[-2, 50]$. It appears that these equations have two points of intersection.

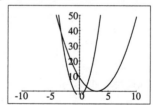

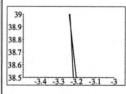

In the viewing rectangle
$[-3.5, -3]$ by $[38.5, 39]$
we find the solution
$x \approx -3.236$.

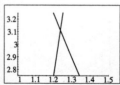

In the viewing rectangle
$[[1, 1.5]$ by $[2.75, 3.25]$
we find the solution
$x \approx 1.236$.

Note: $x = -1 + \sqrt{5} \approx 1.236$ and
$x = -1 - \sqrt{5} \approx -3.236$.

5. Solve the equation graphically.

$$x^3 + 3x^2 - 3 = 0$$

We graph the equation $y = x^3 + 3x^2 - 3$ in the viewing rectangle $[-5, 5]$ by $[-5, 5]$. The solutions occur at the x-intercepts. Using the ZOOM and TRACE, we can approximate the solutions as $x = -2.53$, $x = -1.35$, and $x = 0.88$.

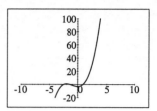

6. Solve $x^3 + 3 = \dfrac{2}{x^2 + 2}$ by setting the equation equal to zero.

$$x^3 + 3 = \frac{2}{x^2 + 2} \qquad \textit{Move all terms to one side of the equation}$$

$$x^3 + 3 - \frac{2}{x^2 + 2} = 0 \qquad \textit{Set the result equal to y}$$

$$y = x^3 + 3 - \frac{2}{x^2 + 2} \qquad \textit{Graph}$$

We graph $y = x^3 + 3 - \dfrac{2}{x^2 + 2}$ in the viewing rectangle $[-5, 5]$ by $[-5, 5]$. The solutions occur at the x-intercepts of the equation. The approximate solution is $x = -1.35$.

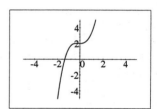

7. Solve $x^3 + 3 = \dfrac{2}{x^2 + 2}$ by graphing each side of the equation.

We graph $y = x^3 + 3$ and $y = \dfrac{2}{x^2 + 2}$ in the viewing rectangle $[-5, 5]$ by $[-5, 5]$. The solutions occur at the intercepts of the two equation. Again the approximate solution is $x = -1.35$.

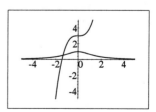

C. We solve inequalities graphically by graphing each side of the inequality in the same viewing rectangle. We determine which graph goes with which equation by evaluating each equation at an x value. Then determine where the graph of one equation lies above the graph of the other.

8. Solve the inequality $\sqrt{x^2 + x + 2} > 3 - x$ graphically.

We set $y_1 = \sqrt{x^2 + x + 2}$ and $y_2 = 3 - x$ in the viewing rectangle $[-5, 5]$ by $[-10, 10]$. From the graph, we see that $y_1 < y_2$ when $x > 1$. So the solution is $(1, \infty)$.

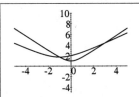

9. Solve the inequality $\sqrt{x^2 + 2x + 4} \geq \sqrt{2x^2 + 1}$ graphically.

We set $y_1 = \sqrt{x^2 + 2x + 4}$ and $y_2 = \sqrt{2x^2 + 1}$ in the viewing rectangle $[-5, 5]$ by $[-5, 10]$. From the graph, we see that $y_1 \geq y_2$ when $-1 \leq x \leq 3$, so the solution is $[-1, 3]$.

10. An umbrella manufacturer estimates that the cost per week of producing x umbrellas is $0.01x^2 + 4x + 800$. How many umbrellas do they need to manufacture each week so that the *average* cost per umbrella is less than \$12?

The average cost per umbrella manufactured during the week is $\dfrac{0.01x^2 + 4x + 800}{x}$. So we need to solve $\dfrac{0.01x^2 + 4x + 800}{x} < 12$. We set $y_1 = \dfrac{0.01x^2 + 4x + 800}{x}$ and $y_2 = 12$ and graph both equations in the viewing rectangle $[0, 800]$ by $[0, 50]$. Note x represents the number of umbrellas manufactured during the week, so $x > 0$. From the viewing rectangle, the average cost is less than \$12 when $117.2 < x < 682.8$. Thus they must manufacture between 118 and 682 umbrellas.

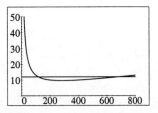

54

Section 1.10 Lines

Key Ideas

A. Slope of a line.

B. Point-slope equation.

C. Slope-intercept equation.

D. Other line equations.

E. Parallel and perpendicular lines.

A. Slope is an extremely important concept in mathematics. Slope is a central focus in the study of calculus. The slope of a nonvertical line that passes through the points $P_1(x_1, y_1)$ and $P_2(x_2, y_2)$ is

$$m = \frac{y_2 - y_1}{x_2 - x_1} = \frac{\text{rise}}{\text{run}} = \frac{\text{change in } y}{\text{change in } x}.$$ The slope of a vertical line is not defined.

1. Find the slope of the line that passes through the points P and Q.

    (a) $P(1, 3)$ and $Q(-2, 9)$

$$m = \frac{9 - 3}{-2 - 1} = \frac{6}{-3} = -2$$

$$\text{or } m = \frac{3 - 9}{1 - (-2)} = \frac{-6}{3} = -2$$

The order the points are taken is not important, but be consistent.

    (b) $P(2, -5)$ and $Q(-3, 4)$

$$m = \frac{4 - (-5)}{-3 - 2} = \frac{9}{-5} = -\frac{9}{5}$$

    (c) $P(-7, -4)$ and $Q(-4, -3)$

$$m = \frac{-3 - (-4)}{-4 - (-7)} = \frac{1}{3}$$

    (d) $P(5, 2)$ and $Q(-3, 2)$

$$m = \frac{2 - 2}{-3 - 5} = \frac{0}{-8} = 0$$

Note: $\dfrac{0}{-8} = 0$ *is defined but* $\dfrac{-8}{0}$ *is not defined; many students get these two concepts confused.*

    (e) $P(4, 2)$ and $Q(4, -1)$

$$m = \frac{-1 - 2}{4 - 4} = \frac{-3}{0} \text{ which is undefined.}$$

So this line has no slope (it's a vertical line).

B. The following is the **point-slope form** of the equation of the line that passes through the point (x_1, y_1) and has slope m: $y - y_1 = m(x - x_1)$. This is a formula you should learn.

2. Find an equation of the line that passes through the point $(2, -1)$ with slope $\frac{1}{3}$. Sketch the line.

Substitute into the equation $y - y_1 = m(x - x_1)$ with $(x_1, y_1) = (2, -1)$ and $m = \frac{1}{3}$.
$$y - (-1) = \tfrac{1}{3}(x - 2)$$
$$y + 1 = \tfrac{1}{3}(x - 2)$$

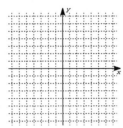

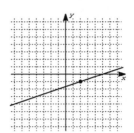

3. Find an equation of the line that passes through the point $(1, 3)$ with slope $-\frac{2}{5}$. Sketch the line.

Substitute into the equation $y - y_1 = m(x - x_1)$ with $(x_1, y_1) = (1, 3)$ and $m = -\frac{2}{5}$.
$$y - 3 = -\tfrac{2}{5}(x - 1)$$

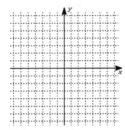

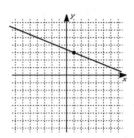

C. The following is the **slope-intercept form** of the equation of a line: $y = mx + b$. This is a special form of the point-slope equation, where the y-intercept b is the point $(0, b)$. Again, this is a formula you should learn.

4. Find an equation of the line with slope -2 and and y-intercept 5.

Substituting into $y = mx + b$ we get:

$y = -2x + 5$.

5. Find an equation of the line with slope $\frac{4}{5}$ and and y-intercept -3.

Substituting into $y = mx + b$ we get:

$y = \tfrac{4}{5}x - 3$.

D. The **general equation** of a line is the equation $ax + by = c$. Every line can be expressed in this form. **Vertical lines** have the form $x = c$ and have *no* slope. **Horizontal lines** have the form $y = d$ and have *zero* slope.

6. Find an equation of the vertical line that passes through the point $(2, -1)$.

$x = 2$

7. Find the slope and y-intercept of the line $3x - 2y = 0$.

Solve for y.
$$3x - 2y = 0$$
$$-2y = -3x$$
$$y = \tfrac{3}{2}x$$
Slope $m = \tfrac{3}{2}$, y-intercept $= 0$

8. Find the slope and y-intercept of the line $2x + 5y = -24$.

Solve for y.
$$2x + 5y = -24$$
$$5y = -2x - 24$$
$$y = \frac{-2x - 24}{5} \qquad \text{\textit{Remember to divide}}$$
$$\textit{everything by } 5$$
$$y = -\tfrac{2}{5}x - \tfrac{24}{5} \qquad \tfrac{a+b}{c} = \tfrac{a}{c} + \tfrac{b}{c}$$
Slope $m = -\tfrac{2}{5}$, y-intercept $= -\tfrac{24}{5}$

E. Two lines are **parallel** if and only if they have the same slope (or both have no slope). Two lines with slopes m_1 and m_2 are **perpendicular** if and only if $m_1 m_2 = -1$, that is, their slopes are negative reciprocals, so $m_2 = -\dfrac{1}{m_1}$. Also, horizontal lines (0 slope) are perpendicular to vertical lines (no slope).

9. Determine if each pair of lines are parallel, perpendicular, or neither.
   (a) $2x + 6y = 7$ and $x + 3y = 9$

Solve each equation for y in order to put each line into slope-intercept form.
$$2x + 6y = 7 \qquad\qquad x + 3y = 9$$
$$6y = -2x + 7 \qquad\qquad 3y = -x + 9$$
$$y = -\tfrac{1}{3}x + \tfrac{7}{6} \qquad\qquad y = -\tfrac{1}{3}x + 3$$
Both lines have the same slope, $-\tfrac{1}{3}$, and different y-intercepts, thus they are parallel.

(b) $5x + 2y = 0$ and $-5x + 2y = 2$

Solve each equation for y in order to put each line into slope-intercept form.

$$5x + 2y = 0 \qquad\qquad -5x + 2y = 2$$
$$2y = -5x \qquad\qquad 2y = 5x + 2$$
$$y = -\tfrac{5}{2}x \qquad\qquad y = \tfrac{5}{2}x + 1$$

Since $\left(-\tfrac{5}{2}\right) \cdot \left(\tfrac{5}{2}\right) \neq -1$ and the slopes are not equal, so the lines are neither parallel nor perpendicular.

(c) $3x - y = 10$ and $x + 3y = 9$

Solve each equation for y in order to put each line into slope-intercept form.

$$3x - y = 10 \qquad\qquad x + 3y = 9$$
$$-y = -3x + 10 \qquad\qquad 3y = -x + 9$$
$$y = 3x - 10 \qquad\qquad y = -\tfrac{1}{3}x + 3$$

Since $3 = \dfrac{-1}{-1/3}$, the lines are perpendicular.

10. Are $A(1, 7)$, $B(7, 9)$, $C(9, 3)$, and $D(3, 1)$ the vertices of a rectangle, or parallelogram, or neither.

Plot the points to find their relative position. Find the slope of the lines passing through each pair of points.

slope of $AB = \frac{9-7}{7-1} = \frac{1}{3}$

slope of $BC = \frac{3-9}{9-7} = -3$

slope of $CD = \frac{1-3}{3-9} = \frac{1}{3}$

slope of $DA = \frac{7-1}{1-3} = -3$

Since AB and CD are parallel (denoted by $AB \parallel CD$) and BC and DA are parallel, $ABCD$ is at least a parallelogram. Since AB is perpendicular to BC (denoted by $AB \perp BC$), this is a rectangle.

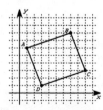

Chapter 2
Functions

Section 2.1 What is a Function?

Key Ideas

A. Functions.

B. Domains and ranges.

A. A **function** f is a rule that assigns to each element x in a set A exactly one element, called $f(x)$, in a set B. That is, $x \in A$ and $f(x) \in B$. Set A is called the **domain** and set B is called the **range**. "$f(x)$" is read as "f of x" or "f at x" and represents the **value of f at x**, or the **image of x under f**. Since $f(x)$ depends on the value x, $f(x)$ is the called the **dependent variable** and x is called the **independent variable**.

1. Express the rule in functional notation.

   (a) Add 6 then divide the sum by 5.

$$f(x) = \frac{x + 6}{5}$$

   (b) Take the square root and subtract 9.

$$f(x) = \sqrt{x} - 9$$

2. Find the values of f defined by the equation
$$f(x) = \frac{3x^2 + x}{2 - x}.$$

   (a) $f(-2)$

   Substituting -2 for x
$$f(-2) = \frac{3(-2)^2 + (-2)}{2 - (-2)} = \frac{12 - 2}{2 + 2} = \frac{10}{4} = \frac{5}{2}.$$

   (b) $f(4)$

   Substituting 4 for x
$$f(4) = \frac{3(4)^2 + (4)}{2 - (4)} = \frac{48 + 4}{2 - 4} = \frac{52}{-2} = -26.$$

   (c) $f(0)$

   Substituting 0 for x
$$f(0) = \frac{3(0)^2 + (0)}{2 - (0)} = \frac{0 - 0}{2 - 0} = \frac{0}{2} = 0.$$

B. If not explicitly stated, the **domain** of a function is the set of all real numbers for which the function makes sense and defines a real number. The domain is the 'input' to the function. The **range** is the set of all possible values, $f(x)$, as x varies throughout the domain of f. The range is sometimes called the **image**, because it is the "image of x under the rule f." The range depends on the domain, so the range is referred to as the *dependent* variable. The range is the 'output' of the function.

There are four ways to describe a specific function: **verbally** - *by a description in words*, **algebraically** - *by an explicit formula*, **visually** - *by a graph*, **numerically** - *by a table of values.*

3. Find the domain of the function $f(x) = \dfrac{4}{x+2}$.

> The only values of x for which f does not make sense occur when the denominator is 0, that is, when $x + 2 = 0$ or $x = -2$. So the domain of f is $(-\infty, -2) \cup (-2, \infty)$. The domain can also be expressed as: $\{x \mid x \neq -2\}$ or $x \neq -2$.

4. Find the domain of the function $f(x) = \dfrac{x-9}{x^3 - 25x}$.

> Again, the only values of x for which f does not make sense occur when the denominator is 0.
> $$x^3 - 25x = 0 \qquad \textit{Factor.}$$
> $$x(x-5)(x+5) = 0$$
> $$x = 0 \qquad x - 5 = 0 \qquad x + 5 = 0$$
> $$x = 5 \qquad x = -5$$
> Domain is $\{x \mid x \neq -5, x \neq 0, x \neq 5\}$.
> The domain can also be expressed as:
> $(-\infty, -5) \cup (-5, 0) \cup (0, 5) \cup (5, \infty)$.

5. Find the domain of the function $f(x) = \sqrt{x^2 - 4}$.

> We are looking for the values of x for which the function makes sense and *defines a real number*. Since we can only find the square root of a nonnegative number, using the methods from Section 1.7, we have
> $$x^2 - 4 \geq 0$$
> $$(x-2)(x+2) \geq 0.$$
>
> | Interval | $(-\infty, -2)$ | $(-2, 2)$ | $(2, \infty)$ |
> |---|---|---|---|
> | Test value | -5 | 0 | 5 |
> | Sign of $x - 2$ | $-$ | $-$ | $+$ |
> | Sign of $x + 2$ | $-$ | $+$ | $+$ |
> | Sign of $x^2 - 4$ | $+$ | $-$ | $+$ |
>
> Since the inequality includes the equality, we check the endpoints of the intervals and see that both $x = -2$ and $x = 2$ are acceptable as input values. Thus the domain is $(-\infty, -2] \cup [2, \infty)$.

Section 2.2   Graphs of Functions

Key Ideas

A. Graphs of linear functions and other functions.
B. Piecewise defined functions.
C. Vertical line test.

A. If f is a function with domain A, its **graph** is the set of ordered pairs, $\{(x, f(x)) \mid x \in A\}$. This means, the graph of f is all the points (x, y) where $y = f(x)$ and x is in the domain of f. A function defined by $f(x) = mx + b$ is called a **linear function** because the graph of the equation $y = mx + b$ is a line with slope m and y-intercept b. When $m = 0$, the linear function becomes the **constant function**, $f(x) = b$.

1. If $f(x) = 3x + 1$ find the domain and range of f and sketch the graph of f.

This is a linear function with slope 3 and y-intercept 1. Generate some points and plot.

x	$f(x)$
-1	-2
0	1
1	4
2	7

Domain: all real numbers.
Range: all real numbers.

Note: The domain for linear functions is all real numbers.

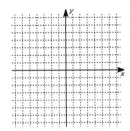

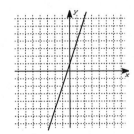

2. If $f(x) = \sqrt{2x + 1}$ find the domain and range of f and sketch the graph of f.

This is not a linear function. We need to find the domain *before* we can generate points to plot.
Domain: $2x + 1 \geq 0 \quad \Leftrightarrow \quad 2x \geq -1 \quad \Leftrightarrow$
$x \geq -\frac{1}{2}$

x	$f(x)$
$-\frac{1}{2}$	0
0	1
$\frac{3}{2}$	2
4	3

Domain: $[-\frac{1}{2}, \infty)$.
Range: $[0, \infty)$.

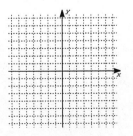

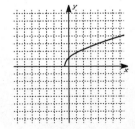

3. If $f(x) = -4$ find the domain and range of f and sketch the graph of f.

This is a constant function, so every input value gives the same output value, in this case -4.

x	$f(x)$
-100	-4
0	-4
5	-4

Domain: all real numbers.
Range: $\{-4\}$.

Note: The domain for constant functions is all real numbers.

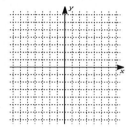

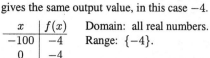

B. A function is called **piecewise defined** if it is defined differently for distinct intervals of its domain.

4. Sketch the graph of the function defined by:
$$f(x) = \begin{cases} x + 1 & \text{if } x < 0 \\ 2x^2 + 1 & \text{if } x \geq 0. \end{cases}$$

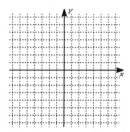

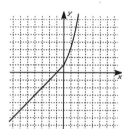

5. Sketch the graph of the function defined by:
$$f(x) = \begin{cases} \sqrt{-5 - 3x} & \text{if } x \leq -2 \\ \sqrt{5 + 2x} & \text{if } -2 < x < 2 \\ -x + 8 & \text{if } 2 \leq x. \end{cases}$$

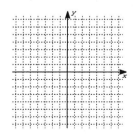

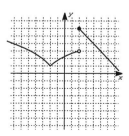

C. The **Vertical Line Test** states that "a curve in the plane is the graph of a function if and only if no vertical line intersects the curve more than once." It works because a function defines a unique value, $f(a)$, so if a vertical line, $x = a$, intersects a graph in more than one place the value of $f(a)$ is not unique. The vertical line test is a very easy test to apply.

6. Use the vertical line test to determine which curves are graphs of functions of x.

    (a)

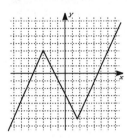

Passes the vertical line test. This is the graph of a function.

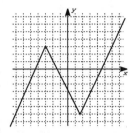

    (b)

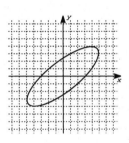

Fails the vertical line test. This is not the graph of a function.

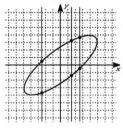

    (c)

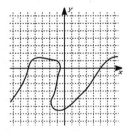

Fails the vertical line test. This is not the graph of a function.

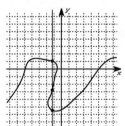

Section 2.3   Applied Functions: Variation

Key Ideas

A. Direct variation.

B. Inverse variation.

C. Joint variation.

A. A *mathematical model* is a function that describes, at least approximately, the dependence of one physical quantity on another physical quantity. In the direct variation model, **y is directly proportional to x** if x and y are related by the equation $y = kx$ for some constant $k \neq 0$. The constant k is called the constant of proportionality.

1. Federal excise tax on gasoline is 13¢ per gallon. Write an equation that directly relates the tax as a function of gallons.

Let $g =$ gallons of gas and let $T =$ the tax. Federal excise tax on gasoline is 13¢ per gallon. Since tax is 13¢ per gallon we have $T = 0.13g$. Here 0.13 is the constant of proportionality.

2. The volume of liquid in a soda can is directly related to the height of the liquid in the can.

  (a) If the can is 12.5 cm tall and contains 355 ml when full, determine the constant of proportionality and write the equation for the variation.

Let $h =$ height of the can, $V =$ volume of the can, and k be the constant of proportionality.
$$V = kh$$
When $h = 12.5$, $V = 355$. Substituting we have
$$355 = 12.5k \quad \Leftrightarrow \quad k = 28.4$$
Thus the model is $V = 28.4h$.

  (b) How much is left in the can when the liquid is 3 cm deep?

Setting $h = 3$ in the equation yields
$$V = 28.4h$$
$$= 28.4(3) = 85.2 \text{ ml.}$$

  (c) How deep is the liquid when there is 300 ml left?

Setting $V = 300$ and solving for h yields
$$V = 28.4h$$
$$300 = 28.4h$$
$$10.56 \text{ cm} = h.$$

B. In another model, **y is inversely proportional to x** if x and y are related by the equation $y = \dfrac{k}{x}$ for some constant $k \neq 0$.

3. Write an equation that relates the width to the length of all rectangles whose area is 4 square units.

Let w = width and let l = length.
Since $A = wl$ we have
$$4 = wl \quad \Leftrightarrow \quad w = \frac{4}{l}$$
Thus width is inversely proportional to length.

C. If one quantity is proportional to two or more quantities, we call this relationship, **jointly proportional** if x, y, and z are related by the equation $z = kxy$ for some constant $k \neq 0$.

4. Glu-lam beams are long beams used in construction. They are made by laminating 2 inch by 8 inch planks of wood together to create a wide beam.

(a) Write an equation that directly relates the volume of the wood used to the length and width of the glu-lam, both given in feet.

Let V = the volume of wood (ft^3);
l = length (in ft) and w = width (in ft).
Since the plank is 8 inches = $\frac{2}{3}$ foot thick.
$$V = \tfrac{2}{3}lw$$

(b) How much wood is in a 2 foot wide, 50 foot long glu-lam beam.

Substituting $w = 2$ and $l = 50$ we find
$$V = \tfrac{2}{3}lw = \tfrac{2}{3}(50)(2) = \tfrac{200}{3} \text{ ft}^3 \approx 66.7 \text{ ft}^3.$$

Section 2.4   Average Rate of Change: Increasing and Decreasing Functions

Key Ideas

A.   Average rate of change.

B.   Increasing and decreasing functions.

A.   The **average rate of change** of the function $y = f(x)$ between $x = a$ and $x = b$ is:

$$\text{average rate of change} = \frac{\text{change in } y}{\text{change in } x} = \frac{f(b) - f(a)}{b - a}$$

The average rate of change is the slope of the **secant line** that joins the points $(a, f(a))$ and $(b, f(b))$.

1.   A function is given. Determine the average rate of change of the function between the given values of the variable.

(a) $f(x) = 5x - 7$; $x = 2, x = 4$

$$\begin{aligned}
\text{average rate} \atop \text{of change} &= \frac{f(4) - f(2)}{4 - 2} \\[1mm]
&= \frac{(5 \cdot 4 - 7) - (5 \cdot 2 - 7)}{2} \\[1mm]
&= \frac{13 - 3}{2} = \frac{10}{2} = 5
\end{aligned}$$

(b) $g(x) = 3x - x^2$; $x = -1, x = 3$

$$\begin{aligned}
\text{average rate} \atop \text{of change} &= \frac{g(3) - g(-1)}{3 - (-1)} \\[1mm]
&= \frac{[3(3) - (3)^2] - [3(-1) - (-1)^2]}{3 + 1} \\[1mm]
&= \frac{(9 - 9) - (-3 - 1)}{4} \\[1mm]
&= \frac{0 - (-4)}{4} = \frac{4}{4} = 1
\end{aligned}$$

(c) $h(x) = x^3 - x$; $x = -1, x = 1$

$$\begin{aligned}
\text{average rate} \atop \text{of change} &= \frac{h(1) - h(-1)}{1 - (-1)} \\[1mm]
&= \frac{[(1)^3 - (1)] - [(-1)^3 - (-1)]}{1 + 1} \\[1mm]
&= \frac{(1 - 1) - (-1 + 1)}{2} = \frac{0 - 0}{2} = 0
\end{aligned}$$

2. A graph of a function is given. Determine the average rate of change of the function between the indicated values of the variable.

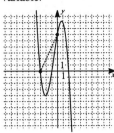

We use the points $(-3, 0)$ and $(0, 6)$.

$$\begin{array}{l} \text{average rate} \\ \text{of change} \end{array} = \frac{6 - 0}{0 - (-3)} = \frac{6}{3} = 2$$

B. Graphs of functions are described by what the function does as x increases over an interval in the domain.

A function f is called **increasing** on an interval I if $f(x_1) < f(x_2)$ whenever $x_1 < x_2$ in I.

A function f is called **decreasing** on an interval I if $f(x_1) > f(x_2)$ whenever $x_1 < x_2$ in I.

3. State the intervals in which the function whose graph is shown below is increasing or decreasing.

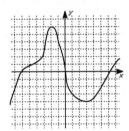

The function is increasing on $x \leq -2$ and on $3 \leq x$.

The function is decreasing on $-2 \leq x \leq 3$.

4. Use a graphing device to graph the function $f(x) = x^3 - x^2 - x + 2$. State approximately the intervals on which f is increasing and on which f is decreasing.

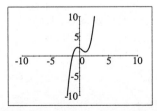

The function is increasing on $x \leq -0.33$ and on $1 \leq x$.

The function is decreasing on $-0.33 \leq x \leq 1$.

Section 2.5   Transformations of Functions
Key Ideas
A. Vertical shifts of graphs.
B. Horizontal shifts of graphs.
C. Vertical stretching, shrinking, and reflecting.
D. Horizontal stretching, shrinking, and reflecting.
E. Symmetry, odd, and even functions.

A. When $c > 0$, the equation $y = f(x) + c$ moves the graph of $y = f(x)$ vertically c units up, and the equation $y = f(x) - c$ moves the graph of $y = f(x)$ vertically c units down.

1. Graph the functions $f(x) = x^3$ and $g(x) = x^3 + 4$ on the same coordinate axis.

First graph $f(x) = x^3$. Then graph $g(x) = x^3 + 4$ by translating the graph of f up 4 units.

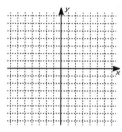

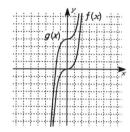

2. Graph the functions $f(x) = |x|$ and $g(x) = |x| - 2$ on the same coordinate axis.

First graph $f(x) = |x|$. Then graph $g(x) = |x| - 2$ by translating the graph of f down 2 units.

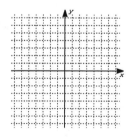

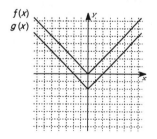

B. When $c > 0$, the equation $y = f(x - c)$ moves the graph of $y = f(x)$ horizontally c units to the right, and the equation $y = f(x + c)$ moves the graph of $y = f(x)$ horizontally c units to the left.

3. Let $f(x) = x^2$. Evaluate the functions $f(x - 3)$ and $f(x + 2)$. Sketch all three on the same coordinate axis.

$f(x - 3) = (x - 3)^2 = x^2 - 6x + 9$
moves $f(x)$ right 3 units.
$f(x + 2) = (x + 2)^2 = x^2 + 4x + 4$
moves $f(x)$ left 2 units.

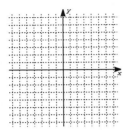

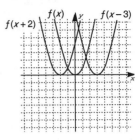

4. Let $f(x) = \dfrac{5}{x^2 + 1}$. Evaluate the functions $f(x - 2)$ and $f(x + 1)$. Sketch all three on the same coordinate axis.

$f(x - 2) = \dfrac{5}{(x - 2)^2 + 1} = \dfrac{5}{x^2 - 4x + 5}$
moves $f(x)$ right 2 units.
$f(x + 1) = \dfrac{5}{(x + 1)^2 + 1} = \dfrac{5}{x^2 + 2x + 2}$
moves $f(x)$ left 1 unit.

Notice that both $f(x - 2)$ and $f(x + 1)$ are much easier to graph using the horizontal shifts.

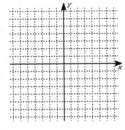

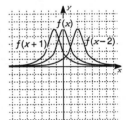

C. For $c > 1$, the equation $y = cf(x)$ *stretches* the graph of $y = f(x)$ vertically by a factor of c. For $0 < c < 1$, the equation $y = cf(x)$ *shrinks* the graph of $y = f(x)$ vertically by a factor of c. And $y = -f(x)$ reflects the graph of $y = f(x)$ about the x-axis.

5. Let $f(x) = 2x - 1$. Find $3f(x)$ and $\frac{1}{2}f(x)$. Sketch all three graphs on the same coordinate axis.

$3f(x) = 3(2x - 1) = 6x - 3$
$\frac{1}{2}f(x) = \frac{1}{2}(2x - 1) = x - \frac{1}{2}$

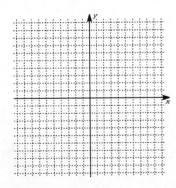

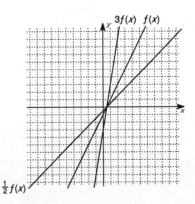

6. Let $f(x) = \sqrt{x} - 1$. Find $2f(x)$ and $-2f(x)$.
   Sketch all three graphs on the same coordinate axis.

$$2f(x) = 2(\sqrt{x} - 1) = 2\sqrt{x} - 2$$
$$-2f(x) = -2(\sqrt{x} - 1) = -2\sqrt{x} + 2$$

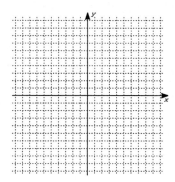

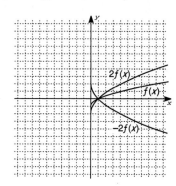

7. Let $f(x) = 3x + \dfrac{2}{x}$. Find $-f(x)$ and $-\frac{1}{2}f(x)$.
   Sketch all three graphs on the same coordinate axis.

$$-f(x) = -\left(3x + \frac{2}{x}\right) = -3x - \frac{2}{x}$$

$$-\tfrac{1}{2}f(x) = -\frac{1}{2}\left(3x + \frac{2}{x}\right) = -\frac{3}{2}x - \frac{1}{x}$$

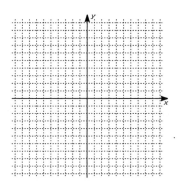

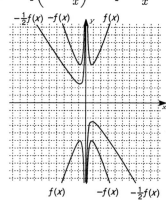

8. Let $f(x) = \dfrac{5}{x^2 + 1}$. Find the functions $\frac{1}{2}f(x)$ and
   $-2f(x)$. Sketch all three graphs on the same
   coordinate axis.

$$\tfrac{1}{2}f(x) = \frac{1}{2}\left(\frac{5}{x^2 + 1}\right) = \frac{5}{2(x^2 + 1)}$$

$$-2f(x) = -2\left(\frac{5}{x^2 + 1}\right) = \frac{-10}{x^2 + 1}$$

Compare these with the transformations in problem 4.

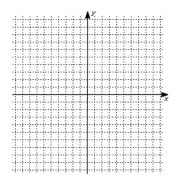

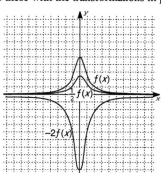

D. For $c > 1$, the equation $y = f(cx)$ *shrinks* the graph of $y = f(x)$ horizontally by a factor of $\frac{1}{c}$. For $0 < c < 1$, the equation $y = f(cx)$ *stretches* the graph of $y = f(x)$ horizontally by a factor of $\frac{1}{c}$ And $y = f(-x)$ reflects the graph of $y = f(x)$ about the y-axis.

9. Let $f(x) = x + 1$. Find $f(3x)$ and $f\left(\frac{1}{2}x\right)$. Sketch all three graphs on the same coordinate axis.

$f(3x) = (3x) + 1 = 3x + 1$

$f\left(\frac{1}{2}x\right) = \left(\frac{1}{2}x\right) + 1 = \frac{1}{2}x + 1$

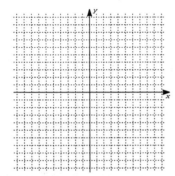

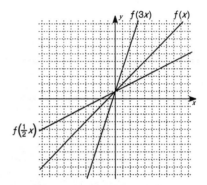

10. Let $f(x) = x^2 + x$. Find $f(2x)$ and $f(-x)$. Sketch all three graphs on the same coordinate axis.

$f(2x) = (2x)^2 + 2x = 4x^2 + 2x$

$f(-x) = (-x)^2 + (-x) = x^2 - x$

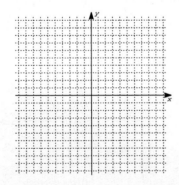

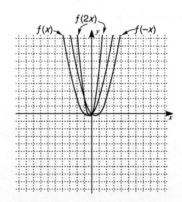

E. Symmetry often helps in graphing a function. An **even function** f is a function for which $f(-x) = f(x)$ for all x in its domain. The graph of an even function is symmetric about the y-axis. An **odd function** f is a function for which $f(-x) = -f(x)$ for all x in its domain. The graph of an odd function is symmetric about the origin. Before graphing, quickly check to see if the function is odd, even, or neither.

11. Check each function to see if it is odd, even, or neither. Then graph.

    (a) $f(x) = x^2 - 4$

$f(x) = x^2 - 4$

$f(-x) = (-x)^2 - 4 = x^2 - 4 = f(x)$

Since $f(-x) = f(x)$, f is even.

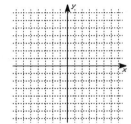

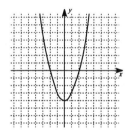

    (b) $f(x) = \dfrac{8x}{x^2 + 1}$

$f(x) = \dfrac{8x}{x^2 + 1}$

$f(-x) = \dfrac{8(-x)}{(-x)^2 + 1} = \dfrac{-8x}{x^2 + 1} = -\dfrac{8x}{x^2 + 1}$

$= -f(x)$

Since $f(-x) = -f(x)$, f is odd.

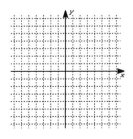

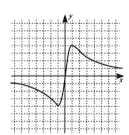

(c) $f(x) = |x|$

$f(x) = |x|$
$f(-x) = |-x| = |x| = f(x)$
Since $f(-x) = f(x)$, f is even.

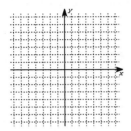

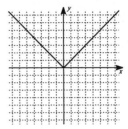

(d) $f(x) = x^3 - x^2 - 8x + 4$

$f(x) = x^3 - x^2 - 8x + 4$
$f(-x) = (-x)^3 - (-x)^2 - 8(-x) + 4$
$\qquad = -x^3 - x^2 + 8x + 4$
And $-f(x) = -x^3 + x^2 + 8x - 4$.
Since $f(-x) \neq -f(x)$ and $f(-x) \neq f(x)$, f is neither odd nor even.

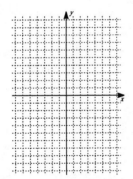

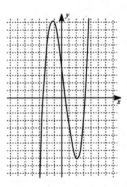

Section 2.6   Extreme Values of Functions

Key Ideas

A. Quadratic functions.

B. Maximum and minimum values.

C. Local extreme values.

A. A **quadratic function** is a function of the form $f(x) = ax^2 + bx + c$, where a, b, and c are real numbers, with $a \neq 0$. It is often easier to graph and work with quadratic functions when we have completed the square and expressed f in the form $f(x) = a(x - h)^2 + k$.

1. Express f in the form $f(x) = a(x - h)^2 + k$ and sketch the graph of f.

  (a) $f(x) = x^2 + 6x$

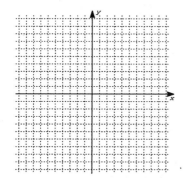

$$f(x) = x^2 + 6x$$
$$= x^2 + 6x + \mathbf{9} - \mathbf{9}$$
$$= (x + 3)^2 - 9$$

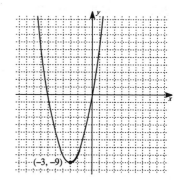

$(-3, -9)$

  (b) $f(x) = x^2 - 4x - 2$

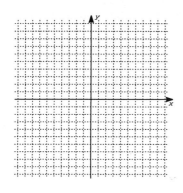

$$f(x) = x^2 - 4x - 2$$
$$= x^2 - 4x + \mathbf{4} - 2 - \mathbf{4}$$
$$= (x - 2)^2 - 6$$

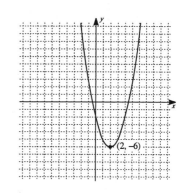

$(2, -6)$

75

(c) $f(x) = 2x^2 - 6x - \frac{5}{2}$

$f(x) = 2x^2 - 6x - \frac{5}{2} = 2(x^2 - 3x) - \frac{5}{2}$

$\qquad = 2\left(x^2 - 3x + \frac{9}{4}\right) - \frac{5}{2} - 2\left(\frac{9}{4}\right)$

$\qquad = 2\left(x - \frac{3}{2}\right)^2 - 7$

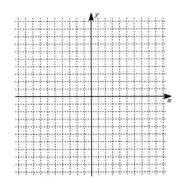

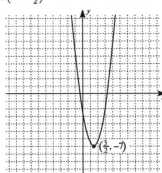

$\left(\frac{3}{2}, -7\right)$

B. When $a > 0$, the parabola (quadratic function), $f(x) = a(x - h)^2 + k$, opens up. The lowest point is the vertex (h, k), so the **minimum value** of the function occurs when $x = h$, and this minimum value is $f(h) = k$. Also $f(x) = a(x - h)^2 + k \geq k$ for all x.

When $a < 0$, the parabola (quadratic function), $f(x) = a(x - h)^2 + k$, opens down. The highest point is the vertex (h, k), so the **maximum value** of the function occurs when $x = h$ and this maximum value is $f(h) = k$. Also $f(x) = a(x - h)^2 + k \leq k$ for all x.

2. Determine the maximum or minimum values of f by completing the square. Find the intercepts and then sketch the graph.

(a) $f(x) = x^2 + 5x + 4$

$f(x) = x^2 + 5x + 4 = (x^2 + 5x) + 4$

$\qquad = \left(x^2 + 5x + \frac{25}{4}\right) + 4 - \frac{25}{4}$

$\qquad = \left(x + \frac{5}{2}\right)^2 - \frac{9}{4}$

Minimum value of $-\frac{9}{4}$ at $x = -\frac{5}{2}$.

To find the x-intercepts set $f = 0$ and solve.

$f(x) = x^2 + 5x + 4 = 0$

$\qquad (x + 1)(x + 4) = 0$

$\qquad x + 1 = 0 \qquad\qquad x + 4 = 0$

$\qquad\qquad x = -1 \qquad\qquad\qquad x = -4$

x-intercepts: -1 and -4

To find the y-intercept evaluate $f(0)$.

y-intercept: $f(0) = 4$

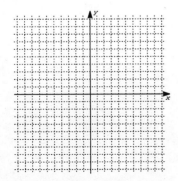

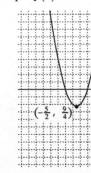

$\left(-\frac{5}{2}, \frac{9}{4}\right)$

(b) $f(x) = 3x^2 - 6x + 4$

$$f(x) = 3x^2 - 6x + 4 = 3(x^2 - 2x) + 4$$
$$= 3(x^2 - 2x + 1) + 4 - \mathbf{3(1)}$$
$$= 3(x - 1)^2 + 1$$

Minimum value of 1 at $x = 1$.
Since $f(x) = 3(x - 1)^2 + 1 \geq 1$, there are no x-intercepts.
y-intercept: $f(0) = 4$

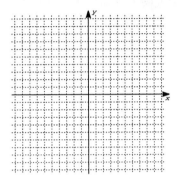

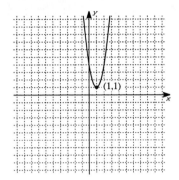

(c) $f(x) = -2x^2 + 5x$

$$f(x) = -2x^2 + 5x = -2\left(x^2 - \tfrac{5}{2}x\right)$$
$$= -2\left(x^2 - \tfrac{5}{2}x + \tfrac{25}{16}\right) + \mathbf{2\left(\tfrac{25}{16}\right)}$$
$$= -2\left(x - \tfrac{5}{4}\right)^2 + \tfrac{25}{8}$$

Maximum value of $\frac{25}{8}$ at $x = \frac{5}{4}$.
To find the x-intercepts set $f = 0$ and solve.
$$f(x) = -2x^2 + 5x = 0$$
$$x(-2x + 5) = 0$$
$$x = 0 \qquad\qquad -2x + 5 = 0$$
$$-2x = -5$$
$$x = \tfrac{5}{2}$$

x-intercepts: 0 and $\frac{5}{2}$ *Don't forget $x = 0$.*
y-intercept: $f(0) = 0$

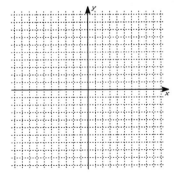

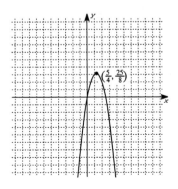

(d) $f(x) = -5x^2 - 10x + 1$

$$f(x) = -5x^2 - 10x + 1 = -5(x^2 + 2x) + 1$$
$$= -5(x^2 - 2x + 1) + 1 + \mathbf{5(1)}$$
$$= -5(x + 1)^2 + 6$$

Maximum value of 6 at $x = -1$.
To find the x-intercepts, set $f = 0$ and solve.
$f(x) = -5x^2 - 10x + 1 = 0$
Since this polynomial cannot be factored with integers, use the quadratic formula.

$$x = \frac{-(-10) \pm \sqrt{(-10)^2 - 4(-5)(1)}}{2(-5)}$$

$$= \frac{10 \pm \sqrt{100 + 20}}{-10} = \frac{10 \pm \sqrt{120}}{-10} = -1 \pm \frac{\sqrt{30}}{5}$$

$$\approx -1 \pm 1.10$$

x-intercepts: -2.10 and 0.10
y-intercept: $f(0) = 1$

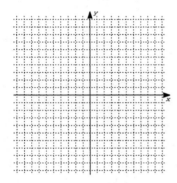

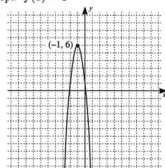

3. A farmer wants to enclose a rectangular field on three sides by a fence and divide it into two smaller rectangular fields by a fence perpendicular to the second side. He has 3000 yards of fencing. Find the dimension of the field so that the total enclosed area is a maximum.

First make a drawing and label the sides.

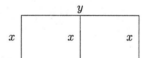

Equating the fencing material, we have the equation $3x + y = 3000$. We want to maximize the area given by the equation $A = xy$. So we solve the first equation for y and substitute into the second equation.
$y = 300 - 3x$
$A = x(3000 - 3x) = -3x^2 + 3000x$
Since the leading coefficient is negative, A will have a maximum. Now complete the square:
$A = -3(x^2 - 1000x)$
$= -3(x^2 - 1000x + 250,000) + 750,000$
$= -3(x - 500)^2 + 750,000$
So A is a maximum of 750,000 sq. yds. when $x = 500$ ids and $y = 3000 - 3(500) = 1500$ yds.

C. A point a is called a **local maximum** of the function f if $f(a) > f(x)$ for all x in some interval containing a. Likewise, a **local minimum** of the function f is a point a where $f(a) < f(x)$ for all x in some interval containing a. When using a graphing device, these local extreme values can be found by looking for the highest (or lowest) point *within* a viewing rectangle.

4. Find the approximate local maximum and minimum values of $f(x) = x^3 - 2x + 4$.

Local minimum: $y \approx 2.911$ at $x \approx 0.816$

$[0, 1]$ by $[2.5, 3]$

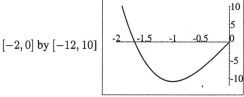

Local maximum: $y \approx 5.089$ at $x \approx -0.816$

$[-1, -0.5]$ by $[5, 5.5]$

5. Find the approximate local maximum and minimum values of $f(x) = x^4 - 4x^3 + 16x$.

Local minimum: $y \approx -11$ at $x \approx -1$

$[-2, 0]$ by $[-12, 10]$

No local maximum.

6. Find the approximate local maximum and minimum values of $f(x) = x^3 + 3x^2 - 2\sqrt{x^2 + 1}$.

Local minimum: $y \approx -2$ at $x \approx 0$
Local maximum: $y \approx -0.18$ at $x \approx -1.66$

$[-12, 12]$ by $[-10, 10]$

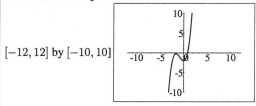

Section 2.7   Modeling with Functions

Key Ideas

A. Guidelines for modeling with functions.

A. Steps of how we model with functions are summarized in the following table.

Guidelines for Modeling with Functions
1. *Express the model in words.* Identify the quantity you want to model and express it as a function of the other quantities in the problem.
2. *Choose the variable.* Identify all the variables used to express the function in Step 1. Assign a symbol, such as x, to one variable and express the variables in terms of this symbol.
3. *Set up the model.* Express the function in the language of algebra, by writing it as a function of the single variable chosen in Step 2.
4. *Express the Model in words.* Identify the quantity you want to model and express it as a function of the other quantities in the problem.
5. *State the answers.* Write your answer as a complete sentence.

1. A child has 15 coins, nickels and dimes. Model the amount of money she has as a function of the number dimes she has? How many dimes must she have in order to have at least $1?

Lets experiment with the problem and make a table for various combinations of nickels and dimes.

# of nickels	15	14	13	12	...
# of dimes	0	1	2	3	...
Total amount	$0.75	$0.80	$0.85	$0.90	...

# of nickels	...	3	2	1	0
# of dimes	...	12	13	14	15
Total amount	...	$1.35	$1.40	$1.45	$1.50

From the table we see she has between 75¢ and $1.50. There are two quantities, here nickels and dimes, but the function we want must depend on one variable only. We choose to have the variable represent the number of dimes

Let x = number of dimes. Since she has 15 coins $15 - x$ = number of nickels. Then using the model

$$Total\ value = 0.05 \left(\begin{array}{c} \text{\# of} \\ \text{nickels} \end{array} \right) + 0.10 \left(\begin{array}{c} \text{\# of} \\ \text{dimes} \end{array} \right)$$

we get $T = 0.05(15 - x) + 0.10x = 0.75 + 0.05x$.

To answer the question "How many dimes must she have in order to have at least $1?" we use the above model, $T = 0.75 + 0.05x$, we solve

$0.75 + 0.05x \geq 1.00 \quad \Leftrightarrow \quad 0.05x \geq 0.25 \quad \Leftrightarrow$

$x \geq 5$. Thus in order for her to have at least $1 she must have at least 5 dimes.

2. A company wants to construct a can whose height is 1.62 times it's radius. Model the volume of the can as a function of it radius. Make a table to show the volume when the radius is 5, 6, 7, 8, 9, 10, 11, and 12 cm.

Let $x =$ the radius of the can.

Then the height of the can $= 1.62x$

Volume = (area of the base) × (height)

$V(x) = (\pi x^2)(1.62x) = 1.62\pi x^3$

radius x	volume $V(x)$
5 cm	636 cm^3
6 cm	1,099 cm^3
7 cm	1,746 cm^3
8 cm	2,606 cm^3
9 cm	3,710 cm^3
10 cm	5,089 cm^3
11 cm	6,774 cm^3
12 cm	8,794 cm^3

3. A concrete sidewalk, 4 inches thick and 3 feet wide is to be constructed using bagged concrete. Each bag contains 1/2 cube feet of concrete. Find a function that relates the number of bags of concrete needed to construct x feet of sidewalk
How many bags are needed to make a 15 foot sidewalk?

Since the bags are given in cubic feet, we convert all measurements, and use the model

Volume = (length) × (width) × (thickness)

$$= x(3)\left(\tfrac{4}{12}\right) = x \text{ cubic feet.}$$

Since each bags contains 1/2 cubic feet of concrete, the function that relates the number of bags of concrete needed to construct x feet of sidewalk is given by $f(x) = 2x$.

Use this model, we need $f(15) = 2 \cdot 15 = 30$ bags of concrete.

4. Find a function that relates the volume of a cube to the surface area.

Let $V =$ the volume of the cube.

Then $\sqrt[3]{V} =$ length of an edge of the cube.

The area of one side of the cube is $\left(\sqrt[3]{V}\right)^2 = \sqrt[3]{V^2}$.

Since a cube has six sides, the surface area S is given by the function $S(V) = 6\sqrt[3]{V^2}$.

Section 2.8   Combining Functions

Key Ideas

A.   Sum of functions.
B.   Difference of functions.
C.   Product of functions.
D.   Quotient of functions.
E.   Composition of functions.

A.   Let $f(x)$ be a function with domain A and let $g(x)$ be a function with domain B. Then the **sum** of the functions f and g is $(f+g)(x) = f(x) + g(x)$ where the domain is $A \cap B$.

1. Let $f(x) = 8x + 3$ and $g(x) = x^2 - 4$. Find $f + g$ and find its domain.

   $$(f+g)(x) = f(x) + g(x)$$
   $$= (8x + 3) + (x^2 - 4)$$
   $$= x^2 + 8x - 1$$

   Domain: Since the domain for both f and g is the real numbers, the domain of $f + g$ is the real numbers.

2. Let $f(x) = -(x-2)^2$ and $g(x) = |x|$. Find $f + g$ and find its domain.

   $$(f+g)(x) = f(x) + g(x)$$
   $$= -(x-2)^2 + |x|$$

   Domain: Since the domain for both f and g is the real numbers, the domain of $f + g$ is the real numbers.

3. Let $f(x) = \sqrt{x + 3}$ and $g(x) = \sqrt{9 - x^2}$. Find $f + g$ and find its domain.

   $$(f+g)(x) = f(x) + g(x)$$
   $$= \sqrt{x + 3} + \sqrt{9 - x^2}$$

   Domain of f: $x + 3 \geq 0$ $\Leftrightarrow$ $x \geq -3$ or $[-3, \infty)$.
   Domain of g: $9 - x^2 \geq 0$ $\Leftrightarrow$
   $(3 - x)(3 + x) \geq 0$. Using earlier methods we find that the domain is $[-3, 3]$.
   Domain of $f + g$ is $[-3, \infty) \cap [-3, 3] = [-3, 3]$.

B. Let $f(x)$ be a function with domain A and let $g(x)$ be a function with domain B. Then the **difference** of the functions f and g is $(f - g)(x) = f(x) - g(x)$ where the domain is $A \cap B$.

4. Let $f(x) = \sqrt{x^2 - 1}$ and $g(x) = \sqrt{16 - x^2}$. Find $f - g$ and it domain.

$(f - g)(x) = f(x) - g(x)$
$$= \sqrt{x^2 - 1} - \sqrt{16 - x^2}$$
Domain of f: $x^2 - 1 \geq 0$ $\Leftrightarrow$
$(x - 1)(x + 1) \geq 0$. Using earlier methods we find $(-\infty, -1] \cup [1, \infty)$.
Domain of g: $16 - x^2 \geq 0$ $\Leftrightarrow$
$(4 - x)(4 + x) \geq 0$. Using earlier methods we find that the domain is $[-4, 4]$.
Domain of $f - g$ is
$((-\infty, -1] \cup [1, \infty)) \cap [-4, 4] = [-4, -1] \cup [1, 4]$.

5. Let $f(x) = \dfrac{2}{x - 3}$ and $g(x) = \dfrac{1}{x + 5}$. Find $f - g$ and it domain.

$(f - g)(x) = f(x) - g(x)$
$$= \frac{2}{x - 3} - \frac{1}{x + 5}$$
Domain of f: $x - 3 \neq 0$ $\Leftrightarrow$ $x \neq 3$
Domain of g: $x + 5 \neq 0$ $\Leftrightarrow$ $x \neq -5$
Domain of $f - g$ is $\{x \mid x \neq 3 \text{ or } x \neq -5\}$.

6. Let $f(x) = x$ and $g(x) = \dfrac{1}{x^2 - 4}$. Find $f - g$ and it domain.

$(f - g)(x) = f(x) - g(x)$
$$= x - \frac{1}{x^2 - 4}$$
Domain of f: All real numbers.
Domain of g: $x^2 - 4 \neq 0$ $\Leftrightarrow$
$(x - 2)(x + 2) \neq 0$ $\Rightarrow$ $x \neq \pm 2$
Domain of $f - g$ is $\{x \mid x \neq \pm 2\}$.

C. Let $f(x)$ be a function with domain A and let $g(x)$ be a function with domain B. Then the **product** of the functions f and g is $(fg)(x) = f(x)g(x)$ where the domain is $A \cap B$.

7. Let $f(x) = x + 2$ and $g(x) = \dfrac{2}{x + 1} - 4$. Find fg and it domain.

$(fg)(x) = f(x)g(x)$
$$= (x + 2)\left(\frac{2}{x + 1} - 4\right)$$
$$= \frac{2x + 4}{x + 1} - 4x - 8$$
Domain of f: All real numbers.
Domain of g: $x + 1 \neq 0$ $\Leftrightarrow$ $x \neq -1$.
Domain of fg is $\{x \mid x \neq -1\}$.

8. Let $f(x) = \sqrt{x + 12}$ and $g(x) = \sqrt{x - 5}$. Find fg and it domain.

$$(fg)(x) = f(x)g(x)$$
$$= \left(\sqrt{x + 12}\right)\left(\sqrt{x - 5}\right)$$
$$= \sqrt{(x + 12)(x - 5)} = \sqrt{x^2 + 7x - 60}$$

Domain of f: $x + 12 \geq 0 \quad \Leftrightarrow \quad x \geq -12$ or $[-12, \infty)$.

Domain of g: $x - 5 \geq 0 \quad \Leftrightarrow \quad x \geq 5$ or $[5, \infty)$.

Domain of fg is $[-12, \infty) \cap [5, \infty) = [5, \infty)$.

D. Let $f(x)$ be a function with domain A and let $g(x)$ be a function with domain B. Then the **quotient** of the functions f and g is $\left(\dfrac{f}{g}\right)(x) = \dfrac{f(x)}{g(x)}$ where the domain is $\{x \in A \cap B \,|\, g(x) \neq 0\}$.

9. Let $f(x) = \sqrt{x - 1}$ and $g(x) = x^2 - 4$.

   (a) Find $\dfrac{f}{g}$ and find its domain.

   $$\frac{f(x)}{g(x)} = \frac{\sqrt{x - 1}}{x^2 - 4}$$

   Domain of numerator: $x - 1 \geq 0 \quad \Leftrightarrow \quad x \geq 1$ or $[1, \infty)$.

   Domain of denominator: $x^2 - 4 \neq 0 \quad \Leftrightarrow$
   $(x - 2)(x + 2) \neq 0 \quad \Rightarrow \quad x \neq \pm 2$.

   Domain of $\dfrac{f}{g}$ is $[1, \infty) \cap \{x \,|\, x \neq \pm 2\} =$
   $[1, 2) \cup (2, \infty)$.

   (b) Find $\dfrac{g}{f}$ and find its domain.

   $$\frac{f(x)}{g(x)} = \frac{x^2 - 4}{\sqrt{x - 1}}$$

   Domain of numerator: all real numbers.

   Domain of denominator: $x - 1 > 0 \quad \Leftrightarrow \quad x > 1$.

   Domain of $\dfrac{g}{f}$ is $(1, \infty)$.

10. Let $f(x) = x^2 - 9$ and $g(x) = x^2 + 9$.

    (a) Find $\dfrac{f}{g}$ and find its domain.

    $$\frac{f(x)}{g(x)} = \frac{x^2 - 9}{x^2 + 9}$$

    Domain of numerator: all real numbers.

    Domain of denominator: since $x^2 + 9$ is always positive, so the domain is all real numbers.

    Domain of $\dfrac{f}{g}$ is all real numbers.

(b) Find $\dfrac{g}{f}$ and find its domain.

$$\frac{f(x)}{g(x)} = \frac{x^2 + 9}{x^2 - 9}$$

Domain of numerator: all real numbers.

Domain of denominator: $x^2 - 9 \neq 0$ $\Leftrightarrow$

$(x - 3)(x + 3) \neq 0$ $\Rightarrow$ $x \neq \pm 3$.

Domain of $\dfrac{g}{f}$ is $\{x \mid x \neq \pm 3\}$.

E. Given two functions f and g, the **composition function** $f \circ g$ is defined by $(f \circ g)(x) = f(g(x))$. We sometimes say that $f \circ g$ is 'f acting on $g(x)$', in other words, x is the input to g and $g(x)$ is the input to f. The domain of $f \circ g$ is that subset of the domain of g where $g(x)$ is in the domain of f. Another way of saying this is that the domain of $f \circ g$ is those points where both $g(x)$ and $f(g(x))$ are defined.

11. Let $f(x) = \sqrt{x - 10}$ and let $g(x) = x + 9$.

   (a) Find $f \circ g$ and find its domain.

$$(f \circ g)(x) = f(g(x)) = f(x + 9)$$
$$= \sqrt{(x + 9) - 10} = \sqrt{x - 1}$$

Domain: $x - 1 \geq 0$ $\Leftrightarrow$ $x \geq 1$ or $[1, \infty)$.

   (b) Find $g \circ f$ and find its domain.

$$(g \circ f)(x) = g(f(x)) = f(\sqrt{x - 10})$$
$$= \sqrt{x - 10} + 9$$

Domain: $x - 10 \geq 0$ $\Leftrightarrow$ $x \geq 10$ or $[10, \infty)$.

Notice that $f \circ g \neq g \circ f$.

12. Let $f(x) = \dfrac{1}{x} - 4$ and let $g(x) = \dfrac{1}{x + 4}$.

   (a) Find $f \circ g$ and find its domain.

$$(f \circ g)(x) = f(g(x)) = f\left(\frac{1}{x + 4}\right)$$
$$= \frac{1}{\left(\dfrac{1}{x + 4}\right)} - 4 = x + 4 - 4 = x$$

Domain: $x + 4 \neq 0$ $\Leftrightarrow$ $x \neq -4$ or $\{x \mid x \neq -4\}$.

Note: $\dfrac{1}{x + 4}$ *is never* 0.

   (b) Find $g \circ f$ and find its domain.

$$(g \circ f)(x) = g(f(x)) = f\left(\frac{1}{x} - 4\right)$$
$$= \frac{1}{\left(\dfrac{1}{x} - 4\right) + 4} = \frac{1}{\dfrac{1}{x}} = x$$

Domain: $x \neq 0$ or $\{x \mid x \neq 0\}$.

13. Let $f(x) = \sqrt{x^2 + x + 1}$ and let $g(x) = x^2 - 1$.

(a) Find $f \circ g$ and find its domain.

$$(f \circ g)(x) = f(g(x)) = f(x^2 - 1)$$
$$= \sqrt{(x^2 - 1)^2 + (x^2 - 1) + 1}$$
$$= \sqrt{x^4 - x^2 + 1}$$

Domain: Since the domain of g is all real numbers and since $x^4 - x^2 + 1 = \left(x^2 - \frac{1}{2}\right)^2 + \frac{3}{4}$ (by completing the square) is always positive, the domain is all real numbers.

(b) Find $g \circ f$ and find its domain.

$$(g \circ f)(x) = g(f(x)) = f\left(\sqrt{x^2 + x + 1}\right)$$
$$= \left(\sqrt{x^2 + x + 1}\right)^2 - 1$$
$$= x^2 + x + 1 - 1 \qquad \textit{See note below.}$$
$$= x^2 + x$$

Domain: Since $x^2 + x + 1 = \left(x + \frac{1}{2}\right)^2 + \frac{3}{4}$, $x^2 + x + 1 > 0$. So the domain is all real numbers.

Note: Since $x^2 + x + 1 > 0$, we have
$$\left(\sqrt{x^2 + x + 1}\right)^2 = x^2 + x + 1.$$

Section 2.9   One-to-One Functions and Their Inverses

Key Ideas

A. One-to-one functions and the horizontal line test.

B. Domain and ranges of inverse functions.

C. Finding the inverse function of a one-to-one function.

D. Graphing the inverse function of a one-to-one function.

A. A function with domain A is called a **one-to-one function** if no two elements of A have the same image; that is, $f(x_1) \neq f(x_2)$ whenever $x_1 \neq x_2$. Another way of stating this is, "If $f(x_1) = f(x_2)$, then $x_1 = x_2$." One test for determining when a function is one-to-one is called the **horizontal line test** which states that a function is one-to-one if and only if *no* horizontal line intersects its graph more than once. Care must be taken to avoid some common errors in algebraically showing a function is one-to-one. While it is true that $a \neq b$ implies that $a + c \neq b + c$, it is <u>not</u> true that $a \neq b$ and $c \neq d$ implies that $a + c \neq b + d$.

1. Show that $f(x) = x^2 + 2x - 3$ is not one-to-one by finding two points $x_1 \neq x_2$ where $f(x_1) = f(x_2)$.

> There are many answers possible. An easy pair of points is found by setting $f(x) = 0$ and factoring.
> $$x^2 + 2x - 3 = (x - 1)(x + 3) = 0$$
> $$x - 1 = 0 \qquad\qquad x + 3 = 0$$
> $$x = 1 \qquad\qquad x = -3$$
> So, $f(1) = f(-3) = 0$, thus f is not one-to-one. Since $f(x) = (x + 1)^2 - 4$ (*found by completing the square*), all solutions to $f(x_1) = f(x_2)$ are of the form $|x_1 + 1| = |x_2 + 1|$.
> Example: $|(1) + 1| = |(-3) + 1| = 2$.

2. Use the horizontal line test to determine if each function is one-to-one.

(a)

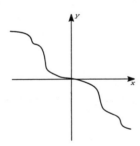

> This graph passes the horizontal line test. This is a one-to-one function.

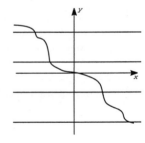

(b)

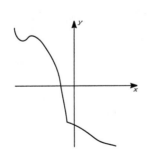

> This graph does *not* pass the horizontal line test. This function is *not* one-to-one.

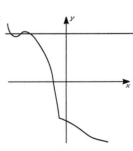

(c)

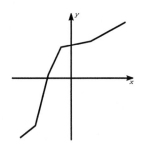

This graph passes the horizontal line test. This is a one-to-one function.

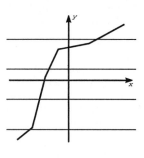

B. Let f be a one-to-one function with domain A and range B. Then its **inverse function f^{-1}** has domain B and range A and is defined by

$$f^{-1}(y) = x \quad \Leftrightarrow \quad f(x) = y \quad \text{for any } y \text{ in } B.$$

It is a common mistake to think of the -1 in f^{-1} as an exponent: $f^{-1}(x) \neq \dfrac{1}{f(x)}$.

3. Suppose f is one-to-one and $f(0) = 3$, $f(1) = 4$, $f(3) = 5$, and $f(4) = 6$.

   (a) Find $f^{-1}(3)$.

   $\qquad\qquad\qquad$ $f^{-1}(3) = 0$ since $f(0) = 3$.

   (b) Find $f^{-1}(4)$.

   $\qquad\qquad\qquad$ $f^{-1}(4) = 1$ since $f(1) = 4$.

C. If f is a one-to-one function, then we find the inverse function by the following procedure.

> 1. Write $y = f(x)$.
> 2. Solve this equation for x in terms of y (if possible).
> 3. Interchange x and y. The resulting function is $y = f^{-1}(x)$.

4. Find the inverse function of $f(x) = \sqrt{2x - 1}$. What is the domain and range of f^{-1}?

$$y = \sqrt{2x - 1}$$
$$y^2 = 2x - 1 \qquad \text{Square both sides.}$$
$$y^2 + 1 = 2x$$
$$\frac{y^2 + 1}{2} = x$$
$$x = \tfrac{1}{2}y^2 + \tfrac{1}{2}$$
$$y = \tfrac{1}{2}x^2 + \tfrac{1}{2} \qquad \text{Interchange } x \text{ and } y.$$
$$f^{-1}(x) = \tfrac{1}{2}x^2 + \tfrac{1}{2}$$

Domain of f^{-1} (= range of f): $[0, \infty)$
Range of f^{-1} (= domain of f): $[\tfrac{1}{2}, \infty)$

Note: Here it is easier to find the range of f in order to find the domain of f^{-1}.

88

5. Find the inverse function of $f(x) = \dfrac{3x - 1}{2 + x}$.

What is the domain and range of f^{-1}?

$$y = \frac{3x - 1}{2 + x}$$

$y(2 + x) = 3x - 1$

$2y + xy = 3x - 1$ *Gather the terms with x*

$xy - 3x = -2y - 1$ *to one side of the equation.*

$x(y - 3) = -(2y + 1)$ *Factor out the x.*

$x = -\dfrac{2y + 1}{y - 3}$

$y = -\dfrac{2x + 1}{x - 3}$ *Interchange x and y.*

$f^{-1}(x) = -\dfrac{2x + 1}{x - 3}$

Domain of f^{-1} (= range of f): $\{x \mid x \neq 3\}$.

Range of f^{-1} (= domain of f): $\{y \mid y \neq -2\}$.

Note: Here it is easier to find the domain of f in order to find the range of f^{-1}.

D. Even though we might not be able to find the formula for f^{-1}, we can still find the graph of f^{-1}. The graph of f^{-1} is obtained by reflecting the graph of f about the line $y = x$. The important thing here is that (a, b) is a point on the graph of f if and only if (b, a) is a point on the graph of f^{-1}.

6. Find the graph of f^{-1} for each graph of f shown below.

(a)

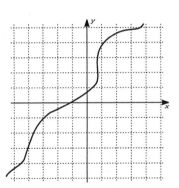

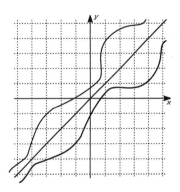

(b)

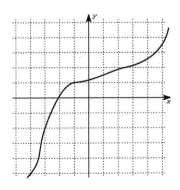

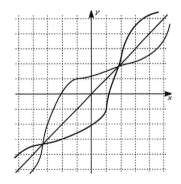

Chapter 3
Polynomial and Rational Functions

Section 3.1 Polynomial Functions and Their Graphs

Key Ideas

A. Polynomial functions.

B. Graphs of polynomials.

C. End behavior of polynomials.

D. Zeros of Polynomials.

E. Intermediate Value Theorem for polynomials.

F. Local extrema of polynomials.

A. A **polynomial of degree n**, P, is a function of the form $P(x) = a_n x^n + a_{n-1} x^{n-1} + \cdots + a_1 x^1 + a_0$ where $a_n, a_{n-1}, a_{n-2}, \ldots, a_1, a_0$ are constants with $a_n \neq 0$ and n is called the **degree**. The numbers $a_n, a_{n-1}, a_{n-2}, \ldots, a_1, a_0$ are called the **coefficients** of the polynomial. The **constant coefficient, a_0**, and the **leading coefficient, a_n**, are important in factoring polynomials and finding zeros. The constant coefficient is sometimes called the constant term.

1. For each polynomial determine the degree, the leading coefficient, and the constant coefficient.

   (a) $P(x) = 8x^4 + 3x^3 - x^2 + 5x - 7$

   Degree: 4
   Leading coefficient: 8
   Constant coefficient: -7

   (b) $P(x) = x^7 - 3x^4 - 3x$

   Degree: 7
   Leading coefficient: 1
   Constant coefficient: 0
   Note: Even though no constant term is shown, it is understood to be zero.

   (c) $P(x) = 6 - 5x - 7x^3 + 2x^4 - 4x^6$

   First write in descending order:
   $P(x) = -4x^6 + 2x^4 - 7x^3 - 5x + 6$
   Degree: 6
   Leading coefficient: -4
   Constant coefficient: 6

   (d) $P(x) = 8x - 17 + 9x^3 + 2x^6 - 3x^2 + x^8$

   First write in descending order:
   $P(x) = x^8 + 2x^6 + 9x^3 - 3x^2 + 8x - 17$
   Degree: 8
   Leading coefficient: 1
   Constant coefficient: -17

B. The graph of a polynomial is always a smooth curve that has no breaks or sharp corners. The graphs of polynomials of degree 0 and degree 1 are **lines** and graphs of polynomials of degree 2 are **parabolas**. Polynomials with only one term are called **monomials**.

91

2. Sketch the graph of the function by translating the graph of an appropriate function of the form $y = x^n$.

(a) $P(x) = -\frac{1}{2}x^2$

This is a reflection about the x-axis and shrinking of the parabola $y = x^2$.

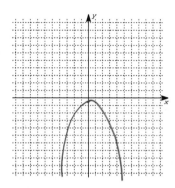

(b) $P(x) = 2(x - 1)^3$

We start with the graph of $y = x^3$, translate it 1 unit to the right, then stretch by a factor of 2.

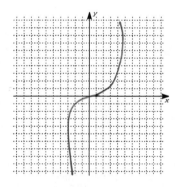

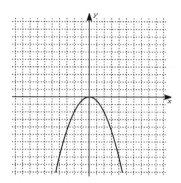

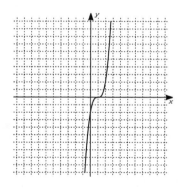

C. The **end behavior** of a function is what happens to the values of the function as $|x|$ becomes large. We use the following notation:

$x \to \infty$ means "x becomes large in the positive direction"

$x \to -\infty$ means "x becomes large in the negative direction"

The end behavior of the polynomial $P(x) = a_n x^n + a_{n-1} x^{n-1} + \cdots + a_1 x^1 + a_0$ is completely determined by the leading term, $a_n x^n$, as shown in the table on the following page.

	n even		n odd	
$a > 0$	$x \to \infty$	$y \to \infty$	$x \to \infty$	$y \to \infty$
	$x \to -\infty$	$y \to \infty$	$x \to -\infty$	$y \to -\infty$
$a < 0$	$x \to \infty$	$y \to -\infty$	$x \to \infty$	$y \to -\infty$
	$x \to -\infty$	$y \to -\infty$	$x \to -\infty$	$y \to \infty$

Use a <u>large</u> viewing rectangle when using a graphing device to find the end behavior. Be sure to verify that what you see agrees with the degree of the polynomial.

3. Describe the end behavior of each polynomial.

(a) $P(x) = -2x^5 + 4x^4 - 2x^3 - 3x$

Since we are only interested in the end behavior, use a large viewing rectangle.

$y \to \infty$ as $x \to -\infty$ and $y \to -\infty$ as $x \to \infty$. Shown is the viewing rectangle $[-20, 20]$ by $[-100, 100]$.

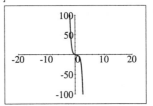

(b) $Q(x) = 0.1x^8 + 5x^7 + x + 10$

Since the degree of $Q(x)$ is 8, we know that both ends of the graph must do the same thing. We see

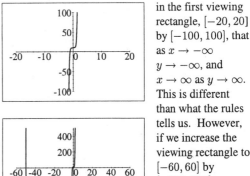

in the first viewing rectangle, $[-20, 20]$ by $[-100, 100]$, that as $x \to -\infty$ $y \to -\infty$, and $x \to \infty$ as $y \to \infty$. This is different than what the rules tells us. However, if we increase the viewing rectangle to $[-60, 60]$ by $[-500, 500]$, we see that in fact as $x \to -\infty$ we have $y \to \infty$, and as $x \to \infty$ we have $y \to \infty$.

D. The graph of the polynomial $y = P(x)$ touches or crosses the x-axis at the point c if and only if $P(c) = 0$. The point c is a **zero** of $P(x)$.

Equivalent Statements	
1. c is a zero of P.	3. $x - c$ is a factor of $P(x)$.
2. $x = c$ is a root of $P(x) = 0$.	4. $x = c$ is a x-intercept of the graph of P.

93

4. Sketch the graph of the polynomial $P(x) = 4x^2 - x^4$.

Start by finding the zeros of $P(x)$ by factoring. This polynomial is easy to factor.
$$P(x) = 4x^2 - x^4 = x^2(4 - x^2)$$
$$= x^2(2 - x)(2 + x)$$

$x^2 = 0$	$2 - x = 0$	$2 + x = 0$
$x = 0$	$2 = x$	$x = -2$

Thus the x-intercepts are $x = 0$, $x = 2$, and $x = -2$.
Since $P(0) = 0$, the y-intercept is $y = 0$.
Find some additional points and sketch.

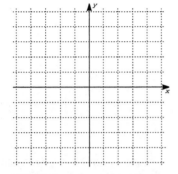

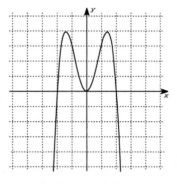

5. Sketch the graph of the polynomial $P(x) = (x + 2)(x - 1)^2$.

Set each factor equal to zero to find the x-intercepts.
$$P(x) = (x + 2)(x - 1)^2$$

$x + 2 = 0$	$x - 1 = 0$
$x = -2$	$x = 1$

Thus the x-intercepts are $x = -2$ and $x = 1$.
Since $P(0) = (2)(1)^2 = 2$, the y-intercept is $y = 2$.
Find some additional points and sketch.

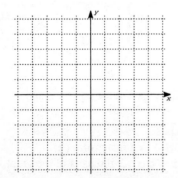

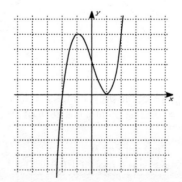

6. Sketch the graph of the polynomial
   $P(x) = (2x - 3)(x + 1)(x + 4)$.

Set each factor equal to zero to find the x-intercepts.
$$P(x) = (2x - 3)(x + 1)(x + 4)$$

$2x - 3 = 0$	$x + 1 = 0$	$x + 4 = 0$
$2x = 3$	$x = -1$	$x = -4$
$x = \frac{3}{2}$		

The x-intercepts are $x = \frac{3}{2}$, $x = -1$, and $x = -4$.
The y-intercept is $P(0) = (-3)(1)(4) = -12$.
Find some additional points and sketch.

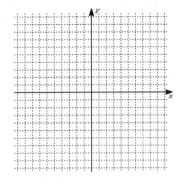

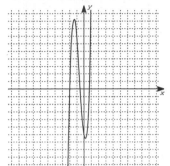

In the graph above, each grid line represents 2 units.

E. The **Intermediate Value Theorem for Polynomials** states that if P is a polynomial function and $P(a)$ and $P(b)$ have opposite signs, then there exists at least one value c between a and b for which $P(c) = 0$. Between successive zeros, the values of the polynomial are either all positive or all negative. To sketch the graph of $y = P(x)$, we first find all the zeros of P. Then choose **test points** between two successive zeros of the graph to determine the sign of $P(x)$. When $P(x)$ is negative, the graph of $y = P(x)$ is below the x-axis. When $P(x)$ is positive, the graph of $y = P(x)$ is above the x-axis. If c is a zero of $P(x)$, and the corresponding factor $(x - c)^m$ occurs *exactly* m times in the factorization of $P(x)$, then the graph crosses the x-axis at c when m is odd and does not cross the x-axis when m is even.

7. Sketch the graph of the polynomial
   $P(x) = (x - 2)^3(x + 1)^2$.

$$P(x) = (x - 2)^3(x + 1)^2$$

$x - 2 = 0$	$x + 1 = 0$
$x = 2$	$x = -1$

The x-intercepts are $x = 2$ and $x = -1$. Since 2 occurs three times, the graph crosses the x-axis at $x = 2$. Since -1 occurs twice, the graph does not cross the x-axis at $x = -1$. $P(0) = (-2)^3(1)^2 = -8$, so the y-intercept is $y = -8$.

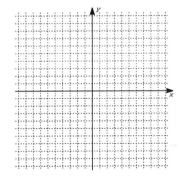

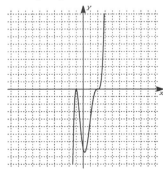

8. Sketch the graph of the polynomial
   $P(x) = x^2(x+1)(x-3)$.

$P(x) = x^2(x+1)(x-3)$

$$x^2 = 0 \qquad x+1 = 0 \qquad x-3 = 0$$
$$x = 0 \qquad x = -1 \qquad x = 3$$

The x-intercepts are $x = 0$, $x = -1$, and $x = 3$.
Since 0 occurs twice, the graph does not cross the x-axis at $x = 0$. Since -1 and 3 each occur only once the graph crosses the x-axis at $x = -1$ and $x = 3$.

$P(0) = (0)^2(1)(-3) = 0$, so the y-intercept is $y = 0$.

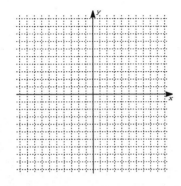

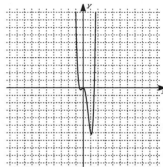

In the graph above, each grid line represents 2 units.

F. The points at which the graph of a function changes from increasing to decreasing or from decreasing to increasing are called **local extrema**. Calculus tells us that a polynomial of degree n can have at most $n - 1$ local extrema. A local extremum is a **local maximum** when it is the highest point *within* some viewing rectangle and a **local minimum** when it is the lowest point *within* some viewing rectangle.

9. Graph the following polynomials using a graphing device. Determine the y-intercept and approximate the x-intercepts. Also approximate the coordinates of all local extrema to 2 decimal places.

   (a) $Q(x) = \frac{1}{4}x^4 - 3x^2 - 3x + 3$

Pick a large viewing rectangle to make sure you find all local extrema. Then reduce the size of the viewing rectangle to find the values to the nearest hundredth.

y-intercept: 3
x-intercepts: 0.62 and 3.79
local maximum: $(-0.52, 3.77)$
local minimum: $(-2.15, 0.92)$ and $(2.67, -13.69)$
$[-5, 5]$ by $[-15, 15]$

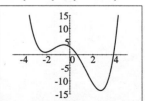

(b) $R(x) = -0.1x^5 + 2x^3 - 12x - 3$

y-intercept: -3

x-intercepts: -0.25

local maximum: $(-1.59, 9.06)$ and $(3.08, -9.24)$

local minimum: $(-3.08, 3.24)$ and $(1.59, -15.06)$

$[-5, 5]$ by $[-20, 20]$

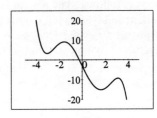

Section 3.2   Dividing Polynomials

Key Ideas

A. Division Algorithm.

B. Synthetic division, Remainder Theorem, and Factor Theorem.

A. Long division for polynomials is a process similar to division for numbers. The terminology of division is as follows. The **divisor** is divided into the quantity called the **dividend**. The result is called the **quotient** and what is left over is the **remainder**. A very useful algorithm is the **Division Algorithm**. If $P(x)$ and $D(x)$ are polynomials, with $D(x) \neq 0$, then there exist unique polynomials $Q(x)$ and $R(x)$ such that: $P(x) = D(x) \cdot Q(x) + R(x)$, where $R(x) = 0$ or the degree of $R(x)$ is less than the degree of $D(x)$. $P(x)$ is the **dividend**, $D(x)$ is the **divisor**, $Q(x)$ is the **quotient**, and $R(x)$ is the **remainder**. The important thing to remember when dividing is to insert $0x^k$ for missing terms.

1. Let $P(x) = x^4 + 4x^3 + 6x^2 - x + 6$ and $D(x) = x^2 + 2x$. Find the polynomials $Q(x)$ and $R(x)$ such that $P(x) = D(x) \cdot Q(x) + R(x)$.

$$
\begin{array}{r}
x^2 \; + 2x + 2 \\
\hline
x^2 + 2x \,)\overline{\, x^4 + 4x^3 + 6x^2 \;\; - x + 6 \,} \\
\underline{x^4 + 2x^3 \qquad\qquad\qquad} \\
2x^3 + 6x^2 \qquad\quad \\
\underline{2x^3 + 4x^2 \qquad\quad} \\
2x^2 \; - x \\
\underline{2x^2 + 4x} \\
-5x + 6
\end{array}
$$

So $x^4 + 4x^3 + 6x^2 - x + 6 =$
$$(x^2 + 2x)(x^2 + 2x + 2) + (-5x + 6)$$

2. Let $P(x) = x^4 + x^3 + 4$ and $D(x) = x^2 - 3$. Find the polynomials $Q(x)$ and $R(x)$ such that $P(x) = D(x) \cdot Q(x) + R(x)$.

$$
\begin{array}{r}
x^2 \; + x + 3 \\
\hline
x^2 - 3 \,)\overline{\, x^4 + x^3 + 0x^2 + 0x \;\; + 4 \,} \\
\underline{x^4 \qquad\quad - 3x^2 \qquad\qquad} \\
x^3 + 3x^2 + 0x \quad\;\; \\
\underline{x^3 \qquad\quad - 3x \quad\;\;} \\
3x^2 + 3x \; + 4 \\
\underline{3x^2 \qquad\quad - 9} \\
3x + 13
\end{array}
$$

So $x^4 + x^3 + 4 = (x^2 - 3)(x^2 + x + 3) +$
$$(3x + 13)$$

B. **Synthetic division** is a very important tool used in factoring, in finding solutions to $P(x) = 0$, and in evaluating polynomials, that is, finding the value of $P(c)$. This is a skill that is developed by practice.

Remember to insert a 0 in the corresponding place when the x^k term is missing. When $D(x) = x - c$, the Division Algorithm becomes the **Remainder Theorem**: $P(x) = (x - c) \cdot Q(x) + r$, where $r = P(c)$. This means that the remainder of $P(x)$ when divided by $(x - c)$ is the same as the functional value at $x = c$. When the remainder is zero, the Remainder Theorem yields the **Factor Theorem**: $P(c) = 0$ if and only if $x - c$ is a factor of $P(x)$.

3. Verify the Remainder Theorem by dividing $P(x) = x^3 + 2x^2 - 10x$ by $x - 3$. Then find $P(3)$.

Using long division we have:

$$
\begin{array}{r}
x^2 + 5x + 5 \\
x - 3 \overline{\smash{)}\ x^3 + 2x^2 - 10x\ + 0} \\
\underline{x^3 - 3x^2} \\
5x^2 - 10x \\
\underline{5x^2 - 15x} \\
5x\ + 0 \\
\underline{5x - 15} \\
15
\end{array}
$$

So $x^3 + 2x^2 - 10x = (x - 3)(x^2 + 5x + 5) + 15$.

Using synthetic division we have:
$x - 3 = 0 \quad \Leftrightarrow \quad x = 3$

$$
\begin{array}{r|rrrr}
3 & 1 & 2 & -10 & 0 \\
  &   & 3 & 15 & 15 \\
\hline
  & 1 & 5 & 5 & 15
\end{array}
$$

We verify the Remainder theorem by finding $P(3)$.
$P(3) = (3)^3 + 2(3)^2 - 10(3) = 27 + 18 - 30$
$\qquad = 15$

4. Verify the Remainder Theorem by dividing $P(x) = x^5 - 3x^3 - 7x^2 - 5x + 2$ by $x - 2$. Then find $P(2)$.

$$
\begin{array}{r|rrrrrr}
2 & 1 & 0 & -3 & -7 & -5 & 2 \\
  &   & 2 & 4 & 2 & -10 & -30 \\
\hline
  & 1 & 2 & 1 & -5 & -15 & -28
\end{array}
$$

So the remainder is -28.
Note: Remember to insert 0 for the coefficient of x^4.
We verify the Remainder theorem by finding $P(2)$.
$P(2) = (2)^5 - 3(2)^3 - 7(2)^2 - 5(2) + 2$
$\qquad = 32 - 24 - 28 - 10 + 2$
$\qquad = -28$

$x - 2 = 0 \quad \Leftrightarrow \quad x = 2$

$$
\begin{array}{r|rrrrrr}
2 & 1 & 0 & -3 & -7 & -5 & 2 \\
  &   & 2 & 4 & 2 & -10 & -28 \\
\hline
  & 1 & 2 & 1 & -5 & -15 & -28
\end{array}
$$

$P(2) = 2^5 - 3(2)^3$

5. Let $P(x) = 6x^3 - x^2 - 31x - 24$. Show that $P(-1) = 0$ and use this fact to factor $P(x)$ completely

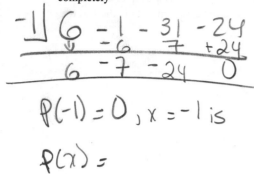

$$-1 \underline{\begin{array}{r} 6 \quad -1 \quad -31 \quad -24 \end{array}}$$

$P(-1) = 0, \; x = -1 \; is$

$P(x) =$

$$-1 \begin{array}{|rrrr} 6 & -1 & -31 & -24 \\ & -6 & 7 & 24 \\ \hline 6 & -7 & -24 & 0 \end{array}$$

So $P(-1) = 0$, $x = -1$ is a zero, and $x + 1$ is a factor. Then
$$\begin{aligned} P(x) &= 6x^3 - x^2 - 31x - 24 \\ &= (x + 1)(6x^2 - 7x - 24) \\ &= (x + 1)(3x - 8)(2x + 3). \end{aligned}$$

6. Let $P(x) = 2x^3 - 7x^2 - 7x + 30$. Show that $P(-2) = 0$ and use this fact to factor $P(x)$ completely

$$-2 \begin{array}{|rrrr} 2 & -7 & -7 & 30 \\ & -4 & 22 & -30 \\ \hline 2 & -11 & 15 & 0 \end{array}$$

So $P(-2) = 0$, $x = -2$ is a zero, and $x + 2$ is a factor.
From the synthetic division table, we have the factorization:
$$\begin{aligned} P(x) &= 2x^3 - 7x^2 - 7x + 30 \\ &= (x + 2)(2x^2 - 11x + 15) \\ &= (x + 2)(2x - 5)(x - 3) \end{aligned}$$

7. Find a polynomial of degree 5 that has zeros at -5, -2, 1, 2, and 3.

$$(x+5)(x+2)(x-1)(x-2)(x-3)$$

$$x-1 \left(x^2 + 2x + 5x + 10 \right)\left(x^2 - 3x - 2x + 6 \right)$$

$$(x-1)$$

Since $P(c) = 0$ if and only if $x - c$ is a factor,
$$\begin{aligned} P(x) &= [x - (-5)][x - (-2)][x - 1][x - 2][x - 3] \\ &= (x + 5)(x + 2)(x - 1)(x - 2)(x - 3) \\ &= x^5 + x^4 - 21x^3 + 11x^2 + 68x - 60. \end{aligned}$$

Section 3.3   Real Zeros of Polynomials
Key Ideas
A. Rational roots.
B. Descartes' Rule of Signs.
C. Upper and lower bounds for roots.
D. Systematic procedure for finding all rational roots.
E. Using graphing devices.

A. If $\dfrac{p}{q}$ is a rational root of the polynomial equation $a_n x^n + a_{n-1} x^{n-1} + \cdots + a_1 x^1 + a_0 = 0$, where

$a_n \neq 0$ (not necessarily $a_n = 1$), $a_0 \neq 0$, and a_i is an integer, then $\dfrac{p}{q}$ is of the form $\pm \dfrac{\text{factor of } a_0}{\text{factor of } a_n}$.

1. What numbers could be the integer zeros of the polynomials?

(a) $P(x) = x^5 + 25x^4 - 2x^3 + 121x^2 - x + 12$

$\underline{\pm 1, 2, 3, 4, 6, 12}$
$\ \ \ \pm 1$

When the leading coefficient is 1, then the roots must be factors of the constant term.
So the possible integer zeros are factors of 12 which are $\pm 1, \pm 2, \pm 3, \pm 4, \pm 6, \pm 12$.

(b) $Q(x) = x^7 - 8x^4 + 9x^2 - 5x + 16$

$\pm \dfrac{1, 2, 4, 8, 16}{1}$

So the possible integer zeros are factors of 16 which are $\pm 1, \pm 2, \pm 4, \pm 8, \pm 16$.

(c) $R(x) = x^6 + 11x^5 + 7x^3 - 6x + 24$

$\underline{\pm 1, 2, 3, 4, 6, 8, 12, 24}$
$\ \ \ \ \ 1$

So the possible integer zeros are factors of 24 which are $\pm 1, \pm 2, \pm 3, \pm 4, \pm 6, \pm 8, \pm 12, \pm 24$.

2. Find all *possible* rational roots of the equation, $3x^3 - 5x^2 - 16x + 12 = 0$ and then solve it completely.

$\dfrac{1, 2, 3, 4, 6, 12}{1, 3}$

$1, 2, 3, 4, 6, 12, \tfrac{1}{3}$

$\pm \dfrac{\text{factor of } 12}{\text{factor of } 3} = \pm 1, \pm 2, \pm 3, \pm 4, \pm 6, \pm 12, \pm \tfrac{1}{3},$

$\pm \tfrac{2}{3}, \pm \tfrac{4}{3}$. We try the positive integers first.

$$\begin{array}{r|rrrr}
1 & 3 & -5 & -16 & 12 \\
  &   & 3 & -2 & -18 \\ \hline
  & 3 & -2 & -18 & -6
\end{array} \qquad
\begin{array}{r|rrrr}
2 & 3 & -5 & -16 & 12 \\
  &   & 6 & 2 & -28 \\ \hline
  & 3 & 1 & -14 & -16
\end{array}$$

$$\begin{array}{r|rrrr}
3 & 3 & -5 & -16 & 12 \\
  &   & 9 & 12 & -12 \\ \hline
  & 3 & 4 & -4 & 0 \quad \text{Yes}
\end{array}$$

Factoring the resulting polynomial yields the last two solutions:

$3x^2 + 4x - 4 = (3x - 2)(x + 2)$

$\begin{aligned} 3x - 2 &= 0 & x + 2 &= 0 \\ 3x &= 2 & x &= -2 \\ x &= \tfrac{2}{3} \end{aligned}$

Solutions: $x = \tfrac{2}{3}$, $x = 3$, and $x = -2$.

3. Find all *possible* rational roots of the equation, $x^4 - x^3 - 11x^2 + 9x + 18 = 0$ and then solve it completely.

$\pm\{1, 2, 3, 4, 6, 18\}$

$$
\begin{array}{r|rrrrr}
1 & 1 & -1 & -11 & +9 & +18 \\
  &   & 1 & 0   & -11 & -20 \\ \hline
  & 1 & 0 & -11 & -20 & -2 \; 16 \; \text{NO}
\end{array}
$$

$$
\begin{array}{r|rrrrr}
2) & 1 & -1 & -11 & 9 & 18 \\
   &   & 2 & 2   & -18 & -18 \\ \hline
3) & 1 & 1 & -9 & -9 & 0 \quad \text{yes} \\
   &   & 3 & 12 & 9 \\ \hline
   & 1 & 4 & +3 & 0
\end{array}
$$

$2, 3,$

$x^2 + 4x + 3$

$(x+1)(x+3)$

$x = -1 \qquad x = -3$

$\pm\dfrac{\text{factor of } 18}{\text{factor of } 1} = \pm1, \pm2, \pm3, \pm6, \pm9, \pm18$

There are 12 possible rational roots, we try the positive integers first.

$$
\begin{array}{r|rrrrr}
1 & 1 & -1 & -11 & 9 & 18 \\
  &   & 1 & 0   & -11 & -2 \\ \hline
  & 1 & 0 & -11 & -2 & 16 \quad \text{No}
\end{array}
$$

$$
\begin{array}{r|rrrrr}
2 & 1 & -1 & -11 & 9 & 18 \\
  &   & 2 & 2   & -18 & -18 \\ \hline
  & 1 & 1 & -9 & -9 & 0 \quad \text{Yes}
\end{array}
$$

Since the resulting polynomial is $x^3 + x^2 - 9x - 9$, $\pm2, \pm6$, and ±18 can be eliminated as possible roots. 3 and 9 are the only positive roots left.

$$
\begin{array}{r|rrrr}
3 & 1 & 1 & -9 & -9 \\
  &   & 3 & 12 & 9 \\ \hline
  & 1 & 4 & 3 & 0 \quad \text{Yes}
\end{array}
$$

Factoring the resulting polynomial yields the last two solutions:
$$x^2 + 4x + 3 = (x+1)(x+3)$$
$$\begin{array}{ll} x + 1 = 0 & x + 3 = 0 \\ x = -1 & x = -3 \end{array}$$

The solutions are $x = -3$, $x = -1$, $x = 2$, and $x = 3$.

4. Find all *possible* rational roots of the equation,
$10x^4 + 11x^3 - 51x^2 - 32x + 20 = 0$ and then solve
it completely.

$1, 2, 4, 5, 10, 20$

$1, 2, 5, 10$

$$\pm\frac{\text{factor of 20}}{\text{factor of 10}} = \pm1, \pm2, \pm4, \pm5, \pm10, \pm20, \pm\frac{1}{2},$$

$$\pm\frac{1}{5}, \pm\frac{1}{10}, \pm\frac{2}{5}, \pm\frac{4}{5}, \pm\frac{5}{2}$$

There are 24 possible rational roots, many
possibilities to try. We try the integers first.

1	10	11	−51	−32	20	
		10	21	−30	−62	
	10	21	−30	−62	−42	No

2	10	11	−51	−32	20	
		20	62	22	−20	
	10	31	11	−10	0	Yes

We try $x = 2$ again in the resulting polynomial
$10x^3 + 31x^2 + 11x - 10$

2	10	31	11	−10	
		20	102	226	
	10	51	113	216	No

*Note: Any number larger than 2 won't work. So
we start trying negative numbers.*

−1	10	31	11	−10	
		−10	−21	10	
	10	21	−10	0	Yes

Factoring the resulting polynomial yields:
$10x^2 + 21x - 10 = (2x + 5)(5x - 2)$

$$2x + 5 = 0 \qquad\qquad 5x - 2 = 0$$
$$2x = -5 \qquad\qquad 5x = 2$$
$$x = -\frac{5}{2} \qquad\qquad x = \frac{2}{5}$$

Solutions: $x = -\frac{5}{2}$, $x = -1$, $x = \frac{2}{5}$, and $x = 2$.

B. Descartes' Rule of Signs states that if $P(x)$ is a polynomial with real coefficients, then

> 1. The number of positive real solutions of $P(x)$ is at most equal to the number of variations in sign in $P(x)$ or is less than that by an even number.
> 2. The number of negative real solutions of $P(x)$ is at most equal to the number of variations in sign in $P(-x)$ or is less than that by an even number.

This is very useful when there is an odd number of variations in sign. In this case you are guaranteed at least one real solution, either negative or positive.

5. Use Descartes' Rule of Signs to help find all rational zeros of the polynomial, $P(x) = x^4 - 2x^3 - x^2 - 4x - 6$, then factor it completely.

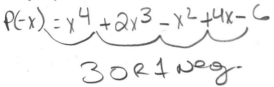

$$P(-x) = x^4 + 2x^3 - x^2 + 4x - 6$$

$$3 \text{ OR } 1 \text{ neg.}$$

$P(x)$ has one variation in sign, so there is one positive real zero.

$P(-x) = x^4 + 2x^3 - x^2 + 4x - 6$ has three variations in sign, so there are either three or one negative real zeros.

These possible zeros are: $\pm 1, \pm 2, \pm 3, \pm 6$.

$$
\begin{array}{r|rrrrr}
-1 & 1 & -2 & -1 & -4 & -6 \\
   &   & -1 & 3 & -2 & 6 \\
\hline
   & 1 & -3 & 2 & -6 & 0 \quad \text{Yes}
\end{array}
$$

Looking at the resulting degree three polynomial,
$Q(x) = x^3 - 3x^2 + 2x - 6$ and
$Q(-x) = -x^3 - 3x^2 - 2x - 6$,
we can see that there is still one positive zero (we cannot increase the number of positive zeros), but no more negative zeros. Trying the possible positive numbers, we get

$$
\begin{array}{r|rrrr}
1 & 1 & -3 & 2 & -6 \\
  &   & 1 & -2 & 0 \\
\hline
  & 1 & -2 & 0 & -6
\end{array}
\qquad
\begin{array}{r|rrrr}
2 & 1 & -3 & 2 & -6 \\
  &   & 2 & -2 & 0 \\
\hline
  & 1 & -1 & 0 & -6
\end{array}
$$

$$
\begin{array}{r|rrrr}
3 & 1 & -3 & 2 & -6 \\
  &   & 3 & 0 & -6 \\
\hline
  & 1 & 0 & 2 & 0 \quad \text{Yes}
\end{array}
$$

Since the result $x^2 + 2$ does not factor over the real numbers, we are finished.

$$
\begin{aligned}
P(x) &= x^4 - 2x^3 - x^2 - 4x - 6 \\
     &= (x+1)(x-3)(x^2+2)
\end{aligned}
$$

C. The number a is a **lower bound** and the number b is an **upper bound** for the roots of a polynomial equation if every real root c of the equation satisfies $a \le c \le b$. Bounds a and b are found in the following way. Let $P(x)$ be a polynomial with real coefficients.

> 1. If we divide $P(x)$ by $x - b$ (where $b > 0$) using synthetic division, and if the row that contains the quotient and remainder has no negative entries, then b is an upper bound for the real roots for $P(x) = 0$.
> 2. If we divide $P(x)$ by $x - a$ (where $a < 0$) using synthetic division, and if the row that contains the quotient and remainder has entries that are alternately nonpositive and nonnegative, then a is a lower bound for the real roots for $P(x) = 0$.

6. Show that all the real roots of the equation $2x^4 + 11x^3 + 4x^2 - 44x - 48 = 0$ lie between -6 and 3.

$$
\begin{array}{r|rrrrr}
-6 & 2 & 11 & 4 & -44 & -48 \\
   &   & -12 & 6 & -60 & 624 \\
\hline
   & 2 & -1 & 10 & -104 & 576
\end{array}
$$

The entries in the last row alternate between nonpositive and nonnegative, therefore -6 is a lower bound.

$$
\begin{array}{r|rrrrr}
3 & 2 & 11 & 4 & -44 & -48 \\
  &   & 6 & 51 & 165 & 363 \\
\hline
  & 2 & 17 & 55 & 121 & 315
\end{array}
$$

Each entry in the last row is positive, therefore 3 is an upper bound.

7. Show that -1 is *not* a lower bound but -2 *is* a lower bound for the roots of the polynomial equation. Then find an integer upper bound for the roots.
$4x^5 - 6x^3 + 7x^2 - 8x - 16 = 0$

$$
\begin{array}{r|rrrrrr}
-1 & 4 & 0 & -6 & 7 & -8 & -16 \\
   &   & -4 & 4 & 2 & -9 & 17 \\
\hline
   & 4 & -4 & -2 & 9 & -17 & 1 \\
   &   & \uparrow & & \uparrow & &
\end{array}
$$

The entries in the last row do *not* alternate in sign, so -1 is not a lower bound.

$$
\begin{array}{r|rrrrrr}
-2 & 4 & 0 & -6 & 7 & -8 & -16 \\
   &   & -8 & 16 & -20 & 26 & -36 \\
\hline
   & 4 & -8 & 10 & -13 & 18 & -52
\end{array}
$$

The entries in the last row alternates in sign, so -2 is a lower bound.

$$
\begin{array}{r|rrrrrr}
1 & 4 & 0 & -6 & 7 & -8 & -16 \\
  &   & 4 & 4 & -2 & 5 & -3 \\
\hline
  & 4 & 4 & -2 & 5 & -3 & -19
\end{array}
$$

$$
\begin{array}{r|rrrrrr}
2 & 4 & 0 & -6 & 7 & -8 & -16 \\
  &   & 8 & 16 & 20 & 54 & 92 \\
\hline
  & 4 & 8 & 10 & 27 & 46 & 76
\end{array}
$$

Each entry in the last row is positive, therefore 2 is an upper bound.

D. Roots of the equation $P(x) = 0$ are the same as the zeros of the polynomial $P(x)$. We summarize the steps in finding the roots of a polynomial $P(x)$:

1. List all possible rational roots.
2. Use Descartes' Rule of Signs to determine the possible number of positive and negative roots.
3. Use synthetic division to test for possible roots (in order). Stop when you have reached an upper or lower bound or when all predicted positive or negative roots have been found.
4. If you find a root, repeat the process with the quotient. Remember you do not have to check roots that did not work before. If you reach a quotient that you can apply the quadratic formula to or that you can factor, then use those techniques to solve.

8. Find all rational zeros of the equation
$x^3 - 3x^2 - 10x + 24 = 0$.

1. List of possible roots:
$\pm 1, \pm 2, \pm 3, \pm 4, \pm 6, \pm 8, \pm 12, \pm 24$.

2. Positive roots: two variation in sign, so there are two or no positive roots.
Negative roots:
$P(-x) = -x^3 - 3x^2 + 10x + 24$,
so there is only one possible negative root.

3.

$$
\begin{array}{r|rrrr}
1 & 1 & -3 & -10 & 24 \\
  &   & 1 & -2 & -12 \\
\hline
  & 1 & -2 & -12 & 12
\end{array}
$$

$$
\begin{array}{r|rrrr}
2 & 1 & -3 & -10 & 24 \\
  &   & 2 & -2 & -24 \\
\hline
  & 1 & -1 & -12 & 0 \quad \text{Yes}
\end{array}
$$

4. We factor the quotient
$x^2 - x - 12 = (x - 4)(x + 3)$

$$x - 4 = 0 \qquad\qquad x + 3 = 0$$
$$x = 4 \qquad\qquad x = -3$$

The zeros are $x = 2$, $x = 4$, and $x = -3$.

9. Find all rational zeros of the equation
$$2x^4 + 11x^3 + 4x^2 - 44x - 48 = 0.$$

1. List of possible roots: $\pm\frac{1}{2}, \pm1, \pm\frac{3}{2}, \pm2, \pm3,$ $\pm4, \pm6, \pm8, \pm12, \pm16, \pm24, \pm48.$

2. Positive roots: one variation in sign, so there is one positive root.
   Negative roots:
   $$P(-x) = 2x^4 - 11x^3 + 4x^2 + 44x - 48,$$
   so there are three or one possible negative roots.

3. Upper and lower bounds, -6 and 3, were established in problem 6. We reduce the possible roots to $\pm\frac{1}{2}, \pm1, \pm\frac{3}{2}, \pm2, -3,$ and -4.

$\frac{1}{2}$	2	11	4	-44	-48
		1	6	5	no chance
	2	12	10	-39

1	2	11	4	-44	-48
		2	13	17	-27
	2	13	17	-27	-75

$\frac{3}{2}$	2	11	4	-44	-48
		3	21	$\frac{75}{2}$
	2	14	25	no chance

2	2	11	4	-44	-48
		4	30	68	48
	2	15	34	24	0 Yes

Since there is only one positive root, 2, we consider the negative roots.

$-\frac{1}{2}$	2	15	34	24
		-1	-7	$-\frac{27}{2}$
	2	14	27	no chance

-1	2	15	34	24
		-2	-13	-21
	2	13	21	3

$-\frac{3}{2}$	2	15	34	24
		-3	-18	-24
	2	12	16	0 Yes

4. We factor the quotient
$$2x^2 + 12x + 16 = 2(x + 4)(x + 2)$$
$$x + 4 = 0 \qquad\qquad x + 2 = 0$$
$$x = -4 \qquad\qquad\quad x = -2.$$

The zeros are $x = -2$, $x = -4$, $x = -\frac{3}{2}$, and $x = 2$.

E. We can use the upper and lower bounds theorems to determine the x-bounds of the viewing rectangle that will contain all solutions to $P(x) = 0$.

10. For each polynomial: (i) Find upper and lower bounds. (ii) Use a graphing device to find all solutions to $P(x) = 0$ to two decimal places.

(a) $P(x) = x^3 + 6x^2 - x - 5$

(i)

$$
\begin{array}{r|rrrr}
1 & 1 & 6 & -1 & -5 \\
  &   & 1 & 7 & 6 \\
\hline
  & 1 & 7 & 6 & 1
\end{array}
$$

So $x = 1$ is an upper bound.

$$
\begin{array}{r|rrrr}
-6 & 1 & 6 & -1 & -5 \\
   &   & -6 & 0 & 6 \\
\hline
   & 1 & 0 & -1 & 1
\end{array}
$$
$\qquad\quad \uparrow \quad \uparrow \quad \uparrow \qquad$ Does not alternate.

$$
\begin{array}{r|rrrr}
-7 & 1 & 6 & -1 & -5 \\
   &   & -7 & 7 & -42 \\
\hline
   & 1 & -1 & 6 & -47
\end{array}
$$

$x = -7$ is a lower bound. Since $P(-6) > 0$ and $P(-7) < 0$, there is a root between -7 and -6.

(ii) We graph $P(x) = x^3 + 6x^2 - x - 5$ in the viewing rectangle $[-7, 1]$ by $[-5, 5]$.

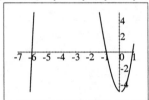

The solutions are $x \approx -6.03$, $x \approx -0.90$, and $x \approx 0.92$.

(b) $P(x) = x^4 + x^3 + 2x^2 - 1$

(i)

$$
\begin{array}{r|rrrrr}
1 & 1 & 1 & 2 & 0 & -1 \\
  &   & 1 & 2 & 4 & 4 \\
\hline
  & 1 & 2 & 4 & 4 & 3
\end{array}
$$

So $x = 1$ is an upper bound.

$$
\begin{array}{r|rrrrr}
-1 & 1 & 1 & 2 & 0 & -1 \\
   &   & -1 & 0 & -2 & 2 \\
\hline
   & 1 & 0 & 2 & -2 & 1
\end{array}
$$

$x = -1$ is a lower bound.
(Note: zero is both nonnegative and nonpositive.)

(ii) We graph $P(x) = x^4 + x^3 + 2x^2 - 1$ in the viewing rectangle $[-1, 1]$ by $[-5, 5]$.

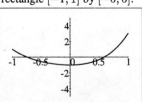

The solutions are $x \approx -0.74$ and $x \approx 0.58$.

Section 3.4   Complex Numbers

Key Ideas
- **A.**   Arithmetic operations on complex numbers.
- **B.**   Dividing complex numbers, the complex conjugate.
- **C.**   Square roots of negative numbers.
- **D.**   Imaginary roots of quadratic equations.

A.   **Complex numbers** are expressions of the form $a + b\,i$, where a and b are real numbers and i is the **imaginary number** defined by the equation $i^2 = -1$. The term a is called the **real part** and the term b is called the **imaginary part**. Two complex numbers are equal if their real parts are equal and their imaginary parts are equal. The number bi, $b \neq 0$, is called a **pure imaginary number**. You add two complex numbers by adding the real part and adding the imaginary part. The distributive property or the FOIL method can be used along with the identity $i^2 = -1$ to multiply two complex numbers.

1. Add or subtract.
   (a) $(3 + 4i) - (4 - 5i)$

$$(3 + 4i) - (4 - 5i) = 3 + 4i - 4 + 5i$$
$$= (3 - 4) + (4 + 5)i$$
$$= -1 + 9i$$

   (b) $\left(\sqrt{3} - i\right) + \left(1 + \sqrt{3}\,i\right)$

$$\left(\sqrt{3} - i\right) + \left(1 + \sqrt{3}\,i\right) = \sqrt{3} - i + 1 + \sqrt{3}\,i$$
$$= \sqrt{3} + 1 + \left(-1 + \sqrt{3}\right)i$$

This is as far as this expression can be simplified.

2. Multiply.
   (a) $(2 - 3i)(7 - 5i)$

We use the distributive property.
$$(2 - 3i)(7 - 5i) = 2(7 - 5i) + (-3i)(7 - 5i)$$
$$= 14 - 10i - 21i + 15i^2$$
$$= 14 - 31i - 15$$
$$= -1 - 31i$$

   (b) $(3 + 2i)(4 + 9i)$

We use the FOIL method.
$$(3 + 2i)(4 + 9i)$$
$$= (3)(4) + (3)(9i) + (2i)(4) + (2i)(9i)$$
$$= 12 + 27i + 8i + 18i^2$$
$$= 12 + 35i + 18(-1)$$
$$= -6 + 35i$$

Note: A different method was used in part (a) and (b) to show how both methods work.

3. Show that $1 + 2i$ is a solution to $x^2 - 2x + 5 = 0$.

Substitute $1 + 2i$ for x and simplify.
$$(1 + 2i)^2 - 2(1 + 2i) + 5$$
$$= [(1)^2 + 2(1)(2i) + (2i)^2] - 2 - 4i + 5$$
$$= 1 + 4i + 4i^2 - 4i + 3$$
$$= 1 - 4 + 3 = 0$$

So $1 + 2i$ is a solution.

B. The difference of squares product $(a - bi)(a + bi) = a^2 - (bi)^2 = a^2 + b^2$ is used to simplify fractional expressions or to divide complex numbers. The numbers $a + bi$ and $a - bi$ are called **complex conjugates.**

4. Simplify $\dfrac{2}{3 - i}$

$$\dfrac{2}{3 - i} = \left(\dfrac{2}{3 - i}\right)\left(\dfrac{3 + i}{3 + i}\right)$$

$$= \dfrac{2(3 + i)}{9 - i^2} = \dfrac{2(3 + i)}{9 + 1}$$

$$= \dfrac{2(3 + i)}{10} = \dfrac{3 + i}{5}$$

$$\text{or} = \tfrac{3}{5} + \tfrac{1}{5} i$$

5. Divide $4 - 3i$ by $3 + 2i$.

$$(4 - 3i) \div (3 + 2i) = \dfrac{4 - 3i}{3 + 2i}$$

$$= \left(\dfrac{4 - 3i}{3 + 2i}\right)\left(\dfrac{3 - 2i}{3 - 2i}\right)$$

$$= \dfrac{12 - 8i - 9i + 6i^2}{9 - 4i^2}$$

$$= \dfrac{12 - 17i - 6}{9 + 4} = \dfrac{6 - 17i}{13}$$

$$\text{or} = \tfrac{6}{13} - \tfrac{17}{13} i$$

C. If $-r < 0$, then the square roots of $-r$, are $i\sqrt{r}$ and $-i\sqrt{r}$, where $i\sqrt{r}$ is called the **principal square root** of $-r$. Remember $\sqrt{a}\sqrt{b} = \sqrt{ab}$ *only* when a and b are *not both negative.* For example: if a and b are both negative then $\sqrt{-3}\sqrt{-6} = i\sqrt{3} \cdot i\sqrt{6} = 3\sqrt{2}\,i^2 = -3\sqrt{2}$, while $\sqrt{(-3)(-6)} = \sqrt{18} = 3\sqrt{2}$.

6. Evaluate.
   (a) $\sqrt{-1}$

$$\sqrt{-1} = i$$

   (b) $\sqrt{-9}$

$$\sqrt{-9} = i\sqrt{9} = 3i$$

   (c) $\sqrt{-12}$

$$\sqrt{-12} = i\sqrt{12} = 2\sqrt{3}\,i$$

   (d) $-\sqrt{-25}$

$$-\sqrt{-25} = -i\sqrt{25} = -5i$$

Note: $\sqrt{}$ *is a grouping symbol, that means "do the work on the inside first."*

110

D. In the quadratic formula, if the discriminant $D = b^2 - 4ac$ is less than 0, then there are no real solutions to $ax^2 + bx + c = 0$. Both solutions will be complex and they will be complex conjugates.

7. Solve $x^2 + 4x + 6 = 0$

$x^2 + 4x + 6 = 0$

Setting $a = 1$, $b = 4$, $c = 6$ and substituting into the quadratic formula:

$$x = \frac{-b \pm \sqrt{b^2 - 4ac}}{2a}$$

$$= \frac{-(4) \pm \sqrt{(4)^2 - 4(1)(6)}}{2(1)}$$

$$= \frac{-4 \pm \sqrt{16 - 24}}{2} = \frac{-4 \pm \sqrt{-8}}{2}$$

$$= \frac{-4 \pm 2\sqrt{2}\,i}{2} = \frac{2(-2 \pm \sqrt{2}\,i)}{2}$$

$$= -2 \pm \sqrt{2}\,i.$$

8. Solve $5x^2 - 6x + 5 = 0$

$5x^2 - 6x + 5 = 0$

Setting $a = 5$, $b = -6$, $c = 5$ and substituting into the quadratic formula:

$$x = \frac{-b \pm \sqrt{b^2 - 4ac}}{2a}$$

$$= \frac{-(-6) \pm \sqrt{(-6)^2 - 4(5)(5)}}{2(5)}$$

$$= \frac{6 \pm \sqrt{36 - 100}}{10} = \frac{6 \pm \sqrt{-64}}{10} = \frac{6 \pm 8\,i}{10}$$

$$= \frac{2(3 \pm 4\,i)}{10} = \frac{3 \pm 4\,i}{5}$$

or $x = \frac{3}{5} \pm \frac{4}{5}\,i.$

Section 3.5　Complex Zeros and the Fundamental Theorem of Algebra

Key Ideas

A. The Fundamental Theorem of Algebra.

B. Conjugate roots.

C. Linear and quadratic factors.

A. This section contains many important theorems and ideas. Notice that these theorems *only* say that these zeros *exists*, not *how* or *where* to find them. Section 3.4 explain *where* to look for the zeros and how to find them. The **Fundamental Theorem of Algebra** states that every polynomial, $P(x) = a_n x^n + a_{n-1} x^{n-1} + \cdots + a_1 x^1 + a_0$, $n \geq 1$, $a_n \neq 0$, with complex coefficients has at least one complex zero. The **Complete Factorization Theorem** states that $P(x)$ can always be factored completely into $P(x) = a(x - c_1)(x - c_2)\cdots(x - c_{n-1})(x - c_n)$ where $a, c_1, c_2, \ldots, c_{n-1}, c_n$ are complex numbers, $a \neq 0$. If the factor $(x - c)$ appears k times in the complete factorization of $P(x)$, then c is said to have **multiplicity k**. The **Zeros Theorem** states that every polynomial of degree $n \geq 1$ has exactly n zeros, provided that a zero of multiplicity k is counted k times.

1. Find all solutions of the equation $x^2 + 36 = 0$.

$$x^2 + 36 = 0 \quad \Leftrightarrow \quad x^2 = -36 \quad \Leftrightarrow$$
$$x = \pm\sqrt{-36} = \pm i\sqrt{36} = \pm 6i$$

2. Find all solutions of the equation $x^2 - 2x + 5 = 0$.

Using the quadratic formula we have:

$$x = \frac{-(-2) \pm \sqrt{(-2)^2 - 4(1)(5)}}{2(1)} = \frac{2 \pm \sqrt{4 - 20}}{2}$$

$$= \frac{2 \pm \sqrt{-16}}{2} = \frac{2 \pm i\sqrt{16}}{2} = \frac{2 \pm 4i}{2}$$

$$= \frac{2(1 \pm 2i)}{2} = 1 \pm 2i.$$

So the solutions are $1 - 2i$ and $1 + 2i$.

3. Find a polynomial that satisfies the given description.

   (a) A polynomial $P(x)$ of degree 4 with zeros -2, -1, 3, and 5, and with constant coefficient -60.

$$P(x) = a[x - (-2)][x - (-1)][x - 3][x - 5]$$
$$= a(x + 2)(x + 1)(x - 3)(x - 5)$$
$$= a(x^2 + 3x + 2)(x^2 - 8x + 15)$$
$$= a(x^4 - 5x^3 - 7x^2 + 29x + 30)$$

Since $-60 = a(30)$, $a = -2$; hence

$$P(x) = -2(x^4 - 5x^3 - 7x^2 + 29x + 30)$$
$$= -2x^4 + 10x^3 + 14x^2 - 58x - 60.$$

(b) A polynomial $Q(x)$ of degree 6 with zeros 0, 1, and -3, where 1 is zero of multiplicity 2 and -3 is zero of multiplicity 3.

$$Q(x) = a[x - (0)][x - (1)]^2[x - (-3)]^3$$
$$= ax(x - 1)^2(x + 3)^3$$
$$= ax(x^2 - 2x + 1)(x^3 + 9x^2 + 27x + 27)$$
$$= ax(x^5 + 7x^4 + 10x^3 - 18x^2 - 27x + 27)$$
$$= a(x^6 + 7x^5 + 10x^4 - 18x^3 - 27x^2 + 27x)$$

Since no information is given about Q other than its zeros, we choose $a = 1$ and we get

$$Q(x) = x^6 + 7x^5 + 10x^4 - 18x^3 - 27x^2 + 27x.$$

(c) A polynomial $R(x)$ of degree 5 with zeros $2 - i$, $2 + i$ and 2, where 2 is zero of multiplicity 3. The leading coefficient is -1.

$$R(x) = a[x - (2 - i)][x - (2 + i)][x - (2)]^3$$
$$= a[(x - 2) + i][(x - 2) - i)][x - (2)]^3 \ ^\#$$
$$= a[(x^2 - 4x + 4) - i^2](x^3 - 6x^2 + 12x - 8)$$
$$= a(x^2 - 4x + 5)(x^3 - 6x^2 + 12x - 8)$$
$$= a(x^5 - 10x^4 + 41x^3 - 86x^2 + 92x - 40)$$

Since the leading term is ax^5 and we are given that the leading coefficient is -1, $a = -1$. So we have

$$R(x) = -x^5 + 10x^4 - 41x^3 + 86x^2 - 92x + 40.$$

 # Note: The difference of squares was used to multiply $[(x - 2) + i][(x - 2) - i] = (x - 2)^2 - i^2.$

B. The **Conjugate Zeros Theorem** tell us that imaginary roots of polynomial equations with real coefficients come in pairs. If the polynomial P has real coefficients, and if the complex number $a + bi$ is a zero of P, then its **complex conjugate** $a - bi$ is also a zero of P.

4. Find a polynomial $P(x)$ of degree 5, with real coefficients, that has zeros $3 - 2i$, 0, 2, and 4.

Since $3 - 2i$ is a zero, so is its conjugate, $3 + 2i$. Again we will assume $a = 1$. So

$$P(x) = x[x - (3 - 2i)][x - (3 + 2i)](x - 2)(x - 4)$$
$$= x[(x - 3) + 2i][(x - 3) - 2i](x - 2)(x - 4)$$
$$= x[(x - 3)^2 - (2i)^2](x^2 - 6x + 8)$$
$$= x(x^2 - 6x + 9 - 4i^2)(x^2 - 6x + 8)$$
$$= x(x^2 - 6x + 13)(x^2 - 6x + 8)$$
$$= x(x^4 - 12x^3 + 57x^2 - 126x + 104)$$
$$= x^5 - 12x^4 + 57x^3 - 126x^2 + 104x.$$

5. $-1 - 3i$ is a zero of the polynomial $P(x) = 4x^5 + 4x^4 + 25x^3 - 56x^2 - 74x - 20$. Find the other zeros.

Since $-1 - 3i$ is a zero of $P(x)$, $-1 + 3i$ is also a zero of $P(x)$.

$$Q(x) = [x - (-1 - 3i)][x - (-1 + 3i)]$$
$$= [(x + 1) + 3i][(x + 1) - 3i]$$
$$= (x + 1)^2 - (3i)^2 = x^2 + 2x + 1 - 9i^2$$
$$= x^2 + 2x + 10$$

Since $Q(x)$ is factor, we use long division to get

$$
\require{enclose}
\begin{array}{r}
4x^3 - 4x^2 - 7x - 2 \\
x^2 + 2x + 10 \enclose{longdiv}{4x^5 + 4x^4 + 25x^3 - 56x^2 - 74x - 20} \\
\end{array}
$$

$$4x^5 + 8x^4 + 40x^3$$
$$-4x^4 - 15x^3 - 56x^2$$
$$-4x^4 - 8x^3 - 40x^2$$
$$-7x^3 - 16x^2 - 74x$$
$$-7x^3 - 14x^2 - 70x$$
$$-2x^2 - 4x - 20$$
$$-2x^2 - 4x - 20$$
$$0$$

Concentrating on the quotient and using the material from the previous sections, we have:

$R(x) = 4x^3 - 4x^2 - 7x - 2$ one change in sign
$R(-x) = -4x^3 - 4x^2 + 7x - 2$ two changes in sign.

So $R(x)$ has exactly one positive root and two or zero negative roots. Possible rational roots are: $\pm\frac{1}{4}, \pm\frac{1}{2},$ $\pm 1, \pm 2$. Checking for the positive root first,

$$
\begin{array}{r|rrrr}
2 & 4 & -4 & -7 & -2 \\
  &   & 8 & 8 & 2 \\
\hline
  & 4 & 4 & 1 & 0. \\
\end{array}
$$

Factoring the resulting polynomial gives us
$4x^2 + 4x + 1 = 0 \quad \Leftrightarrow \quad (2x + 1)^2 = 0 \quad \Rightarrow$
$2x + 1 = 0 \quad \Leftrightarrow \quad 2x = -1 \quad \Leftrightarrow \quad x = -\frac{1}{2}.$
So the zeros of $P(x)$ are: $-1 - 3i, -1 + 3i, 2$, and $-\frac{1}{2}$, where $-\frac{1}{2}$ has multiplicity 2.

C. If we do not use complex numbers, then a polynomial with real coefficients can always be factored into linear and quadratic factors. A quadratic polynomial with no real zeros is called **irreducible** over the real numbers. Such a polynomial cannot be factored without using complex numbers.

6. Let $P(x) = 5x^6 - 80x^2$.

(a) Factor P into linear and irreducible quadratic factors. Identify these factors.

$$P(x) = 5x^6 - 80x^2$$
$$= 5x^2(x^4 - 16)$$
$$= 5x^2(x^2 - 4)(x^2 + 4)$$
$$= 5x^2(x - 2)(x + 2)(x^2 + 4)$$

The linear factors are x, $x - 2$, and $x + 2$. The factor x has multiplicity 2.
The irreducible quadratic factor is $x^2 + 4$.

(b) Factor P completely.

$$P(x) = 5x^6 - 80x^2$$
$$= 5x^2(x - 2)(x + 2)(x^2 + 4)$$
$$= 5x^2(x - 2)(x + 2)(x - 2i)(x + 2i)$$

Section 3.6   Rational Functions

Key Ideas

A. Domain and intercepts of rational functions.
B. Arrow notation and asymptotes.
C. Multiplicity of factors and its effect on asymptotes.
D. Slant asymptotes.
E. Vertical asymptotes and end behavior on graphing devices.

A. A rational function is a function of the form $r(x) = \dfrac{P(x)}{Q(x)}$, where P and Q are polynomials, Q is not the zero polynomial. The domain of $r(x)$ is the set of real numbers where $Q(x) \neq 0$. The x-intercepts are the real numbers where $P(x) = 0$. The y-intercept is $r(0)$, provided 0 is in the domain of $r(x)$.

1. Find the domain, the x-intercepts, and the y-intercept of the function $r(x) = \dfrac{2x^2 + 5x + 2}{x^2 - 11x + 24}$.

$r(x) = \dfrac{2x^2 + 5x + 2}{x^2 - 11x + 24}$

$ = \dfrac{(2x + 1)(x + 2)}{(x - 3)(x - 8)}$ *Factor the numerator and the denominator.*

Domain: $x \neq 3, x \neq 8$ *denominator $\neq 0$*

x-intercepts: $x = -\frac{1}{2}$ and $x = -2$ *numerator $= 0$*

y-intercepts: $r(0) = \frac{2}{24} = \frac{1}{12}$

2. Find the domain, the x-intercepts, and the y-intercept of the function $r(x) = \dfrac{x^3 + 5x^2 - 6x}{x^2 - 9}$.

$r(x) = \dfrac{x^3 + 5x^2 - 6x}{x^2 - 9}$

$ = \dfrac{x(x + 6)(x - 1)}{(x - 3)(x + 3)}$ *Factor the numerator and the denominator.*

Domain: $x \neq 3, x \neq -3$ *denominator $\neq 0$*

x-intercepts: $x = 0, x = -6$ and $x = 1$ *numerator $= 0$*

y-intercepts: $r(0) = \frac{0}{-9} = 0$

B. The arrow notation in the table below is to help determine the behavior of a function around holes in its domain (vertical asymptotes) and end behavior.

Symbol	Meaning
$x \to a^-$	x approaches a from the left or x goes towards a with $x < a$
$x \to a^+$	x approaches a from the right or x goes towards a with $x > a$
$x \to -\infty$	x goes to negative infinity; that is, x decreases without bound
$x \to \infty$	x goes to infinity; that is, x increases without bound

The line $x = a$ is a **vertical asymptote** of the function $y = f(x)$ whenever $y \to -\infty$ or $y \to \infty$ as x approaches a from either the right side or the left side. Vertical asymptotes occur where the denominator equals zero. The line $y = b$ is a **horizontal asymptote** of the function $y = f(x)$ if $y \to b$ as $x \to -\infty$ or as $x \to \infty$. The end behavior of the rational function

$r(x) = \dfrac{a_n x^n + a_{n-1} x^{n-1} + \cdots + a_1 x^1 + a_0}{b_m x^m + b_{m-1} x^{m-1} + \cdots + b_1 x^1 + b_0}$ is completely determined by the leading terms of the numerator and denominator. There are three cases.

> Case 1. $\underline{n < m.}$ In this case, the denominator goes to infinity faster than the numerator does. So $y \to 0$ as $x \to -\infty$ and as $x \to \infty$. So y = 0 is a horizontal asymptote.
>
> Case 2. $\underline{n = m.}$ In this case, $y \to \dfrac{a_n}{b_m}$ as $x \to -\infty$ and as $x \to \infty$. So $y = \dfrac{a_n}{b_m}$ is a horizontal asymptote.
>
> Cases 3. $\underline{n > m.}$ Here the end behavior is the same as the end behavior of $y = \dfrac{a_n}{b_m} x^{n-m}$. (See slant asymptote in 'D' section for more information about this case.)

3. Sketch the graph of the function $y = \dfrac{x^2 - 4}{x^2 - 2x - 3}$.

First factor the numerator and denominator to find the domain, x-intercepts, y-intercepts, and asymptotes.

$$y = \frac{x^2 - 4}{x^2 - 2x - 3} = \frac{(x-2)(x+2)}{(x-3)(x+1)}$$

Domain: $x \neq 3$ and $x \neq -1$
　　Vertical asymptotes: $x = 3$ and $x = -1$

x-intercepts: $x = 2$ and $x = -2$

y-intercept: $y = \frac{-4}{-3} = \frac{4}{3}$

Since $n = m = 2$, both the numerator and the denominator have the same degree so the horizontal asymptote is the ratio of the leading coefficients.
Horizontal asymptote: $y = \frac{1}{1} = 1$

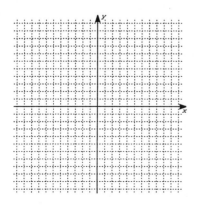

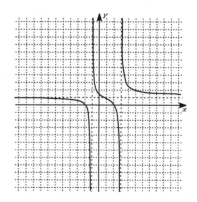

4. Sketch the graph of the function $y = \dfrac{x^3 - 3x^2 + 2x}{x^4 + 1}$.

First factor the numerator and denominator to find The domain, x-intercepts, y-intercepts, and asymptotes.

$$y = \frac{x^3 - 3x^2 + 2x}{x^4 + 1} = \frac{x(x - 2)(x - 1)}{x^4 + 1}$$

Domain: all real numbers since the denominator does not factor. Vertical asymptotes: none

x-intercepts: $x = 2$ and $x = -2$

y-intercept: $y = \frac{0}{1} = 0$

Since $n < m$, horizontal asymptote: $y = 0$.

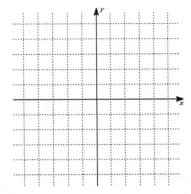

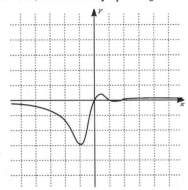

C. In the same way that multiplicity influences how the graph of a polynomial behaves around zeros, multiplicity of the zeros of the numerator and the zeros of the denominator affect the graph of a rational function. If a factor occurs an even number of times, then the graph stays on the same side of the x-axis near the value of x corresponding to that factor. If a factor occurs an odd number of times, then the graph changes sides of the x-axis at that factor.

5. Sketch the graph of the function $y = \dfrac{x^2}{x^2 - 9}$.

$$y = \frac{x^2}{x^2 - 9} = \frac{x^2}{(x - 3)(x + 3)}$$

Domain: $x \neq 3$ and $x \neq -3$

Vertical asymptotes: $x = 3$ and $x = -3$. *Both roots are multiplicity one, so the graph changes sign at each asymptote.*

x-intercepts: $x = 0$. *This is a root of multiplicity two, so the graph does not change sign here.*

y-intercept: $y = 0$.

Horizontal asymptote: $y = 1$.

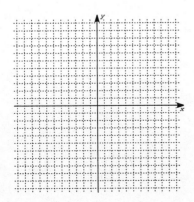

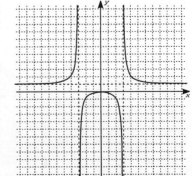

6. Sketch the graph of the function $y = \dfrac{x^2 - 9}{x^2}$.

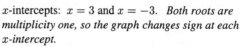

$$y = \frac{x^2 - 9}{x^2} = \frac{(x-3)(x+3)}{x^2}$$

Domain: $x \neq 0$

Vertical asymptotes: $x = 0$. *This is a root of multiplicity two, so the graph does not change sign at this asymptote.*

x-intercepts: $x = 3$ and $x = -3$. *Both roots are multiplicity one, so the graph changes sign at each x-intercept.*

y-intercept: None, $x = 0$ is *not* in the domain.

Horizontal asymptote: $y = 1$

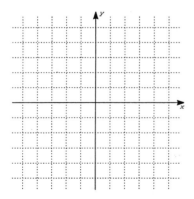

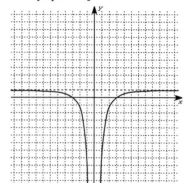

D. When $n > m$ in $r(x) = \dfrac{P(x)}{Q(x)} = \dfrac{a_n x^n + a_{n-1} x^{n-1} + \cdots + a_1 x^1 + a_0}{b_m x^m + b_{m-1} x^{m-1} + \cdots + b_1 x^1 + b_0}$, we use long division to

write $r(x) = D(x) + \dfrac{R(x)}{Q(x)}$, where the degree of $R(x) <$ the degree of $Q(x)$. If $D(x)$ is a polynomial of degree 1, then $D(x)$ is a **slant** or **oblique asymptote**. This means that as $|x|$ gets large, the graph of $y = r(x)$ behaves like the graph of $y = D(x)$.

7. Sketch the graph of the function $y = \dfrac{x^3 - 4x}{x^2 - 1}$.

$$y = \frac{x^3 - 4x}{x^2 - 1} = \frac{x(x-2)(x+2)}{(x-1)(x+1)}$$

Domain: $x \neq -1$ and $x \neq 1$.

Vertical asymptotes: $x = -1$ and $x = 1$. Both multiplicity one.

x-intercepts: $x = 0$, $x = 2$, and $x = -2$. All three are multiplicity one.

y-intercept: $y = 0$

Slant asymptote: $y = x$ since

$$y = \frac{x^3 - 4x}{x^2 - 1}$$

$$= x - \frac{3x}{x^2 - 1}$$

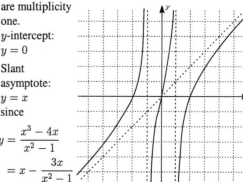

119

E. Some graphing calculators and computer graphing programs do not graph function with vertical asymptotes properly. Sometimes vertical asymptotes will appear as vertical lines. Other problems are caused by viewing rectangles that are too wide. In this case, vertical asymptotes might disappear altogether. However, wide viewing rectangles are useful in showing horizontal asymptotes as well as other end behavior. The key is to use a variety of sizes of viewing rectangles to determine vertical asymptotes as well as end behavior.

8. Graph each rational function in each of the given viewing rectangles. Determine all vertical and horizontal asymptotes from the graphs.

(a) $y = \dfrac{5x + 7}{x - 3}$ $[-5, 5]$ by $[-20, 20]$

$[-5, 5]$ by $[-20, 20]$

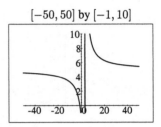

$[-50, 50]$ by $[-1, 10]$

$[-50, 50]$ by $[-1, 10]$

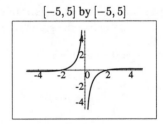

Vertical asymptote: $x = 3$
Horizontal asymptote: $y = 5$

(b) $y = \dfrac{x^2 - 4}{x^3 + 3x}$ $[-5, 5]$ by $[-5, 5]$

$[-5, 5]$ by $[-5, 5]$

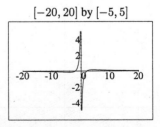

$[-20, 20]$ by $[-5, 5]$

$[-20, 20]$ by $[-5, 5]$

Vertical asymptote: $x = 0$
Horizontal asymptote: $y = 0$

(c) $y = \dfrac{8 + 3x + 2x^2}{9 - x^2}$ $\quad$ $[-5, 5]$ by $[-5, 5]$ $\qquad$ $[-5, 5]$ by $[-5, 5]$

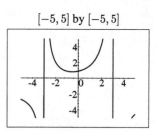

$[-50, 50]$ by $[-5, 5]$ $\qquad$ $[-50, 50]$ by $[-5, 5]$

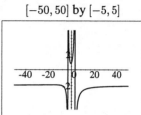

Vertical asymptote: $x = -3$ and $x = 3$
Horizontal asymptote: $y = -2$

9. Graph the rational function in appropriate
viewing rectangles. Determine all vertical and
horizontal asymptotes, x- and y-intercepts and all
local extrema correct to two decimal places.

$$y = \frac{3x^2 + x - 4}{x^2 - 5x + 6}.$$

$[-10, 10]$ by $[-10, 10]$

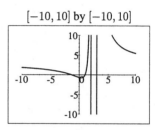

$[-100, 100]$ by $[-5, 5]$

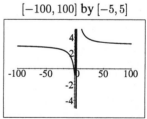

Vertical asymptotes: $x = 2$ and $x = 3$
Horizontal asymptote: $y = 3$
x-intercept: $x = -1.33$ and $x = 1.00$
y-intercept: $y = -0.67$
Extreme point is at $(0.37, -0.75)$.

$[0, 1]$ by $[-2, 0]$

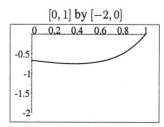

10. Graph each rational function in an appropriate viewing rectangle. Determine all vertical asymptotes. Then graph f and g in a sufficiently large viewing rectangle to show that they have the same end behavior.

(a) $f(x) = \dfrac{x^3 + 7x^2 - 5}{x^2 - 1}$

$g(x) = x + 7$

$[-5, 5]$ by $[-10, 10]$

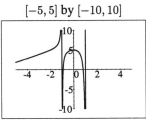

Vertical asymptote: $x = -1$ and $x = 1$

$[-20, 20]$ by $[-20, 20]$

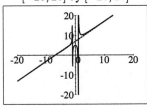

$f(x) = x + 7 + \dfrac{x + 2}{x^2 - 1}$

(b) $f(x) = \dfrac{x^3 - x + 12}{x + 4}$

$g(x) = x^2 - 4x + 15$

$[-10, 10]$ by $[-10, 10]$

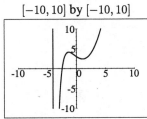

Vertical asymptote: $x = -4$

$[-20, 20]$ by $[-10, 200]$

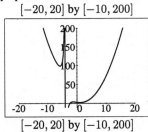

$[-20, 20]$ by $[-10, 200]$

$f(x) = x^2 - 4x + 15 - \dfrac{48}{x + 4}$

122

Chapter 4
Exponential and Logarithmic Functions

Section 4.1   Exponential Functions

Key Ideas

A. Exponential functions and their graphs.

B Natural exponential function.

C. Compound interest.

A. For $a > 0$, the **exponential function with base a** is defined as $f(x) = a^x$ for every real number x. The value of the base determines the general shape of the graph of $f(x) = a^x$. For $0 < a < 1$, the graph of $f(x) = a^x$ decreases rapidly. When $a = 1$, $f(x) = 1^x = 1$ is a constant function. And for $a > 1$, the function $f(x) = a^x$ increases rapidly. Remember

$$a^{-1} = \frac{1}{a}, \text{ so } a^{-x} = \frac{1}{a^x}, a^0 = 1, \text{ and a positive number to any power is always positive.}$$

As a result, the domain of $f(x) = a^x$ is the real numbers and the range is the positive real numbers.

1. Sketch the graphs of $f(x) = 2^x$ and $g(x) = \left(\frac{1}{2}\right)^x$ on the same axis.

Since $\frac{1}{2} = 2^{-1}$, $g(x) = \left(\frac{1}{2}\right)^x = (2^{-1})^x = 2^{-x}$.

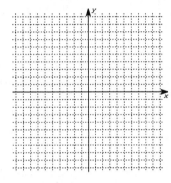

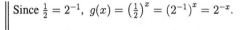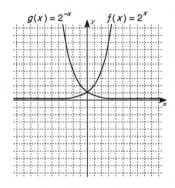

2. Sketch the graphs of $f(x) = 3^x$ and $g(x) = 5^x$ on the same axis.

$f(x) = 3^x$ and $g(x) = 5^x$.

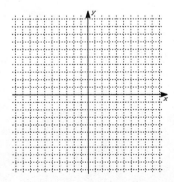

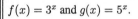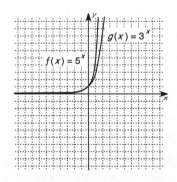

3. Sketch the graphs of $y = 3^x$ and use it to sketch the graphs of $y = -2 + 3^x$ and $y = 4 - 3^x$ all on on the same axis.

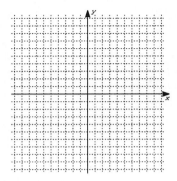

First graph $y = 3^x$.
To graph $y = -2 + 3^x$, start with $y = 3^x$ and shift the graph vertically 2 units down.
To graph $y = 4 - 3^x$, start with $y = 3^x$, first reflect the graph of $y = 3^x$ about the x-axis, and *then* shift the result 4 units up.

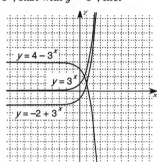

4. Compare the rates of growth of the functions $f(x) = 2.5^x$, $g(x) = x^3$, and $h(x) = \dfrac{2.5^x}{x^3}$.

  (a) Draw all three functions in the following viewing rectangles.

    (i) $[0, 5]$ by $[0, 100]$

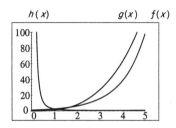

    (ii) $[0, 20]$ by $[0, 2000]$

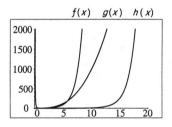

    (iii) $[0, 40]$ by $[0, 20000]$

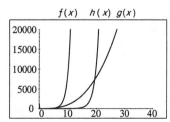

  (b) Comment on the rates of growth.

Although, $g(x)$ starts out larger than $f(x)$ and $h(x)$, these two functions quickly over take and pass $g(x)$.

125

B. As n becomes large, the values of $\left(1 + \dfrac{1}{n}\right)^n$ approaches an irrational number called e, $e \approx 2.71828$.

The function $f(x) = e^x$ is called the **natural exponential function**. Its domain is the real numbers and its range is the positive real numbers.

5. Complete the table and graph the function
$f(x) = -2e^x$.

x	$f(x) = -2e^x$
-3	
-2	
-1	
0	
1	
2	
3	

x	$f(x) = -2e^x$
-3	-0.10
-2	-0.27
-1	-0.74
0	-2
1	-5.44
2	-14.78
3	-40.17

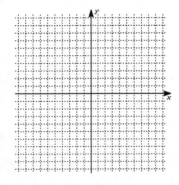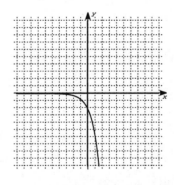

C. If an amount of money, P, called the **principal**, is invested at a rate of r, compounded n compounding periods per year, then the amount of money after t years is given by $A = P\left(1 + \dfrac{r}{n}\right)^{nt}$. The rate $\dfrac{r}{n}$ is the *rate per compounding period* and nt is the number of compounding periods. As the number of compounding periods per year become large, the value of the amount approaches $A = Pe^{rt}$, which is called **continuous compounding of interest**.

6. Compare the amount after 5 years when $1000 is invested in each type of account.

   (a) 10% compounded monthly.

$P = 1000, r = 0.1, n = 12, t = 5$

$A = P\left(1 + \dfrac{r}{n}\right)^{nt}$

$A = 1000\left(1 + \dfrac{0.1}{12}\right)^{(12)(5)}$

$\quad = 1000(1.008333333)^{60}$

$\quad = 1000(1.645308935) = 1645.31$

$A = \$1645.31$

   (b) 9.9% compounded weekly, assuming 52 weeks per year.

$P = 1000, r = 0.099, n = 52, t = 5$

$A = P\left(1 + \dfrac{r}{n}\right)^{nt}$

$A = 1000\left(1 + \dfrac{0.099}{52}\right)^{(52)(5)}$

$\quad = 1000(1.001903846)^{260}$

$\quad = 1000(1.639726394) = 1639.73$

$A = \$1639.73$

   (c) 9.9% compounded continuously.

$P = 1000, r = 0.099, t = 5$

$A = Pe^{rt}$

$A = 1000e^{(0.099)(5)} = 1000e^{0.495}$

$\quad = 1000(1.640498239) = 1640.50$

$A = \$1640.50$

7. How much money needs to be invested in an account paying 7.8% compounded monthly, so that a person will have $5000 in 5 years?

$A = P\left(1 + \dfrac{r}{n}\right)^{rt}$ *Solve for P.*

$P = \dfrac{A}{\left(1 + \dfrac{r}{n}\right)^{rt}}$

Using $A = 5000, r = 0.078, n = 12, t = 5$,

$P = \dfrac{5000}{\left(1 + \dfrac{0.078}{12}\right)^{(12)(5)}} = \dfrac{5000}{(1.0065)^{60}}$

$P = \$3389.56.$

Section 4.2   Logarithmic Functions

Key Ideas

A. What is a logarithm?

B. Graphs of logarithmic functions.

C. Property of Natural and Common logarithms.

A. For $a \neq 1$, the inverse of the exponential function $f(x) = a^x$ is a function called **logarithm**, written as $g(x) = \log_a x$. The function $\log_a x$ is read as 'log base a of x'. So $\log_a x$ is the *exponent* to which the base a must be raised to give x.

Logarithmic Form	Exponential Form
$\log_a x = y$	$a^y = x$

In both forms, a is called the **base** and y is called the **exponent**. It is often useful to switch between the different forms of the definition. Notice that $\log_a 1 = 0$ and $\log_a a = 1$.

1. Express the equation in exponential form.

   (a) $\log_4 64 = 3$

$$\log_4 64 = 3 \quad \Leftrightarrow \quad 4^3 = 64$$

   (b) $\log_5 0.04 = -2$

$$\log_5 0.04 = -2 \quad \Leftrightarrow \quad 5^{-2} = 0.04$$

   (c) $\log_9 243 = \frac{5}{2}$

$$\log_9 243 = \frac{5}{2} \quad \Leftrightarrow \quad 9^{5/2} = 243$$
$$(9^{1/2})^5 = 243$$
$$3^5 = 243$$

2. Express the equation in logarithmic form.

   (a) $6^3 = 216$

$$6^3 = 216 \quad \Leftrightarrow \quad \log_6 216 = 3$$

   (b) $8^{-5/3} = \frac{1}{32}$

$$8^{-5/3} = \frac{1}{32} \quad \Leftrightarrow \quad \log_8\left(\frac{1}{32}\right) = -\frac{5}{3}$$

3. Evaluate.

   (a) $\log_5 25$

                               $\log_5 25 = 2$ since $5^2 = 25$.

   (b) $\log_2\left(\frac{1}{8}\right)$

                               $\log_2\left(\frac{1}{8}\right) = -3$ since $2^{-3} = \frac{1}{8}$.

   (c) $\log_{(1/3)} 27$

                               $\log_{(1/3)} 27 = -3$ since $\left(\frac{1}{3}\right)^{-3} = 3^3 = 27$.

4. Use the definition of the logarithmic function to find x.

   (a) $\log_3 x = 4$

                               $\log_3 x = 4 \quad \Leftrightarrow \quad x = 3^4 = 81$.

   (b) $\log_x 121 = 2$

                               $\log_x 121 = 2 \quad \Leftrightarrow \quad x^2 = 121$

                               Since $x > 0$, we have $x = \sqrt{121} = 11$.

   (c) $\log_{(1/5)} 25 = x$

                               $\log_{(1/5)} 25 = x \quad \Leftrightarrow \quad \left(\frac{1}{5}\right)^x = 25 \quad \Leftrightarrow$

                               $5^{-x} = 25 \quad \Leftrightarrow \quad 5^{-x} = 5^2 \quad \Leftrightarrow \quad -x = 2 \quad \Leftrightarrow$

                               $x = -2$

B. For $a > 0$ and $a \neq 1$, the domain for $f(x) = \log_a x$ is the positive real numbers and the range is the real numbers. As with exponential functions, the graph of $y = \log_a x$ is directly dependent on the value of a. The graphs of $f(x) = \log_a x$ and $f^{-1}(x) = a^x$ for various values of a are shown below.

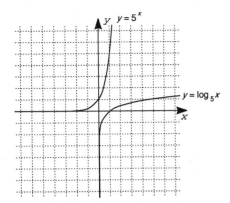

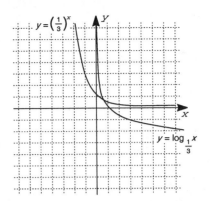

5. Determine the domain and range for $f(x) = \log_3(x - 2)$. Sketch the graph of $y = f(x)$.

Domain: $x - 2 > 0 \iff x > 2$
Range $(-\infty, \infty)$

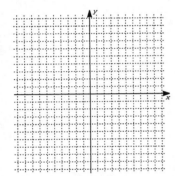

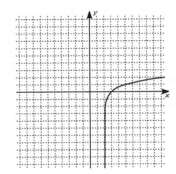

6. Determine the domain and range for $f(x) = \log_5(x^2 + 1)$. Sketch the graph of $y = f(x)$.

Domain: Since $x^2 + 1 > 0$ for all x, the domain is $(-\infty, \infty)$.
Range: Also since $x^2 + 1 \geq 1$,
$\log_5(x^2 + 1) \geq \log_5 1 = 0$, so the range is $[0, \infty)$, the nonnegative real numbers.

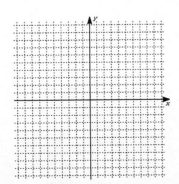

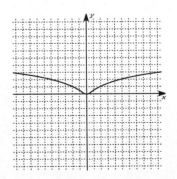

C. The logarithm with base e is called the **natural logarithm** and is given a special name and notation, $\log_e x = \ln x$. The logarithm with base 10 is called the **common logarithm** and given an abbreviated notation, $\log_{10} x = \log x$.

$\ln x = y \quad \Leftrightarrow \quad e^y = x$	and $\qquad \log x = y \quad \Leftrightarrow \quad 10^y = x$

These logarithms have the following properties.

Property	Reason
$\ln 1 = 0$	We must raise e to the power 0 to get 1.
$\ln e = 1$	We must raise e to the power 1 to get e.
$\ln e^x = x$	We must raise e to the power x to get e^x.
$e^{\ln x} = x$	$\ln x$ is the power to which e must be raised to get x.

7. Find the domain of $f(x) = \ln(x+2) + \ln(3-2x)$. Draw the graph of $y = f(x)$ on a calculator and use it to find the asymptotes and local maximums and local minimums.

Since the input to a logarithm function must be positive, we have

$$x + 2 > 0 \qquad \text{and} \qquad 3 - 2x > 0$$
$$x > -2 \qquad\qquad\qquad -2x > -3$$
$$x < \tfrac{3}{2}.$$

So the domain is $-2 < x < \tfrac{3}{2}$. We use this as the x-range of the viewing rectangle.

$$\left[-2, \tfrac{3}{2}\right] \text{ by } [-8, 2]$$

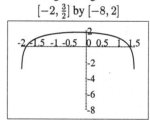

Vertical asymptotes are at $x = -2$ and at $x = \tfrac{3}{2}$.
Local maximum approximately at $(-0.25, 1.812)$;
No local minimum.

Section 4.3   Laws of Logarithms

Key Ideas
A. Basic laws of logarithms.
B. Common errors to avoid!
C. Change of base.

A. Suppose that $x > 0$, $y > 0$, and r is any real number. Then the Laws of Logarithms are:

Rule	Reason
1. $\log_a(xy) = \log_a x + \log_a y$	*The log of a product is the sum of the logs.*
2. $\log_a\left(\dfrac{x}{y}\right) = \log_a x - \log_a y$	*The log of a quotient is the difference of the logs.*
3. $\log_a(x^r) = r\log_a x$	*The log of a power of a number is the exponent times the logarithm of the number.*

1. Use the Laws of Logarithms to rewrite the following expression in a form that contains no logarithms of products, quotients, or powers.

   (a) $\log_2 9x$

   $\log_2 9x = \log_2 9 + \log_2 x$
   The log of a product is the sum of the logs.

   (b) $\log_5\left(\dfrac{2x}{y}\right)$

   $\log_5\left(\dfrac{2x}{y}\right) = \log_5(2x) - \log_5 y$
   The log of a quotient is the difference of the logs.
   $$= \log_5 2 + \log_5 x - \log_5 y$$

   (c) $\log_3(x^4 y^3)$

   $\log_3(x^4 y^3) = \log_3 x^4 + \log_3 y^3$
   The log of a product is the sum of the logs.
   $$= 4\log_3 x + 3\log_3 y$$
   The log of a power of a number is the exponent times the logarithm of the number.

   (d) $\log_4\left(\dfrac{16w^5}{\sqrt[3]{x^2 y}}\right)$

   $\log_4\left(\dfrac{16w^5}{\sqrt[3]{x^2 y}}\right) = \log_4(16w^5) - \log_4\left(\sqrt[3]{x^2 y}\right)$
   $$= \log_4 16 + \log_4 w^5 - \tfrac{1}{3}\log_4(x^2 y)$$
   $$= \log_4 4^2 + \log_4 w^5 - \tfrac{1}{3}(\log_4 x^2 + \log_4 y)$$
   $$= \log_4 4^2 + \log_4 w^5 - \tfrac{1}{3}(2\log_4 x + \log_4 y)$$
   $$= 2 + 5\log_4 w - \tfrac{2}{3}\log_4 x - \tfrac{1}{3}\log_4 y$$

B. There are several common mistakes that you need to avoid, which are the following:

1. $\log_a(x+y) \neq \log_a x + \log_a y$	*Note: $\log_a(x+y)$ cannot be simplified!*
2. $\dfrac{\log_a x}{\log_a y} \neq \log_a x - \log_a y$	*The log of a quotient is not the same as the quotient of the logs.*
3. $(\log_a x)^r \neq r\log_a x$	*Here the exponent is applied to the entire function $\log_a x$, not just the input x.*

2. Use the Laws of Logarithms to rewrite each of the following as a single logarithm.

(a) $4\log_3 x - \frac{3}{7}\log_3 y$

$$4\log_3 x - \frac{3}{7}\log_3 y = \log_3 x^4 - \log_3 \sqrt[7]{y^3}$$
$$= \log_3 \left(\frac{x^4}{\sqrt[7]{y^3}} \right)$$

(b) $2\log_{10} x + 3\log_{10} y - \frac{1}{2}\log_{10} 8$

$$2\log_{10} x + 3\log_{10} y - \frac{1}{2}\log_{10} 8$$
$$= \log_{10} x^2 + \log_{10} y^3 - \log_{10} \sqrt{8}$$
$$= \log_{10} \left(\frac{x^2 y^3}{\sqrt{8}} \right)$$

(c) $4\log_5 x + 2 - \log_5 y - 2\log_5 w$

$$4\log_5 x + 2 - \log_5 y - 2\log_5 w$$
$$= \log_5 x^4 + \log_5 25 - \log_5 y - \log_5 w^2$$
$$= \log_5 (25x^4) - \log_5 (yw^2)$$
$$= \log_5 \left(\frac{25x^4}{yw^2} \right)$$

The hardest part is to convert 2 to the form $\log_5$, since $25 = 5^2$, we have $2 = \log_5 25$.

C. Since most calculators only have the common log and natural log functions, it is important to be able to change the base of a logarithm. The **change of base formula** is $\log_b x = \dfrac{\log_a x}{\log_a b}$. This allows us to evaluate a logarithm of any base (provided the base is greater than 0 and not equal to 1). The change of base formula can also be expressed as $\log_a b \cdot \log_b x = \log_a x$. Of particular importance are:
$\log_b x = \dfrac{\log x}{\log b}$ or $\log_b x = \dfrac{\ln x}{\ln b}$ which allows us to use our calculators to find the logs of uncommon bases.

3. Evaluate $\log_6 7$ to six decimal places.

Using common logs:
$$\log_6 7 = \frac{\log 7}{\log 6} \approx \frac{0.84509804}{0.778151250} = 1.086033$$

Using natural logs:
$$\log_6 7 = \frac{\ln 7}{\ln 6} \approx \frac{1.945910149}{1.791759469} = 1.086033$$

4. Simplify $(\log_3 8)(\log_8 7)$

Apply the change of base formula.
Express all logs to the base 3.
$$(\log_3 8)(\log_8 7) = \log_3 8 \cdot \frac{\log_3 7}{\log_3 8}$$
$$= \log_3 7$$

Section 4.4 Exponential and Logarithmic Equations

Key Ideas

A. Exponential equations.
B. Logarithmic equations.

A. Guidelines for solving exponential equations:

> 1. Isolate the exponential expression on one side of the equation.
> 2. Take the logarithm of each side, then use the Laws of Logarithms to "bring down the exponent."
> 3. Solve for the variable.

1. Solve for x.

   $2^{3x+1} = 32$

   We take the logarithm of each side and use Law 3.

   $2^{3x+1} = 32$

   $\log_2(2^{3x+1}) = \log_2 32$ *Take log of each side.*

   $\log_2(2^{3x+1}) = \log_2 2^5$ *Apply Law 3 (bring down exponent).*

   $(3x+1)\log_2 2 = 5\log_2 2$

   $3x + 1 = 5$ *Definition of $\log_a a = 1$.*

   $3x = 4$

   $x = \frac{4}{3}$

2. Solve $x^5 e^x + 3x^4 e^x = 4x^3 e^x$.

   $x^5 e^x + 3x^4 e^x = 4x^3 e^x$

   $x^5 e^x + 3x^4 e^x - 4x^3 e^x = 0$ *Set equal to 0.*

   $(x^5 + 3x^4 - 4x^3)e^x = 0$ *Factor out the e^x.*

   $x^3(x^2 + 3x - 4)e^x = 0$

   $x^3(x+4)(x-1)e^x = 0$

$x^3 = 0$	$x + 4 = 0$	$x - 1 = 0$
$x = 0$	$x = -4$	$x = 1$

   Solution: $x = 0$, $x = -4$, and $x = 1$.

   Note: $e^x \neq 0$ since $e^x > 0$.

3. Use a graphing calculator to solve $e^{2x} = 5e^x$ to two decimal places.

   First set equal to zero:

   $e^{2x} = 5e^x \quad \Leftrightarrow \quad e^{2x} - 5e^x = 0.$

   Now graph $f(x) = e^{2x} - 5e^x$ and find where it intersects the x-axis.

   $[-2, 2]$ by $[-2, 2]$

   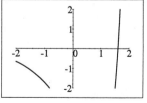

   Solution: $x \approx 1.61$.

   Note: A larger viewing rectangle shows that $f(x)$ approaches the x-axis, but it does not touch the axis.

4. Use a graphing calculator to solve $xe^x - 3x^3 = 0$ to two decimal places.

Graph $f(x) = xe^x - 3x^3$ and find where it intersects the x-axis.

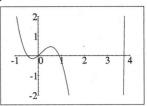

$[-1, 4]$ by $[-2, 2]$

Solution: $x \approx -0.46$, $x = 0$, $x = 0.91$, or $x \approx 3.73$.

5. Solve the equation $4e^{2x-1} = 7$.

$$4e^{2x-1} = 7$$

$$e^{2x-1} = \tfrac{7}{4} \qquad \textit{Take } \ln() \textit{ of both sides.}$$

$$\ln(e^{2x-1}) = 2x - 1 = \ln\left(\tfrac{7}{4}\right)$$

$$2x = 1 + \ln\left(\tfrac{7}{4}\right)$$

$$x = \tfrac{1}{2} + \tfrac{1}{2}\ln\left(\tfrac{7}{4}\right) \approx 1.06$$

6. Solve the equation $e^{2x} - 4e^x + 3 = 0$.

Factor.

$$e^{2x} - 4e^x + 3 = 0$$

$$(e^x - 1)(e^x - 3) = 0 \qquad \textit{Set each factor equal to 0.}$$

$e^x - 1 = 0 \qquad\qquad\qquad e^x - 3 = 0$

$e^x = 1 \qquad\qquad\qquad\qquad e^x = 3$

$x = \ln 1 = 0 \qquad\qquad\qquad x = \ln 3$

So $x = 0$ or $x = \ln 3$.

B. Guidelines for solving logarithmic equations:

1. Isolate the logarithmic term on one side of the equation; you may need to first combine the logarithmic terms.
2. Write the equation in exponential form (or raise the base to each side of the equation).
3. Solve for the variable.

7. Solve for x. $\log_5(62 - x) = 3$

We rewrite the equation in exponential form.
$$\log_5(62 - x) = 3$$
$$62 - x = 5^3 \qquad \text{exponential form}$$
$$62 - x = 125$$
$$-x = 63$$
$$x = -63$$

Check:
$$\log_5[62 - (-63)] = \log_5 125 = \log_5 5^3 = 3. \quad \checkmark$$

8. Solve for x. $\log_3 x + \log_9 x = 6$

We need to convert to a common base. Since $9 = 3^2$, we convert $\log_3 x$ to base 9.
Since $(\log_9 3)(\log_3 x) = \log_9 x$ and $\log_9 3 = \frac{1}{2}$,
we have $\log_3 x = \dfrac{\log_9 x}{\log_9 3} = \dfrac{\log_9 x}{1/2} = 2\log_9 x$.
$$\log_3 x + \log_9 x = 6$$
$$2\log_9 x + \log_9 x = 6$$
$$3\log_9 x = 6$$
$$\log_9 x = 2$$
$$x = 9^2 = 81$$

Check:
$$\log_3 81 = \log_3 3^4 = 4$$
$$\log_9 81 = \log_9 9^2 = 2.$$
So $\log_3 81 + \log_9 81 = 4 + 2 = 6. \quad \checkmark$

9. Solve the equation $\log_8(x+2) + \log_8(x+5) = \frac{2}{3}$.

$$\log_8(x+2) + \log_8(x+5) = \tfrac{2}{3}$$
$$\log_8[(x+2)(x+5)] = \tfrac{2}{3}$$
$$\log_8(x^2 + 7x + 10) = \tfrac{2}{3}$$
$$8^{\log_8(x^2+7x+10)} = 8^{2/3}$$
$$x^2 + 7x + 10 = 8^{2/3} = \left(8^{1/3}\right)^2$$
$$x^2 + 7x + 10 = 2^2 = 4$$
$$x^2 + 7x + 6 = 0$$
$$(x+6)(x+1) = 0$$

$$x + 6 = 0 \qquad\qquad x + 1 = 0$$
$$x = -6 \qquad\qquad x = -1$$

Check $x = -6$:

$\log_8[(-6) + 2] + \log_8[(-6) + 5] \overset{?}{=} \tfrac{2}{3}$

But $\log_8(-4)$ and $\log_8(-1)$ are not defined, so as a result $x = -6$ is not a solution.

Check $x = -1$:

$$\log_8[(-1) + 2] + \log_8[(-1) + 5] \overset{?}{=} \tfrac{2}{3}$$
$$\log_8 1 + \log_8 4 \overset{?}{=} \tfrac{2}{3}$$
$$\log_8 4 = \tfrac{2}{3}$$

since $8^{2/3} = (8^{1/3})^2 = 2^2 = 4$.

Solution is only $x = -1$.

Section 4.5   Modeling with Exponential and Logarithmic Functions

Key Ideas

A. Exponential Growth Models.
B. Compound Interest.
C. Exponential growth and radioactive decay.
D. Newton's Law of Cooling.
E. Logarithmic scales: pH, Richter, decibel scales.

A. A population that experiences **exponential growth** increases according to the formula $n(t) = n_0 e^{rt}$ where n_0 is the initial size of the population, r is the relative rate of growth (expressed as a proportion of the population), t is time, and $n(t)$ is the population at time t.

1. Under ideal conditions a certain bacteria population is known to increase at a relative growth rate of 7.8% per hour. Suppose that there are initially 2000 bacteria.

   (a) What will be the size of the population in 18 hours?

   $n_0 = 2000, r = 0.078, t = 18$

   $n(t) = n_0 e^{rt}$

   $n(18) = 2000 e^{(0.078)(18)} = 2000 e^{1.404} \approx 8143$

   There will be about 8100 bacteria in 18 hours.

   (b) What is the size after 15 days?

   $n_0 = 2000, r = 0.078, t = 15(24) = 360$

   $n(t) = n_0 e^{rt}$

   $n(360) = 2000 e^{(0.078)(360)} = 2000 e^{28.08}$

   $\approx 3.1334 \times 10^{15}$

   There will be about 3.1334×10^{15} bacteria in 15 days.

B. If an amount of money, P, called the **principal**, is invested at a rate of r, compounded n **compounding periods** per year, then the amount of money after t years is given by $A = P\left(1 + \dfrac{r}{n}\right)^{nt}$. The rate $\dfrac{r}{n}$ is the *rate per compounding period* and nt is the number of compounding periods. As the number of compounding periods per year become large, the value of the amount approaches $A = P e^{rt}$, called **continuous compounding of interest**.

2. Karen placed $6000 into a retirement account paying 9% compounded quarterly. How many years will it take for this account to reach $25,000?

$$A = P\left(1 + \frac{r}{n}\right)^{nt}$$

Using $A = 25{,}000$, $P = 6000$, $r = 0.09$, and $n = 4$, we find t.

$$25000 = 6000\left(1 + \frac{0.09}{4}\right)^{4t} = 6000(1.0225)^{4t}$$

$$4.1666667 = 1.0225^{4t}$$

$$\ln(4.1666667) = \ln(1.0225^{4t}) = (4t)\ln 1.0255$$

$$t = \frac{\ln(4.1666667)}{4\ln 1.0255} \approx 16.03 \text{ years}$$

Since interest is paid at the end of each quarter, we move up to the next quarter and t is $16\frac{1}{4}$ years.

3. How much money needs to be invested in an account paying 7.8% compounded monthly, so that a person will have $5000 in 5 years?

Solve for P.

$$A = P\left(1 + \frac{r}{n}\right)^{nt} \quad \Leftrightarrow \quad P = \frac{A}{\left(1 + \dfrac{r}{n}\right)^{nt}}$$

Using $A = 5000$, $r = 0.078$, $n = 12$, and $t = 5$, we find P.

$$P = \frac{5000}{\left(1 + \dfrac{0.078}{12}\right)^{(12)(5)}} = \frac{5000}{(1.0065)^{60}}$$

$$P = \$3389.56$$

C. Logarithms are used to extract the power in exponential functions. This is an especially useful tool when the exponent contains the unknown variable.

4. At a certain time, a bacteria culture has 115,200 bacteria. Four hours later the culture has 1,440,000 bacteria. How many bacteria will be in the population in 15 hours?

$$n(t) = n_0 e^{rt}$$

We use $n_0 = 115{,}200$ and $t = 4$ to find r.

$$n(4) = 1{,}440{,}000 = 115{,}200\, e^{4r}$$

$$e^{4r} = \frac{1{,}440{,}000}{115{,}200} = 12.5$$

$$\ln(e^{4r}) = 4r = \ln 12.5 \approx 2.526$$

$$r \approx 0.631$$

Thus
$$\begin{aligned} n(15) &= 115{,}200 e^{(0.631)(15)} \\ &= 115{,}200 \cdot 12984.12 \\ &= 1{,}495{,}770{,}721 \\ &\approx 1.5 \times 10^9. \end{aligned}$$

D. Newton's Law of Cooling states that the rate of cooling of an object is proportional to the temperature difference between the object and its surroundings, provided that the temperature difference is not too large. It can be shown that $T(t) = T_s + D_0 e^{-kt}$ where T_s is the temperature of the surrounding environment, D_0 is the initial temperature difference, and k is a positive constant that is associated with the cooling object.

5. A bowl of soup that is 115°F is placed in an air-conditioned room at 70°F. After 5 minutes, the temperature of the soup is 100°F.

(a) What is the temperature after 10 minutes?

First find k. Now $T(5) = 100°F$, $T_s = 70°F$, and $D_0 = 115 - 70 = 45°F$. Substituting, we get

$$T(t) = T_s + D_0 e^{-kt}$$
$$100 = 70 + 45e^{-5k}$$
$$30 = 45e^{-5k}$$
$$\tfrac{2}{3} = e^{-5k}$$
$$\ln\left(\tfrac{2}{3}\right) = \ln(e^{-5k}) = -5k$$
$$k = -\tfrac{1}{5} \cdot \ln\left(\tfrac{2}{3}\right) \approx 0.081.$$

So $T(t) = 70 + 45e^{-0.081t}$.

Now $T(10) = 70 + 45e^{-0.081(10)} \approx 90°F$.

(b) How long will it take the soup cool to 80°F?

Find t so that $T(t) = 80$. Substituting into

$$T(t) = 70 + 45e^{-0.081t}$$
$$80 = 70 + 45e^{-0.081t}$$
$$10 = 45e^{-0.081t}$$
$$\tfrac{10}{45} = \tfrac{2}{9} = e^{-0.081t}$$
$$\ln\left(\tfrac{2}{9}\right) = \ln(e^{-0.081t}) = -0.081t$$
$$t = \frac{\ln\left(\tfrac{2}{9}\right)}{-0.081} \approx 19 \text{ minutes}$$

E. Logarithms make quantities which vary over very large ranges more manageable. Three commonly used logarithmic scales are:

6. The hydrogen ion concentration of a shampoo is 8.2×10^{-8}. Find its pH.

$$pH = -\log(8.2 \times 10^{-8})$$
$$= -[\log(8.2) + \log(10^{-8})]$$
$$= -(0.91 - 8) = 7.09$$

7. On October 17, 1989, the Loma Prieta Earthquake rocked the Santa Cruz mountains. Researchers initially measured the earthquake as 6.9 on the Richter scale. Later, the earthquake was reevaluated as 7.1 on the Richter scale. How do the two measures compare in intensity?

We want to compare the intensities of the two earthquakes, so we solve $M = \log I - \log S$ for I.

$$M = \log I - \log S \quad \Leftrightarrow \quad M + \log S = \log I$$
$$10^{(M + \log S)} = 10^{\log I}$$
$$S10^M = I$$

That is, $I = S10^M$

Comparing the two intensities, we have

$$\frac{S10^{7.1}}{S10^{6.9}} = 10^{7.1-6.9} = 10^{0.2} = 1.58.$$

So a 7.1 earthquake is 1.58 times more intense than a 6.9 earthquake.

8. Find the intensity level of a computer speaker system if the system is advertised to deliver $30\ \frac{\text{watts}}{\text{m}^2}$ at a distance of one meter from the speaker.

Using the formula for decibel intensity level, we see that the intensity level is

$$\beta = 10\log\frac{I}{I_0}$$
$$= 10\log\left(\frac{3 \times 10}{10^{-12}}\right)$$
$$= 10\log(3 \times 10^{13})$$
$$= 10 \cdot 13 \cdot \log 3$$
$$\approx 62\,\text{dB}.$$

Chapter 5
Trigonometric Functions of Real Numbers

Section 5.1   The Unit Circle

Key Ideas
A. Unit Circle.
B. Terminal points on the unit circle.
C. Reference numbers.

A. The **unit circle** is the circle of radius 1 centered at the origin in the xy-plane. Its equation is given by $x^2 + y^2 = 1$.

1. Show that each point is on the unit circle.

(a) $\left(-\frac{3}{4}, \frac{\sqrt{7}}{4}\right)$

$\left(-\frac{3}{4}\right)^2 + \left(\frac{\sqrt{7}}{4}\right)^2 = \frac{9}{16} + \frac{7}{16} = \frac{16}{16} = 1$

Since the point satisfies the equation $x^2 + y^2 = 1$, it is on the unit circle.

(b) $\left(-\frac{60}{61}, \frac{11}{61}\right)$

$\left(-\frac{60}{61}\right)^2 + \left(\frac{11}{61}\right)^2 = \frac{3600}{3721} + \frac{121}{3721} = \frac{3721}{3721} = 1$

Since the point satisfies the equation $x^2 + y^2 = 1$, it is on the unit circle.

2. Find the point $P(x, y)$ on the unit circle from the given information.

(a) x-coordinate is $\frac{7}{10}$ and P is in quadrant IV.

$\left(\frac{7}{10}\right)^2 + y^2 = 1 \quad \Leftrightarrow \quad \frac{49}{100} + y^2 = 1 \quad \Leftrightarrow$

$y^2 = 1 - \frac{49}{100} = \frac{51}{100} \quad \Rightarrow \quad y = \pm\frac{\sqrt{51}}{10}$

Since P is in quadrant IV, the y-coordinate is negative. So P is $\left(\frac{7}{10}, -\frac{\sqrt{51}}{10}\right)$.

(b) y-coordinate is $-\frac{9}{41}$ and P is in quadrant III.

$x^2 + \left(-\frac{9}{41}\right)^2 = 1 \quad \Leftrightarrow \quad x^2 + \frac{81}{1681} = 1 \quad \Leftrightarrow$

$x^2 = 1 - \frac{81}{1681} = \frac{1600}{1681} \quad \Rightarrow \quad x = \pm\frac{40}{41}$

Since P is in quadrant III, the x-coordinate is negative. So P is $\left(-\frac{40}{41}, -\frac{9}{41}\right)$.

(c) x-coordinate is negative and the y-coordinate is $\frac{1}{2}$.

$x^2 + \left(\frac{1}{2}\right)^2 = 1 \quad \Leftrightarrow \quad x^2 + \frac{1}{4} = 1 \quad \Leftrightarrow$

$x^2 = 1 - \frac{1}{4} = \frac{3}{4} \quad \Rightarrow \quad y = \pm\frac{\sqrt{3}}{2}$

Since the x-coordinate is negative, P is $\left(-\frac{\sqrt{3}}{2}, \frac{1}{2}\right)$.

B. The **terminal point** determined by the real number t is the point distance t along the unit circle found by starting at the point $(1, 0)$ and moving in a counterclockwise direction when $t > 0$ and in a clockwise direction when $t < 0$.

t	0	$\frac{\pi}{6}$	$\frac{\pi}{4}$	$\frac{\pi}{3}$	$\frac{\pi}{2}$
Terminal point determined by t	$(1, 0)$	$\left(\frac{\sqrt{3}}{2}, \frac{1}{2}\right)$	$\left(\frac{\sqrt{2}}{2}, \frac{\sqrt{2}}{2}\right)$	$\left(\frac{1}{2}, \frac{\sqrt{3}}{2}\right)$	$(0, 1)$

3. Find the terminal point on the unit circle determined by the real number t.

   (a) $t = \frac{3\pi}{2}$

The solution to these problems is found by drawing t on the unit circle.

$(0, -1)$

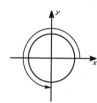

   (b) $t = 5\pi$

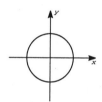

$(-1, 0)$

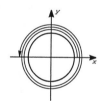

4. Suppose that the terminal point determined by t is the point $\left(-\frac{5}{13}, \frac{12}{13}\right)$ on the unit circle. Find the terminal point determined by

   (a) $\pi - t$

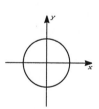

As in the previous problems, it is helpful to draw t on the unit circle.

By drawing the point t and then $\pi - t$ on the same unit circle, we find that t and $\pi - t$ have the same y-coordinate. The terminal point determined by $\pi - t$ lies in quadrant I, so its coordinates are $\left(\frac{5}{13}, \frac{12}{13}\right)$.

   (b) $\pi + t$

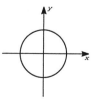

By drawing the point t and then $\pi + t$ on the same unit circle, we find that the terminal point determined by $\pi + t$ is in quadrant IV, so its coordinates are $\left(\frac{5}{13}, -\frac{12}{13}\right)$.

145

C. Let t be a real number. The **reference number**, $\bar{t}$, is the shortest distance along the unit circle between the terminal point determined by t and the x-axis. So $0 \leq \bar{t} \leq \frac{\pi}{2}$.

5. Find the reference number determined by the real number t.

(a) $t = -\frac{3\pi}{4}$

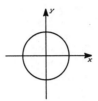

First plot the relative position of t on the unit circle. Then find the reference number.

Since the terminal point determined by t is in quadrant II, $\bar{t} = \frac{\pi}{4}$.

(b) $t = \frac{11\pi}{6}$

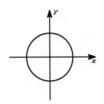

$\bar{t} = \frac{\pi}{6}$

(c) $t = 4$

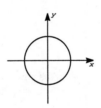

Since $\pi \leq t \leq \frac{3\pi}{2}$, $\bar{t} = 4 - \pi \approx 0.858$.

146

Section 5.2   Trigonometric Functions of Real Numbers

Key Ideas

A. Definition of the six trigonometric functions.

B. Signs and properties of the trigonometric functions.

C. Fundamental identities.

A. Let t be a real number and let $P(x, y)$ be the terminal point on the unit circle determined by t. We define the **trigonometric functions** as:

$$\sin t = y \qquad\qquad \cos t = x \qquad\qquad \tan t = \frac{y}{x} \ \ (x \neq 0)$$

$$\csc t = \frac{1}{y} \ \ (y \neq 0) \qquad\qquad \sec t = \frac{1}{x} \ \ (x \neq 0) \qquad\qquad \cot t = \frac{x}{y} \ \ (y \neq 0)$$

Because the trigonometric functions are defined on the unit circle they are also referred to as the circular functions. The domains of the six trigonometric functions are given in the table to the right.

Function	Domain
sin, cos	all real numbers
tan, sec	all real numbers except $t = \dfrac{\pi}{2} + n\pi$, for any integer n
cot, csc	all real numbers except $t = n\pi$, for any integer n

1. The terminal point $P(x, y)$ determined by t is $\left(\frac{\sqrt{3}}{3}, -\frac{\sqrt{6}}{3} \right)$. Find all six trigonometric functions.

$$\sin t = -\frac{\sqrt{6}}{3}$$

$$\cos t = \frac{\sqrt{3}}{3}$$

$$\tan t = \frac{-\frac{\sqrt{6}}{3}}{\frac{\sqrt{3}}{3}} = -\frac{\sqrt{6}}{\sqrt{3}} = -\sqrt{2}$$

$$\sec t = \frac{1}{\frac{\sqrt{3}}{3}} = \frac{3}{\sqrt{3}} = \sqrt{3}$$

$$\csc t = \frac{1}{-\frac{\sqrt{6}}{3}} = -\frac{3}{\sqrt{6}} = -\frac{\sqrt{6}}{2}$$

$$\cot t = \frac{\frac{\sqrt{3}}{3}}{-\frac{\sqrt{6}}{3}} = -\frac{\sqrt{3}}{\sqrt{6}} = -\frac{\sqrt{2}}{2}$$

2. The terminal point $P(x, y)$ determined by t is $\left(\frac{2}{5}, \frac{\sqrt{21}}{5}\right)$. Find all six trigonometric functions.

$$\sin t = \frac{\sqrt{21}}{5}$$

$$\cos t = \frac{2}{5}$$

$$\tan t = \frac{\frac{\sqrt{21}}{5}}{\frac{2}{5}} = \frac{\sqrt{21}}{2}$$

$$\sec t = \frac{1}{\frac{2}{5}} = \frac{5}{2}$$

$$\csc t = \frac{1}{\frac{\sqrt{21}}{5}} = \frac{5}{\sqrt{21}} = \frac{5\sqrt{21}}{21}$$

$$\cot t = \frac{\frac{2}{5}}{\frac{\sqrt{21}}{5}} = \frac{2}{\sqrt{21}} = \frac{2\sqrt{21}}{21}$$

In Problems 3 and 4, find the approximate value of the given trigonometric function by using (a) the figure and (b) a calculator.

3. $\cos 3.4$

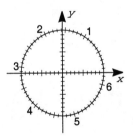

(a) $\cos 3.4 \approx -0.97$

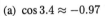
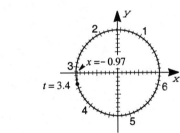

(b) $\cos 3.4 \approx -0.9668$

4. $\tan 5.2$

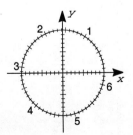

(a) $\tan 5.2 \approx \frac{-0.88}{0.48} \approx -1.83$

(b) $\tan 5.2 \approx -1.8856$

B. To compute other values of the trigonometric functions, we first determine their sign by quadrant. This information is shown in the table to the right. The sign and reference number coupled with the even-odd properties shown below are used to find the trigonometric values for any real number t.

Quadrant	Positive functions	Negative function
I	all	none
II	sin, csc	cos, sec, tan, cot
III	tan, cot	sin, csc, cos, sec
IV	cos, sec	sin, csc, tan, cot

$$\sin(-t) = -\sin t \qquad \cos(-t) = \cos t \qquad \tan(-t) = -\tan t$$
$$\csc(-t) = -\csc t \qquad \sec(-t) = \sec t \qquad \cot(-t) = -\cot t$$

5. Find the sign of the expression if the terminal point determined by t is in the given quadrant.

   (a) $\cos^2 t \sin t$, quadrant II

   In quadrant II, cosine is negative and sine is positive, so the sign of $\cos^2 t \sin t$ is determined by $(-)^2(+)$, which is positive.

   (b) $\sin t \tan t$, quadrant III

   In quadrant III, sine is negative and tangent is positive, so the sign of $\sin t \tan t$ is determined by $(-)(+)$, which is negative.

6. Find each value.

   (a) $\sin \frac{5\pi}{3}$

   The reference number for $\frac{5\pi}{3}$ is $\frac{\pi}{3}$. Since $\frac{5\pi}{3}$ is in quadrant IV, $\sin \frac{5\pi}{3}$ is negative. So $\sin \frac{5\pi}{3} = -\sin \frac{\pi}{3} = -\frac{\sqrt{3}}{2}$.

   (b) $\tan \frac{7\pi}{6}$

   The reference number for $\frac{7\pi}{6}$ is $\frac{\pi}{6}$. Since $\frac{7\pi}{6}$ is in quadrant III, $\tan \frac{7\pi}{6}$ is positive. Thus $\tan \frac{7\pi}{6} = \tan \frac{\pi}{6} = \frac{\sqrt{3}}{3}$.

7. Use the even-odd properties of the trigonometric functions to determine each value.

   (a) $\cos\left(-\frac{\pi}{3}\right)$

   $\cos\left(-\frac{\pi}{3}\right) = \cos \frac{\pi}{3} = \frac{1}{2}$

   (b) $\sin\left(-\frac{\pi}{4}\right)$

   $\sin\left(-\frac{\pi}{4}\right) = -\sin \frac{\pi}{4} = -\frac{\sqrt{2}}{2}$

C. The trigonometric functions are related to each other through equations called **trigonometric identities**. An important class of identities are the **reciprocal identities**:

$$\csc t = \frac{1}{\sin t} \qquad \sec t = \frac{1}{\cos t} \qquad \tan t = \frac{\sin t}{\cos t} \qquad \cot t = \frac{\cos t}{\sin t} \qquad \cot t = \frac{1}{\tan t}$$

Another important class of identities are **Pythagorean identities**:

$$\sin^2 t + \cos^2 t = 1 \qquad\qquad \tan^2 t + 1 = \sec^2 t \qquad\qquad 1 + \cot^2 t = \csc^2 t$$

Note the usual convention of writing $\sin^n t$ for $(\sin t)^n$ for all integers n <u>except $n = -1$</u>, that is, the exponent is placed on the trigonometric function.

8. Find the values of the trigonometric functions of t for the given information.

   (a) $\cos t = \frac{7}{8}$, where t is in quadrant IV

   We find the easy one first: $\sec t = \dfrac{1}{\cos t} = \frac{8}{7}$.

   Next find sine, since t is in quadrant IV, $\sin t < 0$, so

   $$\sin^2 t + \cos^2 t = 1 \quad \Leftrightarrow \quad \sin^2 t + \left(\tfrac{7}{8}\right)^2 = 1 \quad \Leftrightarrow$$

   $$\sin^2 t = 1 - \tfrac{49}{64} = \tfrac{15}{64} \quad \Rightarrow \quad \sin t = -\frac{\sqrt{15}}{8}.$$

   $$\csc t = \frac{1}{\sin t} = -\frac{8}{\sqrt{15}} = -\frac{8\sqrt{15}}{15}$$

   $$\tan t = \frac{\sin t}{\cos t} = \frac{-\sqrt{15}/8}{7/8} = -\frac{\sqrt{15}}{7}$$

   $$\cot t = \frac{1}{\tan t} = -\frac{7}{\sqrt{15}} = -\frac{7\sqrt{15}}{15}$$

   (b) $\csc t = \frac{25}{7}$, where t is in quadrant II

   We find the easy one first: $\sin t = \dfrac{1}{\csc t} = \frac{7}{25}$.

   Next find cosine, since t is in quadrant II, $\cos t < 0$.

   $$\sin^2 t + \cos^2 t = 1 \quad \Leftrightarrow \quad \left(\tfrac{7}{25}\right)^2 + \cos^2 t = 1 \quad \Leftrightarrow$$

   $$\cos^2 t = 1 - \tfrac{49}{625} = \tfrac{576}{625} \quad \Rightarrow \quad \cos t = -\tfrac{24}{25}.$$

   $$\sec t = \frac{1}{\cos t} = -\tfrac{25}{24}$$

   $$\tan t = \frac{\sin t}{\cos t} = \frac{7/25}{-24/25} = -\tfrac{7}{24}$$

   $$\cot t = \frac{1}{\tan t} = -\tfrac{24}{7}$$

(c) $\cot t = 3$, $\cos t < 0$.

Since $\cos t < 0$ and $\cot t > 0$, t is in quadrant III.

We find the easy one first: $\tan t = \dfrac{1}{\cot t} = \tfrac{1}{3}$.

Next we use the Pythagorean identity to find $\csc t$.

$$1 + \cot^2 t = \csc^2 t \quad \Leftrightarrow \quad \csc^2 t = 1 + (3)^2 = 10$$

$\csc t = -\sqrt{10}$ (cosecant is negative in quadrant III).

$$\sin t = \frac{1}{\csc t} = \frac{1}{-\sqrt{10}} = -\frac{\sqrt{10}}{10}$$

We find $\cos t$ using the reciprocal identity,

$$\cot t = \frac{\cos t}{\sin t} \quad \Rightarrow \quad \cos t = \cot t \sin t. \text{ Substituting}$$

we get $\cos t = (3)\left(-\dfrac{\sqrt{10}}{10}\right) = -\dfrac{3\sqrt{10}}{10}$.

$$\sec t = \frac{1}{\cos t} = -\frac{10}{3\sqrt{10}} = -\frac{\sqrt{10}}{3}$$

Section 5.3   Trigonometric Graphs

Key Ideas

A. Periodic function and graphs of the sine and cosine functions.

B. Graphs of transformations of sine and cosine.

C. Using graphing devices in trigonometry.

A. A function f is **periodic** if there is a positive number p such that $f(t + p) = f(t)$. The least positive number p is called the **period** of f and the graph of f on any interval of length p is called **one complete period**. Sine and cosine each have a period of 2π. The graph of the sine function is symmetric with respect to the origin. The graph of the cosine function is symmetric with respect to the y-axis.

1. Sketch the graph of the given function.

   (a) $f(x) = 2\sin x$

The graph of $f(x)$ is the same as the graph of $\sin x$, but multiply the y-coordinate of each point by 2.

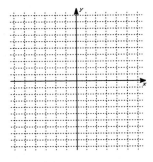

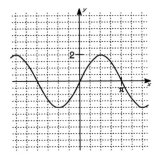

   (b) $g(x) = 3 - \cos x$

The graph of $g(x)$ is the reflection of $\cos x$ shifted up 3 units.

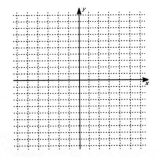

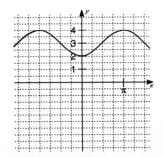

In Problem 2 and 3, determine whether the function whose graph is shown is periodic and, if so, determine the period from the graph.

2.

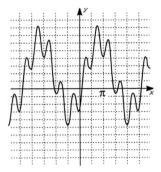

Yes, the function is periodic.
The period is $\frac{5\pi}{2}$.

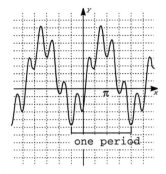

3.

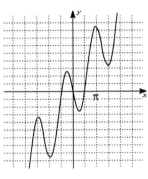

No, the function is not periodic.

B. The graphs of the functions of the form $y = a \sin k(x - b)$ and $y = a \cos k(x - b)$ are simply sine and cosine curves that have been transformed. The number $|a|$ is called the **amplitude** and is the largest value these functions attain. The number $k (k > 0)$ causes the period to change to $\dfrac{2\pi}{k}$. And b is the **phase shift**. When $b > 0$ the graphs are shifted to the right and when $b < 0$ the graphs are shifted to the left.

In Problems 4 and 5, find the amplitude, period, phase shift, and sketch the graph.

4. $y = 3\sin(3x - \pi)$

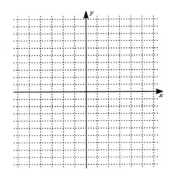

First transform the equation into $y = a \sin k(x - b)$ form.

$y = 3\sin(3x - \pi) = 3\sin 3\left(x - \frac{\pi}{3}\right)$

Amplitude: 3

Period: $\frac{2\pi}{3}$

Phase shift: $\frac{\pi}{3}$

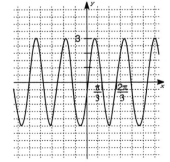

153

5. $y = 3\cos\left(x + \frac{\pi}{2}\right)$

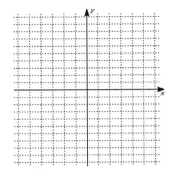

First transform the equation into $y = a\cos k(x - b)$ form.

$y = 3\cos\left(x + \frac{\pi}{2}\right)$

Amplitude: 3

Period: 2π

Phase shift: $-\frac{\pi}{2}$

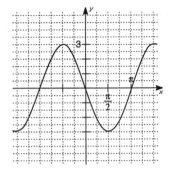

C. As stated before, choosing the appropriate viewing triangle is one of the most important aspects of using a graphing calculator. In general, a viewing rectangle should be only a few periods wide.

6. Use a graphing calculator to draw the graph of the function $f(x) = \sin 30x$ in the following viewing rectangles.

   (a) $[-5, 5]$ by $[-2, 2]$

The solutions shown below are representations of what you should see on your graphing calculator.

Your graphing calculator should give you a graph similar to one of these viewing rectangles.

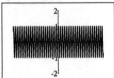

   (b) $[-0.2, 0.2]$ by $[-2, 2]$

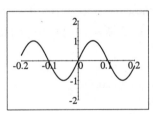

7. Draw the graphs of f, g, and $f + g$ on a common viewing rectangle.

   $f(x) = 2x$ and $g(x) = -\cos x$

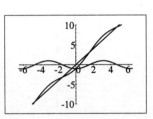

Viewing rectangle $[-6.5, 6.5]$ by $[-10, 10]$.

8. Graph the functions $y = \sqrt{x^2 + 1}$, $y = -\sqrt{x^2 + 1}$, and $y = \sqrt{x^2 + 1}\,(\sin x)$ on a common viewing rectangle. How are they related?

The functions $y = \sqrt{x^2 + 1}$ and $y = -\sqrt{x^2 + 1}$ form a sleeve for the function $y = \sqrt{x^2 + 1}\,(\sin x)$ whose graph lies between these functions.

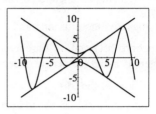

Viewing rectangle $[-10, 10]$ by $[-10, 10]$.

9. (a) Use a graphing device to graph the function.
   (b) Determine from the graph whether the function is periodic and, if so, determine the period.

$$y = \frac{1}{\sin^2 x + 1}$$

(a)

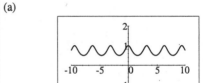

Viewing rectangle: $[-10, 10]$ by $[-2, 2]$.

(b) Yes, this function is periodic. The period is π.

Section 5.4   More Trigonometric Graphs

Key Ideas

A. Periods and domains of the tangent, cotangent, secant, and cosecant functions.

B. Graphs of functions involving tangent and cotangent.

C. Graphs involving secant and cosecant.

A. Recall that the functions $\tan t$ and $\sec t$ are not defined when $t = \dfrac{\pi}{2} + n\pi$, where n is an integer. And the functions $\cot t$ and $\csc t$ are not defined at $t = n\pi$, where n is an integer. The period of tangent and cotangent is π and the period of secant and cosecant is 2π.

B. The functions of the form $y = a \tan k(x - b)$ and $y = a \cot k(x - b)$ are transformations of the tangent and cotangent functions. As in Section 5.3, b is the phase shift, however, now the period is $\dfrac{\pi}{k}$. To sketch a complete period, it is convenient to select an interval between vertical asymptotes.

1. Find the phase shift and period. Sketch at least 3 periods of the given function.

(a) $y = \frac{1}{4}\cot x$

Phase shift: none

Period: π

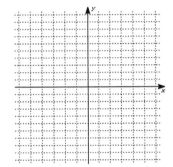

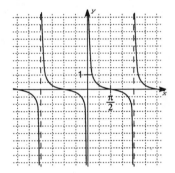

(b) $y = \tan 3x$

Phase shift: none

Period: $\frac{\pi}{3}$

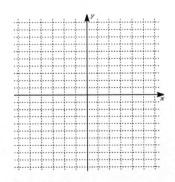

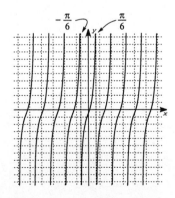

156

C. The functions of the form $y = a \sec k(x - b)$ and $y = a \csc k(x - b)$ are transformations of the secant and cosecant functions. Here b is the phase shift and the period is $\dfrac{2\pi}{k}$. As in other trigonometric functions, it is useful to first determine the phase shift and period before sketching the graph.

2. Find the phase shift and period. Sketch at least one period of the given function.

(a) $y = \frac{1}{4} \csc\left(\frac{x}{2}\right)$

Phase shift: none

Period: $\dfrac{2\pi}{\frac{1}{2}} = 4\pi$

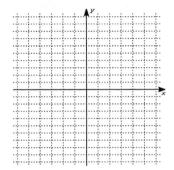

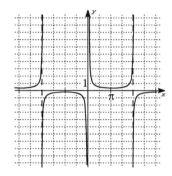

(b) $y = 3 \sec\left(x + \frac{\pi}{3}\right)$

Phase shift: $-\frac{\pi}{3}$

Period: 2π

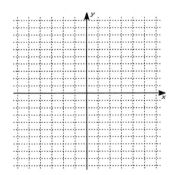

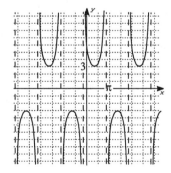

157

(c) $y = -2\csc(3x)$

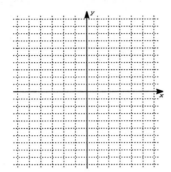

Phase shift: none

Period: $\frac{2\pi}{3}$

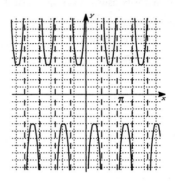

Chapter 6
Trigonometric Functions of Angles

Section 6.1 Angle Measure

Key Ideas

A. Degrees and radian measure.
B. Angles in standard position.
C. Length of a circular arc and area of a sector
d. Linear speed and angular speed.

A. The **measure** of an angle is the amount of rotation about the vertex required to rotate one side to the other side. One unit of measure for angles is the **degree**. One degree is $1/360$ of a complete rotation. Another unit of measure is the **radian**, the amount an angle opens measured along the arc of a circle of radius 1, its center at the vertex of the angle. The two measures are related by the equations:

$$180° = \pi \, \text{rad} \qquad 1 \, \text{rad} = \left(\frac{180}{\pi}\right)° \qquad 1° = \frac{\pi}{180} \, \text{rad}$$

Convert *degrees to radians* by multiplying by $\frac{\pi}{180}$.

Convert *radians to degrees* by multiplying by $\frac{180}{\pi}$.

1. Find the radian measure of an angle with the given degree measure.

   (a) 75°

   Multiply by $\frac{\pi}{180}$. We have
   $$75° = 75\left(\frac{\pi}{180}\right)\text{rad} = \frac{5\pi}{12} \, \text{rad}.$$

   (b) 300°

   Multiply by $\frac{\pi}{180}$. We have
   $$300° = 300\left(\frac{\pi}{180}\right)\text{rad} = \frac{5\pi}{3} \, \text{rad}.$$

2. Find the degree measure of the angle with the given radian measure.

   (a) $\frac{5\pi}{9}$

   Multiply by $\frac{180}{\pi}$. We have
   $$\frac{5\pi}{9} = \frac{5\pi}{9}\left(\frac{180}{\pi}\right) = 100°.$$

   (b) 5

   Multiply by $\frac{180}{\pi}$. We have
   $$5 = 5\left(\frac{180}{\pi}\right) = \left(\frac{900}{\pi}\right)° = 286.5°.$$

B. An angle is in **standard position** if it is drawn in the xy-plane with its vertex at the origin and its initial side on the positive x-axis. Two angles in standard position are **coterminal** if their other sides coincide. The measures of coterminal angles differ by an integer multiple of $360°$ or 2π.

3. The measures of two angles in standard position are given. Determine whether the angles are coterminal.

   (a) $-693°$ and $27°$

   Do these angles differ by a multiple of $360°$?
   $|-693 - 27| = 720$ and $720 = 2 \cdot 360$.
   So these angles are coterminal.

   (b) $\frac{17\pi}{12}$ and $\frac{-19\pi}{12}$

   Do these angles differ by a multiple of 2π?
   $\left|\frac{17\pi}{12} - \frac{-19\pi}{12}\right| = \frac{36\pi}{12} = 3\pi \neq k \cdot 2\pi$, for any integer k.
   So these angles are NOT coterminal.

4. Find angles that are coterminal with the given angle.

   (a) $47°$

   Here the key point is that two coterminal angles differ by $360°$. So the angles we seek are:
   $47° + n360° = 407°, 767°, 1127°, 1487°, \ldots$
   and
   $47° - n360° = -313°, -673°, -1033°, -1393°, \ldots$
   where n is any integer.

   (b) $\frac{11\pi}{6}$

   In this case, we use radians and now the key point is that two coterminal angles differ by 2π. So the angles angles we seek are:
   $\frac{11\pi}{6} + 2n\pi = \frac{23\pi}{6}, \frac{35\pi}{6}, \frac{47\pi}{6}, \frac{59\pi}{6}, \ldots$
   and
   $\frac{11\pi}{6} - 2n\pi = \frac{-\pi}{6}, \frac{-13\pi}{6}, \frac{-25\pi}{6}, \frac{-37\pi}{6}, \ldots$
   where n is any integer.

5. Find an angle between $0°$ and $360°$ that are coterminal with the given angle.

   (a) $992°$

   Add or subtract $360°$ until the angle is between $0°$ and $360°$.
   Since $2 \cdot 360 = 720 < 992 < 1080 = 3 \cdot 360$, we subtract $2 \cdot 360$ from 992.
   $992 - 720 = 272°$

   (b) $-75°$

   Since $-1 \cdot 360 = -360 < -75 < 0 = 0 \cdot 360$, we add 360 from -75.
   $-75 + 360 = 285°$

6. Find an angle between 0 and 2π that is coterminal with the given angle.

   (a) $\frac{-11\pi}{6}$

   Add or subtract 2π until the angle is between 0 and 2π.

   Since $-2\pi < \frac{-11\pi}{6} < 0$, we add 2π and get:

   $\frac{-11\pi}{6} + 2\pi = \frac{-11\pi}{6} + \frac{12\pi}{6} = \frac{\pi}{6}$.

   (b) $\frac{37\pi}{2}$

   Since $\frac{37\pi}{2} = 18.5\pi$, we subtract $9 \cdot 2\pi = 18\pi$.

   So $\frac{37\pi}{2} - 18\pi = \frac{\pi}{2}$.

C. In a circle of radius r, the length s of the arc shown and the area A of the sector shown are related to the central angle (given in radians) by the formulas:

$$s = r\theta \qquad\qquad A = \tfrac{1}{2}r^2\theta$$

In these applications, it is important to first convert the angle measure to radians.

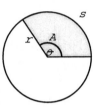

7. Find the area A of the sector and length s of the arc in each figure.

   (a)

   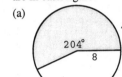

   First express $204°$ in terms of radians.

   So $\theta = 204° = 204 \cdot \frac{\pi}{180} = \frac{17\pi}{15}$.

   In the figure r is 8, so

   $s = r\theta = 8\left(\frac{17\pi}{15}\right) = \frac{136\pi}{15} \approx 28.48$ and

   $A = \tfrac{1}{2}r^2\theta = \tfrac{1}{2}(8^2)\left(\frac{17\pi}{15}\right)$

   $= \frac{544\pi}{15} \approx 113.94$.

   (b)

   From the figure, $\theta = 2\pi - \frac{\pi}{6} = \frac{11\pi}{6}$ and $r = 5$.

   So $s = r\theta = 5\left(\frac{11\pi}{6}\right) = \frac{55\pi}{6} \approx 28.80$ and

   $A = \tfrac{1}{2}r^2\theta = \tfrac{1}{2}(5^2)\left(\frac{11\pi}{6}\right)$

   $= \frac{275\pi}{12} \approx 71.99$.

D. Suppose a point moves along a circle, then there are two ways to describe the motion of point, linear speed and angular speed. **Linear speed** is the rate at which the distance traveled is changing and **angular speed** is the rate at which the central angle changes.

$$\text{Angular Speed} \quad \omega = \frac{\theta}{t} = \frac{\text{radians}}{\text{time}} \qquad\qquad \text{Linear speed} \quad v = \frac{s}{t} = \frac{\text{distance}}{\text{time}}$$

These two descriptions can be related by the equation $v = r\omega$. Note that angular speed does not depend on the radius of the circle, only the angle θ.

8. The carousel at a seaside amusement park revolves five times a minute.

(a) What is the angular speed of this carousel?

Using $\omega = \dfrac{\theta}{t}$ we get

$$\omega = \frac{5 \cdot 2\pi}{1} = 10\pi \text{ rad./min.}$$

(b) How fast (in mi/h) is a person riding a horse near the center, 12 feet from the center traveling?

We first find the linear speed in feet/min and then convert to mi/h. Using $v = r\omega$ with $r = 12$ ft. and $\omega = 10\pi$ rad. (from part (a)) we get:

$v = (12)(10\pi) = 120\pi \approx 376.99$ ft/min.

Convert to mi/h we have:

$$v = 376.99 \cdot \frac{60 \text{ min.}}{1 \text{ hr.}} \cdot \frac{1 \text{ mile}}{5280 \text{ft}} = 4.28 \text{ mi/h.}$$

(c) How fast (in mi/h) is a person riding a horse on the edge, 25 feet from the center traveling?

Again we first find the linear speed in feet/min and then convert to mi/h. Using $v = r\omega$ with $r = 25$ ft. and $\omega = 10\pi$ rad. (from part (a)) we get:

$v = (25)(10\pi) = 250\pi \approx 785.40$ ft/min.

Convert to mi/h we have:

$$v = 785.40 \cdot \frac{60 \text{ min.}}{1 \text{ hr.}} \cdot \frac{1 \text{ mile}}{5280 \text{ft}} = 8.92 \text{ mi/h.}$$

9. A hillside restaurant runs a cable car from the parking lot to the restaurant The cable for the cable car is driven by a one foot diameter wheel attached to a motor at the base of the hill. How fast must the wheel rotate in order for the cable car to make the 300 foot trip in 2 minutes?

We solve $v = r\omega$ for ω to get $\omega = \dfrac{v}{r}$.

Substituting $r = \frac{1}{2}$ and $v = \frac{300}{2} = 150$ ft/min.

we have $\omega = \dfrac{150}{1/2} = 300$ rad./min. Since there are

2π radian in each revolution, the wheel must rotate

$\frac{300}{2\pi} \approx 47.75$ times a minute.

Section 6.2 Trigonometry of Right Triangles

Key Ideas
A. Trigonometric ratios.
B. Special triangles.
C. Application of right triangles.

A. Trigonometric values can be described as ratios of the sides of a right triangle.

$$\sin\theta = \frac{\text{opposite}}{\text{hypotenuse}} \qquad \csc\theta = \frac{\text{hypotenuse}}{\text{opposite}}$$

$$\cos\theta = \frac{\text{adjacent}}{\text{hypotenuse}} \qquad \sec\theta = \frac{\text{hypotenuse}}{\text{adjacent}}$$

$$\tan\theta = \frac{\text{opposite}}{\text{adjacent}} \qquad \cot\theta = \frac{\text{adjacent}}{\text{opposite}}$$

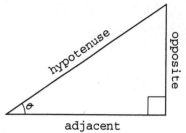

Since any two right triangles with angle θ are similar, these ratios are the same, regardless of the size of the triangle.

1. Find the values of the six trigonometric ratios of the angle θ in the triangle shown.

(a)

$$\sin\theta = \frac{7}{\sqrt{113}} \qquad\qquad \csc\theta = \frac{\sqrt{113}}{7}$$

$$\cos\theta = \frac{8}{\sqrt{113}} \qquad\qquad \sec\theta = \frac{\sqrt{113}}{8}$$

$$\tan\theta = \frac{7}{8} \qquad\qquad \cot\theta = \frac{8}{7}$$

(b)

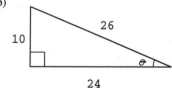

$$\sin\theta = \frac{5}{13} \qquad\qquad \csc\theta = \frac{13}{5}$$

$$\cos\theta = \frac{12}{13} \qquad\qquad \sec\theta = \frac{13}{12}$$

$$\tan\theta = \frac{5}{12} \qquad\qquad \cot\theta = \frac{12}{5}$$

(c)

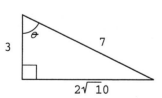

$$\sin\theta = \frac{2\sqrt{10}}{7} \qquad\qquad \csc\theta = \frac{7}{2\sqrt{10}}$$

$$\cos\theta = \frac{3}{7} \qquad\qquad \sec\theta = \frac{7}{3}$$

$$\tan\theta = \frac{2\sqrt{10}}{3} \qquad\qquad \cot\theta = \frac{3}{2\sqrt{10}}$$

2. If $\tan \theta = \frac{12}{5}$ find the other five trigonometric ratios.

Draw a right triangle and label one angle θ.
Since $\tan \theta = \frac{12}{5} = \frac{\text{opposite}}{\text{adjacent}}$, put 12 on the side *opposite* to θ and 5 on the side *adjacent* θ.
The hypotenuse is found by using Pythagorean Theorem.

$$\sqrt{12^2 + 5^2} = \sqrt{144 + 25}$$
$$= \sqrt{169} = 13$$

$\sin \theta = \frac{12}{13}$ $\qquad$ $\cos \theta = \frac{5}{13}$

$\cot \theta = \frac{5}{12}$ $\qquad$ $\sec \theta = \frac{13}{5}$

$\csc \theta = \frac{13}{12}$

3. If $\csc \theta = 3$ find the other five trigonometric ratios.

Draw a right triangle and label one angle θ. Since $\csc \theta = 3 = \frac{3}{1} = \frac{\text{hypotenuse}}{\text{opposite}}$, we put 1 on the side *opposite* to θ and 3 on the *hypotenuse*. Next, find the side *adjacent* to θ by using the Pythagorean

Theorem. $1^2 + x^2 = 3^2$ $\Leftrightarrow$ $1 + x^2 = 9$ $\Leftrightarrow$

$x^2 = 9 - 1 = 8$ $\Rightarrow$ $x = \sqrt{8} = 2\sqrt{2}$

$\sin \theta = \frac{1}{3}$ $\qquad\qquad$ $\cos \theta = \frac{2\sqrt{2}}{3}$

$\tan \theta = \frac{1}{2\sqrt{2}} = \frac{\sqrt{2}}{4}$ $\qquad$ $\cot \theta = \frac{2\sqrt{2}}{1} = 2\sqrt{2}$

$\sec \theta = \frac{3}{2\sqrt{2}} = \frac{3\sqrt{2}}{4}$.

B. 30°- 60° right triangles and 45° right triangles have ratios that can be calculated directly from the Pythagorean Theorem. These ratios appear frequently in the chapters of this textbook and in calculus textbooks, so it is helpful to learn these values.

t deg	t rad	$\sin t$	$\cos t$	$\tan t$	$\csc t$	$\sec t$	$\cot t$
30°	$\dfrac{\pi}{6}$	$\dfrac{1}{2}$	$\dfrac{\sqrt{3}}{2}$	$\dfrac{\sqrt{3}}{3}$	2	$\dfrac{2\sqrt{3}}{3}$	$\sqrt{3}$
45°	$\dfrac{\pi}{4}$	$\dfrac{\sqrt{2}}{2}$	$\dfrac{\sqrt{2}}{2}$	1	$\sqrt{2}$	$\sqrt{2}$	1
60°	$\dfrac{\pi}{3}$	$\dfrac{\sqrt{3}}{2}$	$\dfrac{1}{2}$	$\sqrt{3}$	$\dfrac{2\sqrt{3}}{3}$	2	$\dfrac{\sqrt{3}}{3}$

4. Find the length of the side labeled x correct to six.
   decimal places.

(a)

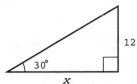

The sides x and 12 are related by the tangent

function, so $\tan 30° = \frac{12}{x}$ and

$x = \frac{12}{\tan 30°} = 12 \cot 30° = 12\sqrt{3} \approx 20.784610.$

(b)

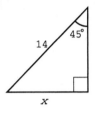

The sides x and 14 are related by the sine function,

so $\sin 45° = \frac{x}{14}$ and

$x = 14 \sin 45° = 14 \cdot \frac{\sqrt{2}}{2} \approx 9.899495.$

C. A triangle consists of three angles and three sides. To **solve a triangle** means to determine all of its
parts from the information known about the triangle. Applications that involve solving triangles
frequently use the following terminology:

line of sight	line from the observer's eye to the object;
angle of elevation	if the object is above the horizontal, then this is the angle between the line of sight and the horizontal;
angle of depression	if the object is below the horizontal, then this is the angle between the line of sight and the horizontal.

5. Express x and y in term of θ.

(a)

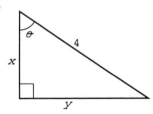

x and θ are related by $\cos \theta = \frac{x}{4}$ so

$x = 4 \cos \theta.$

y and θ are related by $\sin \theta = \frac{y}{4}$ so

$y = 4 \sin \theta.$

(b)

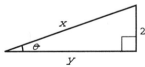

x and θ are related by $\sin \theta = \frac{2}{x}$ so

$x = \frac{2}{\sin \theta} = 2 \csc \theta.$

y and θ are related by $\cot \theta = \frac{y}{2}$ so

$y = 2 \cot \theta.$

6. A builder needs to determine the height of a house from a nearby hill. Using surveying equipment and a tape measure, she determines the horizontal, the angle of elevation, and the angle of depression as well as the distance from where these measurements where taken to the base of the house. This information is shown in the figure below. Determine the height of the house.

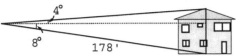

To find the height of the house we need to find the sides labeled x and y.

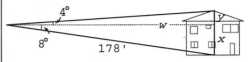

We find x using the sine function:

$$\frac{x}{178} = \sin 8° \quad \Leftrightarrow \quad x = 178 \sin 8° = 24.8 \text{ feet.}$$

To find y we first must find w. We find w using cosine:

$$\frac{w}{178} = \cos 8° \quad \Leftrightarrow \quad w = 178 \cos 8° = 176.3 \text{ feet.}$$

Next find y using w and tangent:

$$\frac{y}{w} = \tan 4° \quad \Leftrightarrow$$

$$y = w \tan 4° \approx 176.3 \tan 4° \approx 12.3 \text{ feet.}$$

Adding x and y , the height of the house is $24.8 + 12.3 = 37.1$ feet.

Section 6.3   Trigonometric Functions of Angles

Key Ideas

A. Definition of the trigonometric functions of angles.
B. Values of the trigonometric functions of any angle.
C. Fundamental Identities.
D. Areas of triangles.

A. Let θ be an angle in standard position and $P(x, y)$ be a point on the terminal side. If $r = \sqrt{x^2 + y^2}$ is the distance from the origin to the point $P(x, y)$, then

$$\sin \theta = \frac{y}{r} \qquad\qquad \cos \theta = \frac{x}{r} \qquad\qquad \tan \theta = \frac{y}{x} \; (x \neq 0)$$

$$\csc \theta = \frac{r}{y} \; (y \neq 0) \qquad \sec \theta = \frac{r}{x} \; (x \neq 0) \qquad \cot \theta = \frac{x}{y} \; (y \neq 0)$$

The definition is similar to the definitions in Chapter 5, the key difference in this case, is that the input to each trigonometric function is an angle whose measure is a real number.

B. Finding the value of a trigonometric function often involves using an acute **reference angle** and using the quadrant to get the sign of the trigonometric function. These signs are given in the table to the right.

Quadrant	Positive functions	Negative functions
I	all	none
II	sin, csc	cos, sec, tan, cot
III	tan, cot	sin, csc, cos, sec
IV	cos, sec	sin, csc, tan, cot

1. Find the exact value of each trigonometric function.

(a) $\sin \frac{7\pi}{4}$

Since $\frac{7\pi}{4}$ is in the fourth quadrant, the reference angle is $2\pi - \frac{7\pi}{4} = \frac{\pi}{4}$, and the sine function is negative.

Since $\sin \frac{\pi}{4} = \frac{\sqrt{2}}{2}$, we have $\sin \frac{7\pi}{4} = -\frac{\sqrt{2}}{2}$.

(b) $\tan \frac{7\pi}{6}$

Since $\frac{7\pi}{6}$ is in the third quadrant, the reference angle is $\frac{7\pi}{6} - \pi = \frac{\pi}{6}$, and the tangent function is positive.

Since $\tan \frac{\pi}{6} = \frac{\sqrt{3}}{3}$, we have $\tan \frac{7\pi}{6} = \frac{\sqrt{3}}{3}$.

(c) $\cos\left(-\frac{3\pi}{4}\right)$

Since $-\frac{3\pi}{4}$ is in the third quadrant, the reference angle is $-\frac{3\pi}{4} + \pi = \frac{\pi}{4}$, and the cosine function is negative.

Since $\cos \frac{\pi}{4} = \frac{\sqrt{2}}{2}$, we have $\cos\left(-\frac{3\pi}{4}\right) = -\frac{\sqrt{2}}{2}$.

(d) $\sec \frac{79\pi}{6}$

We first remove multiples of 2π. Since
$\frac{79\pi}{6} = \frac{72\pi}{6} + \frac{7\pi}{6} = 12\pi + \frac{7\pi}{6}$, we have
$\sec \frac{79\pi}{6} = \sec \frac{7\pi}{6}$. Since $\frac{7\pi}{6}$ is in the third quadrant,
the reference angle is $\frac{7\pi}{6} - \pi = \frac{\pi}{6}$, and the
secant function is negative.
Since $\sec \frac{\pi}{6} = 2$, we have $\sec \frac{79\pi}{6} = -2$.

2. Find the sign of each expression if θ is in the given
    quadrant.

(a) $\sec^2\theta \tan\theta$ Quadrant II

In quadrant II, both $\sec\theta$ and $\tan\theta$ are negative,
so the sign of $\sec^2\theta \tan\theta$ is determined by
$(-)^2(-)$ which is negative.

(b) $\sin\theta \tan\theta$ Quadrant III

In quadrant III, $\sin\theta$ is negative and $\tan\theta$ is positive,
so the sign of $\sin\theta \tan\theta$ is determined by $(-)(+)$
which is negative.

(c) $\dfrac{\cos\theta \sin^3\theta}{\tan^2\theta}$ Quadrant III

In quadrant III, $\cos\theta$ is negative, $\sin\theta$ is negative,
and $\tan\theta$ is positive, so the sign of $\dfrac{\cos\theta \sin^3\theta}{\tan^2\theta}$
is determined by $\dfrac{(-)(-)^3}{(+)^2}$, which is positive.

C. The trigonometric functions are related to each other through equations called **trigonometric identities**. An important class of identities are the **reciprocal identities**:

$$\csc t = \frac{1}{\sin t} \qquad \sec t = \frac{1}{\cos t} \qquad \cot t = \frac{1}{\tan t} \qquad \tan t = \frac{\sin t}{\cos t} \qquad \cot t = \frac{\cos t}{\sin t}$$

Another important class of identities are *the* **Pythagorean identities**:

$$\sin^2 t + \cos^2 t = 1 \qquad \tan^2 t + 1 = \sec^2 t \qquad 1 + \cot^2 t = \csc^2 t$$

Note the usual convention of writing $\sin^n t$ for $(\sin t)^n$ for all integers n <u>except $n = -1$</u>, that is, the exponent is placed on the trigonometric function.

3. Express the first trigonometric function in terms of the second function for θ in the given quadrant.

(a) $\tan \theta$ in terms of $\sin \theta$, Quadrant IV

$\sin \theta$ and $\tan \theta$ appear in several identities but none involving only $\sin \theta$ and $\tan \theta$. The identity we use here is $\tan \theta = \dfrac{\sin \theta}{\cos \theta}$ and realize that we can express $\cos \theta$ as function of $\sin \theta$ by using the identity $\cos^2\theta + \sin^2\theta = 1$. So, $\cos^2\theta = 1 - \sin^2\theta$ and $\cos \theta = \pm\sqrt{1 - \sin^2\theta}$. Since $\cos \theta$ is positive in quadrant IV we have $\cos \theta = \sqrt{1 - \sin^2\theta}$.

Thus $\tan \theta = \dfrac{\sin \theta}{\cos \theta} = \dfrac{\sin \theta}{\sqrt{1 - \sin^2\theta}}$.

(b) $\tan \theta$ in terms of $\sec \theta$, Quadrant III

The functions $\sec \theta$ and $\tan \theta$ are related by the identity $\tan^2\theta + 1 = \sec^2\theta$. So, $\tan^2\theta = \sec^2\theta - 1$ and $\tan \theta = \pm\sqrt{\sec^2 t - 1}$. Since $\tan \theta$ is positive in quadrant III, we have $\tan \theta = \sqrt{\sec^2 t - 1}$.

D. The area $\mathcal{A}$ of a triangle with sides of lengths a and b and included angle θ is

$$\mathcal{A} = \frac{1}{2}ab \sin \theta$$

4. Find the area of a triangle with sides of lengths 17 and 16 and included angle is 157°.

Apply the formula $\mathcal{A} = \frac{1}{2}ab \sin \theta$ with $a = 17$, $b = 16$, and $\theta = 157°$.
$\mathcal{A} = \frac{1}{2}(17)(16) \sin 157° \approx 53.14$.

5. Find the area of an isosceles triangle with a pair of sides each of length 235 and base angles of 22°.

Before we can apply the formula we first need to find the measure of the included angle. Find the included angle using the fact that the sum of the angles of any triangle is 180°.

$\theta = 180 - 22 - 22 = 136°$

$\mathcal{A} = \frac{1}{2}ab \sin \theta$

$\quad = \frac{1}{2}(22)(22) \sin 136°$

$\quad \approx 168.11$

6. The area of a triangle is 75 sq. ft. and two of the sides of the triangle have lengths 16 ft and 20 ft. Find the angle included by these two sides.

We use $\mathcal{A} = \frac{1}{2}ab \sin \theta$ and solve for θ.

$75 = \frac{1}{2}(16)(20) \sin \theta$

$75 = 160 \sin \theta$

$\sin \theta \approx 0.46875$

$\theta \approx 28°$

Since 28° could be just the reference angle, another solution is $\theta = 180° - 28° = 152°$.

7. The radius of the circle is 70 feet. Find the area of the shaded region.

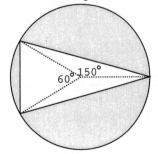

Since the radius is 70 ft, the area of the circle is $\pi(70)^2 \approx 15393.80$ sq. ft. Find the missing central angle: $360 - (150 + 60) = 150$.

Each of the triangles in the figure is an isosceles triangle. Using the formula for area, $\mathcal{A} = \frac{1}{2}ab \sin \theta$, with $a = b = 70$ and $\theta_1 = 60°$, $\theta_2 = 150°$, and $\theta_3 = 150°$.

$\mathcal{A}_1 = \frac{1}{2}(70)(70) \sin 60° \approx 2121.76$,

$\mathcal{A}_2 = \frac{1}{2}(70)(70) \sin 150° = 1225$, and

$\mathcal{A}_3 = \mathcal{A}_2$.

So the area of the shaded region is

$\mathcal{A} \approx 15393.80 - (2121.76 + 1225 + 1225)$

$\quad \approx 10822.04$ sq. ft.

Section 6.4   The Law of Sines

Key Ideas

A. Law of Sines.

B. Solving triangles.

A. The **Law of Sines** says that in any triangle, the length of the sides are proportional to the sines of the corresponding opposite angles.

> In triangle ABC, we have $\dfrac{\sin A}{a} = \dfrac{\sin B}{b} = \dfrac{\sin C}{c}$

The Law of Sines is used to solve a triangle when two sides and an angle opposite it are known or when two angles and a side is known.

1. Solve the triangle.

(a)

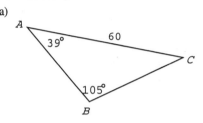

Here we are given the following information:

$\angle A = 39°$, $\angle B = 105°$, and $b = 60$.

We are asked to find: $\angle C$, a, and c.

We find $\angle C$ using the fact that the sum of the angles of a triangle add up to $180°$.

$\angle C = 180° - (39° + 105°) = 36°$

Since we have b and $\angle B$, we use it to find a and c.

$$\frac{\sin 39°}{a} = \frac{\sin 105°}{60} \quad \Leftrightarrow \quad a = \frac{60 \sin 39°}{\sin 105°} \approx 39$$

$$\frac{\sin 36°}{c} = \frac{\sin 105°}{60} \quad \Leftrightarrow \quad c = \frac{60 \sin 36°}{\sin 105°} \approx 36.5$$

(b)

Here we are given the following information:

$\angle A = 38°$, $\angle B = 75°$, and $c = 67$.

We are asked to find: $\angle C$, a, and b.

We find $\angle C$ using the fact that the sum of the angles of a triangle add up to $180°$.

$\angle C = 180° - (38° + 75°) = 67°$

Now that we have c and $\angle C$, we use the law of sines to find a and b.

$$\frac{\sin 38°}{a} = \frac{\sin 67°}{25} \quad \Leftrightarrow \quad a = \frac{25 \sin 38°}{\sin 67°} \approx 16.7$$

$$\frac{\sin 75°}{b} = \frac{\sin 67°}{25} \quad \Leftrightarrow \quad b = \frac{25 \sin 75°}{\sin 67°} \approx 26.2$$

B. A triangle is determined by any three of it parts as long as at least one of these parts is a side. There are four possible cases:

Case 1. <u>Side-Angle-Angle.</u> One side and two angles. (Remember, given any two angles in a triangle, the third angle can be determined.)

Case 2. <u>Side-Side-Angle.</u> Two sides and the angle opposite one of those sides. This case is sometimes called the **ambiguous case** because under some circumstances the triangle is not unique, that is, two triangles are possible. (See problem 3(b) on the next page.)

Case 3. <u>Side-Angle-Side.</u> Two sides and the included angle. Since we don't have a side and its opposite angle, we cannot use the Law of Sines to solve these triangles.

Case 4. <u>Side-Side-Side.</u> Three sides. As in case 3, we cannot use the Law of Sines to solve these triangles.

2. Sketch each triangle, then solve the triangle using the Law of Sines.

(a) $\angle A = 64°$, $\angle B = 64°$, $c = 15$

This is ASA. Draw side c and then the two angles. Their intersection is angle C. We find $\angle C$ and use it to find the sides a and b.

$$\angle C = 180° - (64° + 64°)$$
$$= 52°$$

$$\frac{\sin 64°}{a} = \frac{\sin 52°}{15} \quad \Leftrightarrow$$

$$a = \frac{15 \sin 64°}{\sin 52°} \approx 17.1$$

Since $\angle A = \angle B$, $\triangle ABC$ is an isosceles triangle and $a = b = 17.1$.

(b) $\angle B = 73°$, $\angle C = 28°$, $a = 258$

This is ASA. Draw side c and then the two angles. Their intersection is angle A. We find $\angle A$ and use it to find the sides b and c.

$$\angle A = 180° - (73° + 28°) = 79°$$

$$\frac{\sin 73°}{b} = \frac{\sin 79°}{258} \quad \Leftrightarrow \quad b = \frac{258 \sin 73°}{\sin 79°} \approx 251.3$$

$$\frac{\sin 28°}{c} = \frac{\sin 79°}{258} \quad \Leftrightarrow \quad c = \frac{258 \sin 28°}{\sin 79°} \approx 123.4$$

3. Use the Law of Sines to solve for all possible triangles.

(a) $a = 73$, $b = 58$, $\angle A = 128°$

Since $\angle A > 90°$, at most one triangle is possible.

Find $\angle B$.

$$\frac{\sin B}{58} = \frac{\sin 128°}{73} \quad \Leftrightarrow$$

$$\sin B = \frac{58 \sin 128°}{73} \approx 0.6261 \quad \Rightarrow$$

$$\angle B \approx 38.8°$$

Next find $\angle C$.

$$\angle C \approx 180° - (128° + 38.8°) = 13.2°$$

Now that we have $\angle C$, we can find c.

$$\frac{\sin 13.2°}{c} = \frac{\sin 128°}{73} \quad \Leftrightarrow$$

$$c = \frac{73 \sin 13.2°}{\sin 128°} \approx 21.2$$

(b) $a = 56$, $c = 71$, $\angle A = 45°$

Since $\angle A < 90°$, two triangles are possible.

Find $\angle C$.

$$\frac{\sin C}{71} = \frac{\sin 45°}{56} \quad \Leftrightarrow$$

$$\sin C = \frac{71 \sin 45°}{56} \approx 0.8965 \quad \Rightarrow$$

$\angle C \approx 63.7°$. Since $63.7°$ is the reference angle, another possible value for $\angle C$ is

$$180° - 63.7° = 116.3°.$$

Case 1: $\angle C = 63.7°$

Find $\angle B$. $\angle B \approx 180° - (45° + 63.7°) = 71.3°$

Next we use $\angle B$ to find b.

$$\frac{\sin 71.3°}{b} = \frac{\sin 45°}{56} \quad \Leftrightarrow \quad b = \frac{56 \sin 71.3°}{\sin 45°} \approx 75.0$$

Case 2: $\angle C = 116.3°$

Find $\angle B$: $\angle B \approx 180° - (45° + 116.3°) = 18.7°$.

Next we use $\angle B$ to find b.

$$\frac{\sin 18.7°}{b} = \frac{\sin 45°}{56} \quad \Leftrightarrow \quad b = \frac{56 \sin 18.7°}{\sin 45°} \approx 25.4$$

4. To measure the height of a bridge, a surveyor measures the angle of elevation from two points 20 feet apart (on a direct line from the center of the bridge). This information is shown in the figure below. Find the height of the bridge.

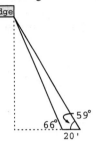

First label the vertices. If we knew CB, then we could use $\sin B = \dfrac{CD}{CB}$ to find the height CD.

Find the length of CB.

Use $\angle ACB$, $\angle CAB$, and AB in the Law of Sines to find CB.

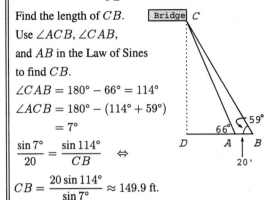

$\angle CAB = 180° - 66° = 114°$

$\angle ACB = 180° - (114° + 59°)$

$\quad = 7°$

$\dfrac{\sin 7°}{20} = \dfrac{\sin 114°}{CB} \quad \Leftrightarrow$

$CB = \dfrac{20\sin 114°}{\sin 7°} \approx 149.9 \text{ ft.}$

Now that we have CB we can find the height.

$\sin 59° = \dfrac{CD}{149.9}$ so $CD = 149.9\sin 59° \approx 128.5 \text{ ft.}$

The bridge height is 128.5 ft.

5. A ski resort wishes to install a new tram from the lodge to the top of the mountain. A surveyor measures the angle of elevation from the base of the tram and from the base of where the mountain becomes steep. She also finds the distance between these sites. Find the distance between the lodge and the top of the mountain.

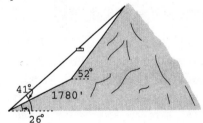

First label the vertices and the line as shown. In order to use the Law of Sines to find AB, we need to find $\angle B$ and $\angle C$.

$\angle A = 41° - 26° = 15°$.

$\angle C = \angle BCD + \angle DCA$

$\angle BCD = 180° - 52° = 128°$

$\angle DCA = 26°$ (Alternate interior angles are equal)

$\angle C = 128° + 26°$

$\quad = 154°$

Finally, we find $\angle B$.

$\angle B = 180° - (154° + 15°) = 11°$.

We now use the Law of Sines to find AB.

$\dfrac{\sin C}{AB} = \dfrac{\sin B}{AC} \quad \Leftrightarrow \quad \dfrac{\sin 154°}{AB} = \dfrac{\sin 11°}{1780} \quad \Rightarrow$

$AB = 1780\,\dfrac{\sin 154°}{\sin 11°} \approx 4089.4 \text{ ft.}$

Section 6.5   The Law of Cosines

Key Ideas

A. Law of Cosines.

B. Area of Triangles and Heron Formula.

A. The Law of Sines cannot be used directly to solve a triangle when we know two sides and the angle between them or when we know all three sides. In these cases the **Law of Cosines** is used.

> In any triangle ABC, we have:
> $$a^2 = b^2 + c^2 - 2bc \cos A$$
> $$b^2 = a^2 + c^2 - 2ac \cos B$$
> $$c^2 = a^2 + b^2 - 2ab \cos C.$$

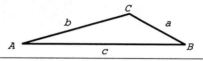

1. The sides of a triangle are $a = 9$, $b = 10$, and $c = 11$. Find the angles of the triangle.

Solve the equation $a^2 = b^2 + c^2 - 2bc \cos A$ for $\cos A$.

$$2bc \cos A = b^2 + c^2 - a^2$$

$$\cos A = \frac{b^2 + c^2 - a^2}{2bc}$$

So, $\cos A = \frac{10^2 + 11^2 - 9^2}{2(10)(11)} = 0.6364$ and

$\underline{\angle A = 50.5°}$.

Similarly,

$$\cos B = \frac{a^2 + c^2 - b^2}{2ac} = \frac{9^2 + 11^2 - 10^2}{2(9)(11)}$$

$$= 0.5152$$

$\underline{\angle B = 59°}$,

$$\cos C = \frac{a^2 + b^2 - c^2}{2ab} = \frac{9^2 + 10^2 - 11^2}{2(9)(10)}$$

$$= 0.3333$$

$\underline{\angle C = 70.5°}$.

Check: $50.5° + 59° + 70.5° = 180°$.

2. $a = 19$, $b = 34$, and $C = 75°$. Solve triangle ABC.

We need to find angles A and B, and side c.

Use **Law of Cosines** to find c.

$c^2 = a^2 + b^2 - 2ab \cos C$

$\quad = 19^2 + 34^2 - 2(19)(34) \cos 75°$

$\quad = 361 + 1156 - 1292(0.2588) = 1182.61$

$\underline{c \approx 34.39}$

Since the Law of Sines is easier to use (less terms) we use it to find $\angle A$.

$$\frac{\sin C}{c} = \frac{\sin A}{a} \quad \Leftrightarrow$$

$$\sin A = a\frac{\sin C}{c} = (19)\frac{\sin 75°}{34.39} \approx 0.5337$$

$\underline{\angle A \approx 32.25°}$

$\underline{\angle B \approx 180° - (75° + 32.25°) = 72.75°}$

3. A homeowner needs to replace the light above a staircase. The dimensions are shown in the figure to the right. How long does the ladder need to be?

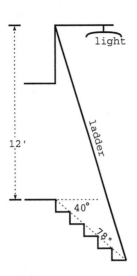

First label the vertices of the triangle. Before the Law of Cosines can be used, we must first convert side c from feet to inches or side b from inches to feet. Since ladders are sold by the foot, we convert side b to feet. $78" = 6.5'$.

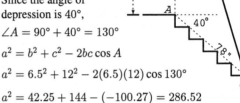

Next, find $\angle A$. Since the angle of depression is $40°$,

$\angle A = 90° + 40° = 130°$

$a^2 = b^2 + c^2 - 2bc \cos A$

$a^2 = 6.5^2 + 12^2 - 2(6.5)(12) \cos 130°$

$a^2 = 42.25 + 144 - (-100.27) = 286.52$

$a \approx 16.9$.

So the homeowner needs a ladder that is 17 feet long.

B. The **semiperimeter**, s, of a triangle is half the perimeter of the triangle.

> The area $\mathcal{A}$ of triangle ABC is given by **Heron's Formula:**
> $$\mathcal{A} = \sqrt{s(s-a)(s-b)(s-c)}, \quad \text{where } s = \frac{1}{2}(a+b+c)$$

4. Find the area of the triangle with the given sides.

(a) $a = 45$, $b = 28$, and $c = 57$.

First find the semiperimeter.

$s = \frac{1}{2}(45 + 28 + 57) = \frac{1}{2} \cdot 130 = 65$

$\mathcal{A} = \sqrt{65(65 - 45)(65 - 28)(65 - 57)}$

$= \sqrt{384800}$

≈ 620.32 sq. units

(b) $a = 23$, $b = 37$, and $c = 21$.

First find the semiperimeter.

$s = \frac{1}{2}(23 + 37 + 21) = \frac{1}{2} \cdot 81 = 40.5$

$\mathcal{A} = \sqrt{40.5(40.5 - 23)(40.5 - 37)(40.5 - 21)}$

$= \sqrt{48372.1875}$

≈ 219.94 sq. units

5. An irregular lot has dimensions shown in the figure below. Find its area.

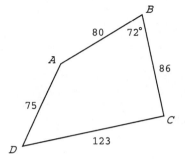

We draw the diagonal AC and use the Law of Cosines to find its length. Then find the area of the triangles ABC and ADC.

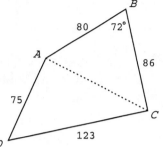

$$(AC)^2 = (AB)^2 + (BC)^2 - 2(AB)(BC)\cos B$$
$$= 80^2 + 86^2 - 2(80)(86)\cos 72°$$
$$\approx 6400 + 7396 - 4252.07$$
$$\approx 9543.93$$
$$AC \approx 97.7$$

Find the area of triangle ABC.
We use the formula for the area of a triangle from Section 6.3.
$$\mathcal{A}_1 = \tfrac{1}{2} \cdot a \cdot c \cdot \sin B$$
$$= \tfrac{1}{2}(80)(86)\sin 72°$$
$$= 3271.63$$

To find the area of triangle ADC, we first find the semiperimeter, s.
$$s = \tfrac{1}{2}(75 + 123 + 97.7) = \tfrac{1}{2} \cdot 295.7 = 147.85$$
Then using Heron's formula, the area is
$$\mathcal{A}_2 = \sqrt{147.85(72.85)(24.85)(50.15)}$$
$$= \sqrt{13422957.51}$$
$$\approx 3663.74.$$

Therefore the area of the lot is approximately
$3271.63 + 3663.74 = 6935.37$ sq. ft.

Chapter 7
Analytic Trigonometry

Section 7.1   Trigonometric Identities

Key Ideas

A. Identities.

B. Proving trigonometric identities.

A. Identities enable us to write the same expression in different way. **Trigonometric identities** are equations that are true for all values for which both sides are defined. Sine and cosine are called **cofunctions**. Similarity, tangent and cotangent are cofunctions, as are secant and cosecant. Cofunction identities show another way these functions are interrelated. The cofunction identities are shown in the table below. Some of the important classes of trigonometric identities are:

Reciprocal Identities	$\csc t = \dfrac{1}{\sin t} \qquad \sec t = \dfrac{1}{\cos t} \qquad \cot t = \dfrac{1}{\tan t}$ $\tan t = \dfrac{\sin t}{\cos t} \qquad \cot t = \dfrac{\cos t}{\sin t}$
Pythagorean Identities	$\sin^2 t + \cos^2 t = 1 \qquad \tan^2 t + 1 = \sec^2 t$ $1 + \cot^2 t = \csc^2 t$
Even-odd Identities	$\sin(-t) = -\sin t \qquad \cos(-t) = \cos t$ $\tan(-t) = -\tan t \qquad \csc(-t) = -\csc t$ $\sec(-t) = \sec t \qquad \cot(-t) = -\cot t$
Cofunction Identities	$\sin\left(\frac{\pi}{2} - u\right) = \cos u \qquad \cos\left(\frac{\pi}{2} - u\right) = \sin u$ $\tan\left(\frac{\pi}{2} - u\right) = \cot u \qquad \cot\left(\frac{\pi}{2} - u\right) = \tan u$ $\sec\left(\frac{\pi}{2} - u\right) = \csc u \qquad \csc\left(\frac{\pi}{2} - u\right) = \sec u$

1. Write following trigonometric expression in terms of sine and cosine, and then simplify.

$$\sin\theta \sec\theta + \cos\theta \csc\theta$$

$\sin\theta \sec\theta + \cos\theta \csc\theta$

$$= \sin\theta \, \frac{1}{\cos\theta} + \cos\theta \, \frac{1}{\sin\theta}$$

$$= \frac{\sin\theta}{\cos\theta} + \frac{\cos\theta}{\sin\theta}$$

$$= \frac{\sin\theta}{\cos\theta} \cdot \frac{\sin\theta}{\sin\theta} + \frac{\cos\theta}{\sin\theta} \cdot \frac{\cos\theta}{\cos\theta} \cdot$$

$$= \frac{\sin^2\theta}{\cos\theta \sin\theta} + \frac{\cos^2\theta}{\sin\theta \cos\theta}$$

$$= \frac{\sin^2\theta + \cos^2\theta}{\cos\theta \sin\theta}$$

$$= \frac{1}{\cos\theta \sin\theta}$$

$$= \sec\theta \csc\theta$$

2. Write the following trigonometric expression in terms of sine and cosine, and then simplify.

$$\frac{1}{\cot x - \csc x}$$

$$\frac{1}{\cot x - \csc x} = \frac{1}{\dfrac{\cos x}{\sin x} - \dfrac{1}{\sin x}}$$

$$= \frac{1}{\dfrac{\cos x - 1}{\sin x}} = \frac{\sin x}{\cos x - 1}$$

3. Simplify the trigonometric expression.
   (a) $(\tan^2\alpha - 1)(\sec^2\alpha + 1)$

$$(\tan^2\alpha - 1)(\sec^2\alpha + 1)$$
$$= \tan^2\alpha \sec^2\alpha + \tan^2\alpha - \sec^2\alpha - 1$$
$$= \tan^2\alpha \sec^2\alpha + \tan^2\alpha - (\tan^2\alpha + 1) - 1$$
$$= \tan^2\alpha \sec^2\alpha + \tan^2\alpha - \tan^2\alpha - 1 - 1$$
$$= \tan^2\alpha \sec^2\alpha - 2$$

   (b) $\dfrac{1}{1 + \cos\beta}$

Here we wish to *clean up* the denominator, that is, write the denominator as a *single* term (if possible).

$$\frac{1}{1 + \cos\beta} = \frac{1}{1 + \cos\beta} \cdot \frac{1 - \cos\beta}{1 - \cos\beta}$$

$$= \frac{1 - \cos\beta}{1 - \cos^2\beta} = \frac{1 - \cos\beta}{\sin^2\beta}$$

$$= \frac{1}{\sin^2\beta} - \frac{\cos\beta}{\sin^2\beta}$$

$$= \frac{1}{\sin^2\beta} - \frac{1}{\sin\beta} \cdot \frac{\cos\beta}{\sin\beta}$$

$$= \csc^2\beta - \csc\beta \cot\beta$$

4. Prove that the equation is an identity.

$$\cos\left(\tfrac{\pi}{4} + x\right) = \sin\left(\tfrac{\pi}{4} - x\right)$$

To show that this equation is an identity, we use the cofunction identity and the following substitution

Let $u = \tfrac{\pi}{4} + x$. Then

$$\tfrac{\pi}{2} - u = \tfrac{\pi}{2} - \left(\tfrac{\pi}{4} + x\right) = \tfrac{\pi}{2} - \tfrac{\pi}{4} - x = \tfrac{\pi}{4} - x.$$

Substituting we get

$$\cos\left(\tfrac{\pi}{4} + x\right) = \cos u = \sin\left(\tfrac{\pi}{2} - u\right)$$

$$= \sin\left(\tfrac{\pi}{4} - x\right).$$

B. To show that an equation is not an identity, all you need to do is to show that the equation is false for *some choice* of variable(s). This *choice* is called a **counterexample**. To prove that an equation is an identity, we must start with one side of the equation and transform it into the other side. At each step, a trigonometric identity or an algebraic identity is used. When stuck as to what to do, start by rewriting all functions in terms of sine and cosine. *In general, it is usually easier to change the more complicated side of the equation into the simpler side.* It is very important to understand that when *proving a trigonometric identity, we are not solving an equation*, so we do *not* perform the same operation on both sides of the equation.

5. Verify each identity.

(a) $\sin w + \cos w = \dfrac{1 + \cot w}{\csc w}$

$$\sin w + \cos w = \dfrac{1 + \cot w}{\csc w}$$

Since the right side is the more complicated side, we transform it into the left side. To get an idea as to what to do, first express the right side in terms of sine and cosine.

$$\sin w + \cos w = \dfrac{1 + \dfrac{\cos w}{\sin w}}{\dfrac{1}{\sin w}}$$

From this, we see that both the numerator and the denominator have a common denominator, $\sin w$. Multiplying top and bottom of the left side by this common denominator we get:

$$= \dfrac{1 + \dfrac{\cos w}{\sin w}}{\dfrac{1}{\sin w}} \cdot \dfrac{\sin w}{\sin w}$$

$$= \dfrac{\sin w + \cos w}{1} = \sin w + \cos w.$$

(b) $\csc^4 \gamma - \cot^4 \gamma = \csc^2 \gamma + \cot^2 \gamma$

Transform the left side into the right side. FACTOR.

$\text{LHS} = \csc^4 \gamma - \cot^4 \gamma$

$$= (\csc^2 \gamma - \cot^2 \gamma)(\csc^2 \gamma + \cot^2 \gamma)$$

The Pythagorean identity $1 + \cot^2 \gamma = \csc^2 \gamma$ lets us use the substitution $1 = \csc^2 \gamma - \cot^2 \gamma$.

$$= (1)(\csc^2 \gamma + \cot^2 \gamma)$$

$$= \csc^2 \gamma + \cot^2 \gamma = \text{RHS}$$

(c) $\sec y - \cos y = \sin y \tan y$

Transform the left side into the right side.

$$\text{LHS} = \sec y - \cos y = \dfrac{1}{\cos y} - \cos y$$

$$= \dfrac{1}{\cos y} - \cos y \cdot \dfrac{\cos y}{\cos y}$$

$$= \dfrac{1 - \cos^2 y}{\cos y}$$

$$= \dfrac{\sin^2 y}{\cos y} = \sin y \cdot \dfrac{\sin y}{\cos y}$$

$$= \sin y \tan y = \text{RHS}$$

Note: To start we expressed the left side in terms of cosine, then got a common denominator.

6. Make the indicated trigonometric substitution and simplify.

(a) $\dfrac{x}{(1-x^2)^{3/2}}$;

$x = \sin\theta \qquad \left(-\frac{\pi}{2} < \theta < \frac{\pi}{2}\right)$

$$\dfrac{x}{(1-x^2)^{3/2}} = \dfrac{\sin\theta}{(1-\sin^2\theta)^{3/2}}$$

$$= \dfrac{\sin\theta}{(\cos^2\theta)^{3/2}} = \dfrac{\sin\theta}{\cos^3\theta}$$

$$= \dfrac{\sin\theta}{\cos\theta}\cdot\dfrac{1}{\cos^2\theta}$$

$$= \tan\theta\,\sec^2\theta$$

(b) $\dfrac{\sqrt{1+x^2}}{x}$;

$x = \tan\theta \qquad \left(-\frac{\pi}{2} \le \theta \le \frac{\pi}{2}\right)$

$$\dfrac{\sqrt{1+x^2}}{x} = \dfrac{\sqrt{1+\tan^2\theta}}{\tan\theta}$$

$$= \dfrac{\sqrt{\sec^2\theta}}{\tan\theta} = \dfrac{\sec\theta}{\tan\theta}$$

Note: sec θ *is positive in the given interval.*
Since both numerator and denominator have the common denominator cos θ,

$$\dfrac{\sec\theta}{\tan\theta} = \dfrac{\sec\theta}{\tan\theta}\cdot\dfrac{\cos\theta}{\cos\theta} = \dfrac{1}{\sin\theta} = \csc\theta.$$

Under the given substitution, $\dfrac{\sqrt{1+x^2}}{x} = \csc\theta.$

(c) $\dfrac{x^2}{\sqrt{x^2-a^2}}$;

$x = a\sec\theta,\, a > 0 \qquad \left(0 \le \theta < \frac{\pi}{2}\right)$

$$\dfrac{x^2}{\sqrt{x^2-a^2}} = \dfrac{(a\sec\theta)^2}{\sqrt{(a\sec\theta)^2-a^2}}$$

$$= \dfrac{a^2\sec^2\theta}{\sqrt{a^2\sec^2\theta-a^2}}$$

$$= \dfrac{a^2\sec^2\theta}{\sqrt{a^2(\sec^2\theta-1)}}$$

$$= \dfrac{a^2\sec^2\theta}{\sqrt{a^2\tan^2\theta}}$$

$$= \dfrac{a^2\sec^2\theta}{a\tan\theta} = a\sec\theta\cdot\dfrac{a\sec\theta}{a\tan\theta}$$

$$= a\sec\theta\,\csc\theta$$

Note: tan θ is positive in the given interval.

7. Show that each equation is not an identity.

(a) $\dfrac{\cos\theta+1}{\sin\theta}=\dfrac{\sin\theta}{\cos\theta-1}$

We show that $\theta=\frac{\pi}{3}$ gives us a counterexample.

$$\frac{\cos\theta+1}{\sin\theta}=\frac{\cos\frac{\pi}{3}+1}{\sin\frac{\pi}{3}}=\frac{\frac{1}{2}+1}{\frac{\sqrt{3}}{2}}=\sqrt{3}$$

$$\frac{\sin\theta}{\cos\theta-1}=\frac{\sin\frac{\pi}{3}}{\cos\frac{\pi}{3}-1}=\frac{\frac{\sqrt{3}}{2}}{\frac{1}{2}-1}=-\sqrt{3}$$

(b) $\dfrac{\sin\gamma\cos\gamma}{\sin\gamma+\cos\gamma}=\cot\gamma+\tan\gamma$

We show that $\gamma=\frac{3\pi}{4}$ gives us a counterexample.

$$\frac{\sin\gamma\cos\gamma}{\sin\gamma+\cos\gamma}=\frac{\sin\frac{3\pi}{4}\cos\frac{3\pi}{4}}{\sin\frac{3\pi}{4}+\cos\frac{3\pi}{4}}$$

$$=\frac{\frac{\sqrt{2}}{2}\cdot\left(-\frac{\sqrt{2}}{2}\right)}{\frac{\sqrt{2}}{2}+\left(-\frac{\sqrt{2}}{2}\right)}$$

$$=\frac{-\frac{1}{2}}{0}\ \text{which is not defined.}$$

$$\cot\gamma+\tan\gamma=\cot\frac{3\pi}{4}+\tan\frac{3\pi}{4}$$

$$=-1+(-1)$$

$$=-2$$

Section 7.2   Addition and Subtraction Formulas

Key Ideas
A. Addition and subtraction formulas.
B. Sums of sines and cosines.

A. The identities for the sum and difference for sine and cosine are shown in the table below. These identities are used to increase the number of angles for which we know the exact value of sine and cosine. These identities are very useful in calculus.

$\sin(s + t) = \sin s \cos t + \cos s \sin t$	$\sin(s - t) = \sin s \cos t - \cos s \sin t$
$\cos(s + t) = \cos s \cos t - \sin s \sin t$	$\cos(s - t) = \cos s \cos t + \sin s \sin t$
$\tan(s + t) = \dfrac{\tan s + \tan t}{1 - \tan s \tan t}$	$\tan(s - t) = \dfrac{\tan s - \tan t}{1 + \tan s \tan t}$

1. Find the exact value of each expression.

   (a) $\sin 75°$

   We must first express $75°$ in terms of angles for which we know the exact values of sine and cosine. Since $75° = 45° + 30°$ and we know the exact values for these two angles, we use these two angles.
   $\sin 75° = \sin(45° + 30°)$
   $= \sin 45° \cos 30° + \cos 45° \sin 30°$
   $= \frac{\sqrt{2}}{2} \cdot \frac{\sqrt{3}}{2} + \frac{\sqrt{2}}{2} \cdot \frac{1}{2} = \frac{\sqrt{6}+\sqrt{2}}{4}$

   (b) $\cos 15°$

   We must first express $15°$ in terms of angles for which we know the exact values of sine and cosine. Since $15° = 45° - 30°$ and we know the exact values for these two angles, we use these two angles.
   $\cos 15° = \cos(45° - 30°)$
   $= \cos 45° \cos 30° + \sin 45° \sin 30°$
   $= \frac{\sqrt{2}}{2} \cdot \frac{\sqrt{3}}{2} + \frac{\sqrt{2}}{2} \cdot \frac{1}{2} = \frac{\sqrt{6}+\sqrt{2}}{4}$
   Yes, $\sin 75° = \cos 15°$ *(see the cofunction identities in Section 7.1).*

2. Express $\sin 3x$ in terms of $\sin x$ and $\cos x$.

   Since $3x = 2x + x$, we first apply the sine addition formula.
   $\sin 3x = \sin(2x + x)$
   $= \sin 2x \cos x + \cos 2x \sin x$
   Next we apply the addition formulas for $2x = x + x$.
   $= (\sin x \cos x + \cos x \sin x) \cos x +$
   $\qquad\qquad (\cos x \cos x - \sin x \sin x) \sin x$
   $= 2\sin x \cos^2 x + \sin x \cos^2 x - \sin^3 x$
   $= 3\sin x \cos^2 x - \sin^3 x$

3. If α and β are acute angles such that $\cos \alpha = \frac{4}{5}$ and $\sin \beta = \frac{5}{13}$, find $\sin(\alpha + \beta)$, $\sin(\alpha - \beta)$, $\cos(\alpha + \beta)$, and $\cos(\alpha - \beta)$.

Before we can find these values, we need to determine $\sin \alpha$ and $\cos \beta$. Since α and β are acute angles, we use right triangles and place the values we know. We then use the Pythagorean Theorem to find the values of the missing sides.

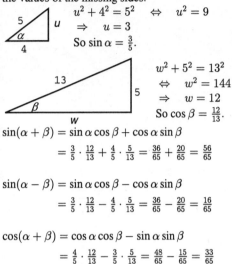

$$u^2 + 4^2 = 5^2 \quad \Leftrightarrow \quad u^2 = 9$$
$$\Rightarrow \quad u = 3$$
So $\sin \alpha = \frac{3}{5}$.

$$w^2 + 5^2 = 13^2$$
$$\Leftrightarrow \quad w^2 = 144$$
$$\Rightarrow \quad w = 12$$
So $\cos \beta = \frac{12}{13}$.

$$\sin(\alpha + \beta) = \sin \alpha \cos \beta + \cos \alpha \sin \beta$$
$$= \tfrac{3}{5} \cdot \tfrac{12}{13} + \tfrac{4}{5} \cdot \tfrac{5}{13} = \tfrac{36}{65} + \tfrac{20}{65} = \tfrac{56}{65}$$

$$\sin(\alpha - \beta) = \sin \alpha \cos \beta - \cos \alpha \sin \beta$$
$$= \tfrac{3}{5} \cdot \tfrac{12}{13} - \tfrac{4}{5} \cdot \tfrac{5}{13} = \tfrac{36}{65} - \tfrac{20}{65} = \tfrac{16}{65}$$

$$\cos(\alpha + \beta) = \cos \alpha \cos \beta - \sin \alpha \sin \beta$$
$$= \tfrac{4}{5} \cdot \tfrac{12}{13} - \tfrac{3}{5} \cdot \tfrac{5}{13} = \tfrac{48}{65} - \tfrac{15}{65} = \tfrac{33}{65}$$

$$\cos(\alpha - \beta) = \cos \alpha \cos \beta + \sin \alpha \sin \beta$$
$$= \tfrac{4}{5} \cdot \tfrac{12}{13} + \tfrac{3}{5} \cdot \tfrac{5}{13} = \tfrac{48}{65} + \tfrac{15}{65} = \tfrac{63}{65}$$

4. Prove the identity.
$$\frac{\sin(x + y)}{\cos x \cos y} = \tan x + \tan y$$

Expand the left side by using the addition formula for sine.

$$\text{LHS} = \frac{\sin(x + y)}{\cos x \cos y}$$

$$= \frac{\sin x \cos y + \cos x \sin y}{\cos x \cos y}$$

$$= \frac{\sin x \cos y}{\cos x \cos y} + \frac{\cos x \sin y}{\cos x \cos y}$$

$$= \frac{\sin x}{\cos x} + \frac{\sin y}{\cos y}$$

$$= \tan x + \tan y = \text{RHS}$$

5. Verify that the period of tangent is π by applying the addition formula.

$$\tan(x + \pi) = \tan x$$

Apply the tangent addition formula.

$$\tan(x + \pi) = \frac{\tan x + \tan \pi}{1 - \tan x \tan \pi}$$

$$= \frac{\tan x + 0}{1 - (\tan x)(0)}$$

$$= \frac{\tan x}{1} = \tan x$$

B. The last theorem in this section shows how expressions of the form $A \sin x + B \cos x$ can be expressed as a single trigonometric function.

$$A \sin x + B \cos x = k\sin(x + \phi)$$

where $k = \sqrt{A^2 + B^2}$ and ϕ satisfies $\cos \phi = \dfrac{A}{\sqrt{A^2 + B^2}}$ and $\sin \phi = \dfrac{B}{\sqrt{A^2 + B^2}}$

6. Write the expression $\frac{1}{\sqrt{3}} \sin x + \cos x$ in terms of sine only.

Here we have $A = \frac{1}{\sqrt{3}}$ and $B = 1$, so

$$k = \sqrt{\left(\tfrac{1}{\sqrt{3}}\right)^2 + 1^2}$$

$$= \sqrt{\tfrac{1}{3} + 1} = \sqrt{\tfrac{4}{3}} = \tfrac{2}{\sqrt{3}}.$$

So $\cos \phi = \dfrac{\frac{1}{\sqrt{3}}}{\frac{2}{\sqrt{3}}} = \dfrac{1}{2}$ and

$$\sin \phi = \dfrac{1}{\frac{2}{\sqrt{3}}} = \dfrac{\sqrt{3}}{2}.$$

Therefore $\phi = \frac{\pi}{3}$ and

$$\tfrac{1}{\sqrt{3}} \sin x + \cos x = \tfrac{2}{\sqrt{3}} \sin\left(x + \tfrac{\pi}{3}\right).$$

7. Write the following expression in terms of sine only.
$$-5 \sin 2x + 5 \cos 2x$$

Here $A = -5$ and $B = 5$, so

$$k = \sqrt{(-5)^2 + (5)^2} = \sqrt{25 + 25} = \sqrt{50}$$

$$= 5\sqrt{2}.$$

$$\cos \phi = \tfrac{-5}{5\sqrt{2}} = -\tfrac{1}{\sqrt{2}} = -\tfrac{\sqrt{2}}{2} \text{ and}$$

$$\sin \phi = \tfrac{5}{5\sqrt{2}} = \tfrac{1}{\sqrt{2}} = \tfrac{\sqrt{2}}{2}.$$

Thus $\phi = \frac{3}{4}\pi$ and

$$-5 \sin 2x + 5 \cos 2x = 5\sqrt{2} \sin\left(2x + \tfrac{3}{4}\pi\right)$$

Section 7.3   Double-Angle, Half-Angle, and Product-Sum Formulas

Key Ideas

A. Double-angle formulas.

B. Half-angle formulas.

C. Product-to-sum formulas.

D. Sum-to-product formulas.

A. These double-angle formulas are special cases of the addition formulas. These formulas allow us to express the trigonometric values of these double-angles in terms of the single angle.

$$\sin 2x = 2\sin x \cos x$$

$$\cos 2x = \cos^2 x - \sin^2 x \qquad \cos 2x = 1 - 2\sin^2 x \qquad \cos 2x = 2\cos^2 x - 1$$

$$\tan 2x = \frac{2\tan x}{1 - \tan^2 x}$$

Notice that there are three variations of the double angle formula for cosine.

1. Find $\sin 2x$, $\cos 2x$, and $\tan 2x$ from the given information.

   (a) $\cos x = -\frac{12}{13}$, x in quadrant III

First find $\sin x$. Since x is in quadrant III, $\sin x < 0$, so applying the Pythagorean Identity:

$$\sin x = -\sqrt{1 - \cos^2 x} = -\sqrt{1 - \left(-\frac{12}{13}\right)^2}$$

$$= -\sqrt{\frac{169}{169} - \frac{144}{169}} = -\sqrt{\frac{25}{169}} = -\frac{5}{13}.$$

So $\tan x = \dfrac{\sin x}{\cos x} = \dfrac{-5/13}{-12/13} = \dfrac{5}{12}.$

Then

$$\sin 2x = 2\sin x \cos x = 2\left(-\frac{5}{13}\right)\left(-\frac{12}{13}\right) = \frac{120}{169}$$

$$\cos 2x = 2\cos^2 x - 1 = 2\left(-\frac{12}{13}\right)^2 - 1$$

$$= \frac{288}{169} - \frac{169}{169} = \frac{119}{169}$$

$$\tan 2x = \frac{2\tan x}{1 - \tan^2 x} = \frac{2\left(\frac{5}{12}\right)}{1 - \left(\frac{5}{12}\right)^2} = \frac{\frac{10}{12}}{1 - \frac{25}{144}}$$

$$= \frac{120}{144 - 25} = \frac{120}{119}.$$

(b) $\sin \theta = \frac{8}{17}$, θ in quadrant II

First find $\cos \theta$. Since θ is in quadrant II,
$\cos \theta < 0$, so applying the Pythagorean Identity:

$$\cos \theta = -\sqrt{1 - \sin^2\theta} = -\sqrt{1 - \left(\frac{8}{17}\right)^2}$$

$$= -\sqrt{\frac{289}{289} - \frac{64}{289}} = -\sqrt{\frac{225}{289}} = -\frac{15}{17}.$$

So $\tan \theta = \dfrac{\sin \theta}{\cos \theta} = \dfrac{8/17}{-15/17} = -\dfrac{8}{15}.$

Then
$$\sin 2\theta = 2 \sin \theta \cos \theta = 2\left(\frac{8}{17}\right)\left(-\frac{15}{17}\right) = -\frac{240}{289}$$

$$\cos 2\theta = 2 \cos^2\theta - 1 = 2\left(-\frac{15}{17}\right)^2 - 1$$

$$= \frac{450}{289} - \frac{289}{289} = \frac{161}{289}$$

$$\tan 2\theta = \frac{2 \tan \theta}{1 - \tan^2\theta} = \frac{2\left(-\frac{8}{15}\right)}{1 - \left(-\frac{8}{15}\right)^2} = \frac{-\frac{16}{15}}{1 - \frac{64}{225}}$$

$$= \frac{-240}{225-64} = -\frac{240}{161}.$$

2. Prove the equation is an identity.
$(\sin \alpha - \cos \alpha)^2 = 1 - \sin 2\alpha$

Expand the left side:

$\text{LHS} = (\sin \alpha - \cos \alpha)^2$

$$= \sin^2\alpha - 2\sin\alpha \cos \alpha + \cos^2\alpha.$$

Rearrange the order:

$$= \sin^2\alpha + \cos^2\alpha - 2\sin\alpha \cos \alpha.$$

Making use of the Pythagorean Identity and the double-angle formula;

$$= 1 - \sin 2\alpha = \text{RHS}.$$

3. Prove the equation is an identity.
$$\cot x = \frac{\sin 2x}{1 - \cos 2x}$$

Use the double-angle formulas to expand the right side. Since the denominator of $\cot x$ is $\sin x$, we choose the double-angle formula for cosine that contains only $\sin x$.

$$\text{RHS} = \frac{\sin 2x}{1 - \cos 2x} = \frac{2 \sin x \cos x}{1 - (1 - 2\sin^2x)}$$

$$= \frac{2\sin x \cos x}{1 - 1 + 2\sin^2x} = \frac{2\sin x \cos x}{2\sin^2x}$$

$$= \frac{\cos x}{\sin x} = \cot x = \text{LHS}$$

B. The double-angle formulas for $\cos 2x$ are used to establish the formulas for lowering powers.

$$\sin^2 x = \frac{1 - \cos 2x}{2} \qquad \cos^2 x = \frac{1 + \cos 2x}{2} \qquad \tan^2 x = \frac{1 - \cos 2x}{1 + \cos 2x}$$

When u is substituted for $2x$ we get the **half-angle formulas**.

$$\sin \frac{u}{2} = \pm \sqrt{\frac{1 - \cos u}{2}} \qquad\qquad \cos \frac{u}{2} = \pm \sqrt{\frac{1 + \cos u}{2}}$$

$$\tan \frac{u}{2} = \pm \sqrt{\frac{1 - \cos u}{1 + \cos u}} \qquad \tan \frac{u}{2} = \frac{1 - \cos u}{\sin u} \qquad \tan \frac{u}{2} = \frac{\sin u}{1 + \cos u}$$

The half-angle formulas for tangent are used in calculus to establish many useful substitutions.

4. Use an appropriate half-angle formula to find the exact value of $\cos 11.25°$.

Since $11.25° = \left(\frac{1}{4}\right)45° = \left(\frac{1}{2}\right)\left(\frac{1}{2}\right)45°$, we first find

$$\cos 22.5° = \cos \frac{45°}{2} = \sqrt{\frac{1 + \cos 45°}{2}}$$

$$= \sqrt{\frac{1 + \frac{\sqrt{2}}{2}}{2}} = \sqrt{\frac{\frac{2}{2} + \frac{\sqrt{2}}{2}}{2}}$$

$$= \sqrt{\frac{2 + \sqrt{2}}{4}} = \frac{\sqrt{2 + \sqrt{2}}}{2}.$$

So,

$$\cos 11.25° = \cos \frac{22.5°}{2} = \sqrt{\frac{1 + \cos 22.5°}{2}}$$

$$= \sqrt{\frac{1 + \frac{\sqrt{2 + \sqrt{2}}}{2}}{2}}$$

$$= \sqrt{\frac{\frac{2}{2} + \frac{\sqrt{2 + \sqrt{2}}}{2}}{2}}$$

$$= \sqrt{\frac{2 + \sqrt{2 + \sqrt{2}}}{4}}$$

$$= \frac{\sqrt{2 + \sqrt{2 + \sqrt{2}}}}{2}.$$

C. These **product-to-sum identities** are useful in calculus. These are used to write products of sines and cosines as sums. These identities are derived by using the addition and subtraction identities.

$\sin u \cos v = \frac{1}{2}[\sin(u+v) + \sin(u-v)]$	$\cos u \sin v = \frac{1}{2}[\sin(u+v) - \sin(u-v)]$
$\cos u \cos v = \frac{1}{2}[\cos(u+v) + \cos(u-v)]$	$\sin u \sin v = \frac{1}{2}[\cos(u-v) - \cos(u+v)]$

5. Write the product as a sum.

(a) $\sin 3x \sin x$

$$\sin 3x \sin x = \frac{1}{2}[\cos(3x - x) - \cos(3x + x)]$$
$$= \frac{1}{2}(\cos 2x - \cos 4x)$$

(b) $\cos 5x \sin 2x$

$$\cos 5x \sin 2x = \frac{1}{2}[\sin(5x + 2x) - \sin(5x - 2x)]$$
$$= \frac{1}{2}(\sin 7x - \sin 3x)$$

6. Find the exact value of each product.

(a) $\tan 82.5° \tan 52.5°$

$$\tan 82.5° \tan 52.5° = \frac{\sin 82.5°}{\cos 82.5°} \cdot \frac{\sin 52.5°}{\cos 52.5°}$$
$$= \frac{\sin 82.5° \sin 52.5°}{\cos 82.5° \cos 52.5°}$$
$$= \frac{\frac{1}{2}[\cos(82.5° - 52.5°) - \cos(82.5° + 52.5°]}{\frac{1}{2}[\cos(82.5° + 52.5°) + \cos(82.5° - 52.5°)]}$$
$$= \frac{\frac{1}{2}(\cos 30° - \cos 135°)}{\frac{1}{2}(\cos 135° + \cos 30°)}$$
$$= \frac{\frac{1}{2}\left(\frac{\sqrt{3}}{2} - \frac{-\sqrt{2}}{2}\right)}{\frac{1}{2}\left(\frac{-\sqrt{2}}{2} + \frac{\sqrt{3}}{2}\right)}$$
$$= \frac{\sqrt{3} + \sqrt{2}}{-\sqrt{2} + \sqrt{3}} = \frac{\sqrt{3} + \sqrt{2}}{\sqrt{3} - \sqrt{2}}$$
$$= \frac{\sqrt{3} + \sqrt{2}}{\sqrt{3} - \sqrt{2}} \cdot \frac{\sqrt{3} + \sqrt{2}}{\sqrt{3} + \sqrt{2}}$$
$$= \frac{3 + 2\sqrt{6} + 2}{3 - 2} = \frac{5 + 2\sqrt{6}}{1}$$
$$= 5 + 2\sqrt{6}$$

(b) $\sin 97.5° \cos 37.5°$

$\sin 97.5° \cos 37.5° =$

$$\tfrac{1}{2}[\sin(97.5° + 37.5°) + \sin(97.5° - 37.5°)]$$

$$= \tfrac{1}{2}(\sin 135° + \sin 60°)$$

$$= \tfrac{1}{2}\left(\tfrac{\sqrt{2}}{2} + \tfrac{\sqrt{3}}{2}\right)$$

$$= \tfrac{\sqrt{2}+\sqrt{3}}{4}$$

D. These **sum-to-product identities** are useful in calculus and are derived from the addition and subtraction identities. These identities are used to write sums of sines and cosines as products which provides a very useful tool for solving equations by factoring.

$$\sin u + \sin v = 2\sin\frac{u+v}{2}\cos\frac{u-v}{2} \qquad \sin u - \sin v = 2\cos\frac{u+v}{2}\sin\frac{u-v}{2}$$

$$\cos u + \cos v = 2\cos\frac{u+v}{2}\cos\frac{u-v}{2} \qquad \cos u - \cos v = -2\sin\frac{u+v}{2}\sin\frac{u-v}{2}$$

7. Write the sum as a product.

   (a) $\sin 3x - \sin x$

   $\sin 3x - \sin x = 2\cos\dfrac{3x+x}{2}\sin\dfrac{3x-x}{2}$

   $= 2\cos 2x \sin x$

   (b) $\cos 4x + \cos 2x + \cos x$

   First express $\cos 4x + \cos 2x$ as a product.

   $\cos 4x + \cos 2x = 2\cos\dfrac{4x+2x}{2}\cos\dfrac{4x-2x}{2}$

   $= 2\cos 3x \cos x$

   Apply this to the original equation.

   $\cos 4x + \cos 2x + \cos x = (\cos 4x + \cos 2x) + \cos x$

   $= 2\cos 3x \cos x + \cos x$

   $= \cos x(2\cos 3x + 1)$

   Note: We chose to use the first two terms of the expression since any product found using $\cos x$ and another term would have resulted in a cosine of some fraction of x, which would not match the third term.

Section 7.4   Inverse Trigonometric Functions

Key Ideas

A. The inverse sine function.
B. The inverse cosine function.
C. The inverse tangent function.
D. The other inverse trigonometric function.

A. The properties of inverse functions were discussed in Section 2.9. Since sine is not a one-to-one function, we restrict the domain of sine to $\left[-\frac{\pi}{2}, \frac{\pi}{2}\right]$ to obtain a one-to-one function.

> The **inverse sine function, $\sin^{-1}$**, is the function with domain $[-1, 1]$ and range $\left[-\frac{\pi}{2}, \frac{\pi}{2}\right]$ defined by
>
> $$\sin^{-1} x = y \qquad \Leftrightarrow \qquad \sin y = x.$$

The inverse sine function is also called **arcsine** and is denoted by the symbol **arcsin**. An important property is that cosine is always positive in the range of arcsine.

1. Find the exact value of the expression $\sin^{-1}\left(-\frac{\sqrt{3}}{2}\right)$.

The number in the interval $\left[-\frac{\pi}{2}, \frac{\pi}{2}\right]$ whose sine is $-\frac{\sqrt{3}}{2}$ is $-\frac{\pi}{3}$. So $\sin^{-1}\left(-\frac{\sqrt{3}}{2}\right) = -\frac{\pi}{3}$.

2. Find the approximate value of $\sin^{-1}(0.95)$.

We use the calculator to approximate this value. Place the calculator in radian mode and then use the $\boxed{\text{INV}}\,\boxed{\text{SIN}}$, or $\boxed{\text{ARCSIN}^{-1}}$, or $\boxed{\text{ARCSIN}}$ key to get $\sin^{-1}(0.95) \approx 1.2532$.

3. Find the exact value of $\cos\left(\sin^{-1}\left(\frac{1}{3}\right)\right)$ and $\tan\left(\sin^{-1}\left(\frac{1}{3}\right)\right)$.

Let $u = \sin^{-1}\left(\frac{1}{3}\right)$, so $\sin u = \frac{1}{3}$. Since $u \in \left[-\frac{\pi}{2}, \frac{\pi}{2}\right]$ we must have $\cos u > 0$.
Using the fundamental Pythagorean identity we get $\cos u = \sqrt{1 - \sin^2 u}$. So we have

$$\cos\left(\sin^{-1}\left(\tfrac{1}{3}\right)\right) = \sqrt{1 - \sin^2\left(\sin^{-1}\left(\tfrac{1}{3}\right)\right)}$$

$$= \sqrt{1 - \left(\tfrac{1}{3}\right)^2} = \sqrt{1 - \tfrac{1}{9}}$$

$$= \sqrt{\tfrac{8}{9}} = \tfrac{2\sqrt{2}}{3}.$$

Since $\tan u = \dfrac{\sin u}{\cos u}$,

$$\tan\left(\sin^{-1}\left(\tfrac{1}{3}\right)\right) = \frac{\sin\left(\sin^{-1}\left(\tfrac{1}{3}\right)\right)}{\cos\left(\sin^{-1}\left(\tfrac{1}{3}\right)\right)}$$

$$= \frac{\frac{1}{3}}{\frac{2\sqrt{2}}{3}} = \frac{1}{2\sqrt{2}} = \frac{\sqrt{2}}{4}.$$

B. Again, since cosine is not a one-to-one function, we restrict the domain of cosine to obtain a one-to-one function.

> The **inverse cosine function, $\cos^{-1}$**, is the function with domain $[-1, 1]$ and range $[0, \pi]$ defined by
>
> $$\cos^{-1} x = y \qquad \Leftrightarrow \qquad \cos y = x.$$

The inverse cosine function is also called **arccosine** and is denote by **arccos**. An important property of the range of arccosine is that sine is always positive in this interval.

4. Find the exact value of each expression.

(a) $\cos^{-1}\left(-\frac{\sqrt{2}}{2}\right)$

The number in the interval $[0, \pi]$ whose cosine is

$-\frac{\sqrt{2}}{2}$ is $\frac{3\pi}{4}$. So $\cos^{-1}\left(-\frac{\sqrt{2}}{2}\right) = \frac{3\pi}{4}$.

(b) $\tan\left(\cos^{-1}\left(-\frac{5}{13}\right)\right)$

Let $u = \cos^{-1}\left(-\frac{5}{13}\right)$, so $\cos u = -\frac{5}{13}$. Since $u \in [0, \pi]$, $\sin u > 0$.

$$\sin u = \sqrt{1 - \cos^2 u} = \sqrt{1 - \left(-\frac{5}{13}\right)^2}$$

$$= \sqrt{1 - \frac{25}{169}} = \sqrt{\frac{144}{169}}$$

$$= \frac{12}{13}$$

Since $\tan u = \dfrac{\sin u}{\cos u}$,

$$\tan\left(\cos^{-1}\left(-\tfrac{5}{13}\right)\right) = \frac{\sin\left(\cos^{-1}\left(-\frac{5}{13}\right)\right)}{\cos\left(\cos^{-1}\left(-\frac{5}{13}\right)\right)}$$

$$= \frac{\frac{12}{13}}{-\frac{5}{13}} = -\frac{12}{5}.$$

5. A 85 foot rope is tied 10 feet from the base of a building (see figure). Find the measure of the angle θ that the rope makes with the ground.

85

θ

10

Since $\cos \theta = \frac{10}{85}$,

$\theta = \cos^{-1}\left(\frac{10}{85}\right) = 1.4529$ radians (about $83.24°$).

C. The **inverse tangent function, tan**$^{-1}$, is the function with domain $(-\infty, \infty)$ and range $\left(-\frac{\pi}{2}, \frac{\pi}{2}\right)$ defined by $\quad \mathbf{tan^{-1}}\boldsymbol{x} = \boldsymbol{y} \quad \Leftrightarrow \quad \mathbf{tan}\,\boldsymbol{y} = \boldsymbol{x}.$

The inverse tangent function is also called **arctangent** and is denoted by **arctan**. Because secant and tangent are related by the fundamental identity $\tan^2 x + 1 = \sec^2 x$, we choose to restrict the range of $\tan^{-1} x$ (domain of $\tan x$) to the interval where secant is positive.

6. Find the value of each expression.

(a) $\tan^{-1}\left(-\frac{\sqrt{3}}{3}\right)$

Since $\tan\left(-\frac{\pi}{6}\right) = -\frac{\sqrt{3}}{3}$ and $-\frac{\pi}{2} < -\frac{\pi}{6} < \frac{\pi}{2}$, we have

$\tan^{-1}\left(-\frac{\sqrt{3}}{3}\right) = -\frac{\pi}{6}.$

(b) $\tan^{-1}\left(\tan \frac{7\pi}{4}\right)$

Since $\tan \frac{7\pi}{4} = -1$, we have

$\tan^{-1}\left(\tan \frac{7\pi}{4}\right) = \tan^{-1}(-1) = -\frac{\pi}{4}.$

Note: $\tan^{-1}\left(\tan \frac{7\pi}{4}\right) \neq \frac{7\pi}{4}$ *since* $\frac{7\pi}{4}$ *is not in the range of* $\tan^{-1}$.

(c) $\tan^{-1}(50)$

We use a calculator to find this value, so $\tan^{-1}(50) = 1.55080.$

Remember to first put your calculator into radian mode.

7. A brace is made to support the frame of a gate. The measurements of the frame are shown in the figure. Find the measure of the angle θ.

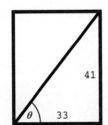

Here we use the relation $\tan \theta = \dfrac{\text{opposite}}{\text{adjacent}} = \frac{41}{33}.$

So $\theta = \tan^{-1}\left(\frac{41}{33}\right) \approx \tan^{-1}(1.242424)$

≈ 0.89308 radians.

This problem could have been worked in degrees, in which case, $\theta \approx \tan^{-1}(1.242424) \approx 51.17°.$

D. The table below shows the defining equations for all six inverse trigonometric functions and the corresponding domains and ranges. Included in the table are the fundamental identities that are used to determine the range of the inverse trigonometric function.

Defining Equations	Inv. trig domain (Trig func. range)	Inv. trig range (Trig func. Domain)	Fundamental identity
$\sin^{-1}x = y \quad \Leftrightarrow \quad \sin y = x$	$[-1, 1]$	$\left[-\frac{\pi}{2}, \frac{\pi}{2}\right]$	$\sin^2\theta + \cos^2\theta = 1$ $\cos\theta = \sqrt{1 - \sin^2\theta}$
$\cos^{-1}x = y \quad \Leftrightarrow \quad \cos y = x$	$[-1, 1]$	$[0, \pi]$	$\sin^2\theta + \cos^2\theta = 1$ $\sin\theta = \sqrt{1 - \cos^2\theta}$
$\tan^{-1}x = y \quad \Leftrightarrow \quad \tan y = x$	$(-\infty, \infty)$	$\left(-\frac{\pi}{2}, \frac{\pi}{2}\right)$	$\tan^2\theta + 1 = \sec^2\theta$ $\sec\theta = \sqrt{\tan^2\theta + 1}$
$\sec^{-1}x = y \quad \Leftrightarrow \quad \sec y = x$	$(-\infty, -1] \cup [1, \infty)$	$[0, \frac{\pi}{2}) \cup [\pi, \frac{3\pi}{2})$	$\tan^2\theta + 1 = \sec^2\theta$ $\tan\theta = \sqrt{\sec^2\theta - 1}$
$\cot^{-1}x = y \quad \Leftrightarrow \quad \cot y = x$	$(-\infty, \infty)$	$(0, \pi)$	$1 + \cot^2\theta = \csc^2\theta$ $\csc\theta = \sqrt{1 + \cot^2\theta}$
$\csc^{-1}x = y \quad \Leftrightarrow \quad \csc y = x$	$(-\infty, -1] \cup [1, \infty)$	$(0, \frac{\pi}{2}] \cup (\pi, \frac{3\pi}{2}]$	$1 + \cot^2\theta = \csc^2\theta$ $\cot\theta = \sqrt{\csc^2\theta - 1}$

8. Rewrite the expression $\tan(\sin^{-1}x)$ as an algebraic function in x. Sketch a right triangle which shows the relationship between the trigonometric functions.

Let $\sin^{-1}x = \theta$. Then $\sin\theta = x$ and $\tan(\sin^{-1}x) = \tan\theta$.

Since $\sin\theta = \dfrac{\text{opposite}}{\text{hypotenuse}}$ we have

$\sin\theta = x = \dfrac{x}{1} = \dfrac{\text{opposite}}{\text{hypotenuse}}$.

Place this information on the right triangle and solve for the remaining side using the Pythagorean theorem.

Since adjacent $= \sqrt{1 - x^2}$, we have

$\cos\theta = \dfrac{\sqrt{1 - x^2}}{1} = \sqrt{1 - x^2}.$

And $\tan(\sin^{-1}x) = \tan\theta = \dfrac{x}{\sqrt{1 - x^2}}.$

Note: $\cos\theta$ *is positive throughout the range of* $\sin^{-1}x$.

197

9. Rewrite the expression $\cos(2\tan^{-1}x)$ as an algebraic function in x. Sketch a right triangle that shows the relationship between the trigonometric functions.

Let $\tan^{-1}x = \theta$. Then $\tan\theta = x$ and

$\cos(2\tan^{-1}x) = \cos 2\theta$. Since $\tan\theta = \dfrac{\text{opposite}}{\text{adjacent}}$ we

have $\tan\theta = x = \dfrac{x}{1} = \dfrac{\text{opposite}}{\text{adjacent}}$.

Label the angle θ, the opposite side x, and the adjacent side 1. The hypotenuse can be found using the Pythagorean Theorem. Thus the

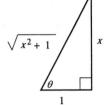

hypotenuse $= \sqrt{x^2 + 1}$ and $\cos(\tan^{-1}x) = \frac{1}{\sqrt{x^2+1}}$.

We now use the double angle identity,

$\cos 2\theta = 2\cos^2\theta - 1$. So

$\cos(2\tan^{-1}x) = 2\cos^2(\tan^{-1}x) - 1$

$$= 2\left(\frac{1}{\sqrt{x^2+1}}\right)^2 - 1$$

$$= 2\left(\frac{1}{x^2+1}\right) - 1$$

$$= \frac{2}{x^2+1} - \frac{x^2+1}{x^2+1}$$

$$= \frac{1-x^2}{x^2+1}.$$

Notes: (1) We selected the double angle identity which only contained cosine;
(2) $\cos\theta$ is positive throughout the range of $\tan^{-1}x$.

Section 7.5   Trigonometric Equations

Key Ideas

A. Trigonometric equations.

B. Using a graphing calculator to solve equations involving both algebraic and trigonometric expressions.

A. A **trigonometric equation** is an equation that contains trigonometric functions. *In general*, if a trigonometric equation has one solution, then it will have infinitely many solutions. First find all the solutions in an *appropriate* interval (usually one period) and then use the period of the trigonometric function to determine the other solutions.

1. Find all solutions of the equation.

   (a) $2 \sin x \cos x = \sin x$

As with polynomial equations, trigonometric equations are solved by moving all terms to one side and factoring.

$2 \sin x \cos x = \sin x$

$2 \sin x \cos x - \sin x = 0$

$\sin x (2 \cos x - 1) = 0$

Set each factor equal to zero; solve. Find all solutions in one period.

$$\sin x = 0 \qquad \text{or} \qquad 2 \cos x - 1 = 0$$
$$x = 0,\ \pi \qquad\qquad\qquad 2 \cos x = 1$$
$$\cos x = \tfrac{1}{2}$$
$$x = \tfrac{\pi}{3},\ \tfrac{5\pi}{3}$$

Because the period of both functions is 2π, we obtain the solutions:

$x = 0 + 2k\pi = 2k\pi$

$x = \pi + 2k\pi = (2k + 1)\pi$

$x = \tfrac{\pi}{3} + 2k\pi$

$x = \tfrac{5\pi}{3} + 2k\pi$

for any integer k.

   (b) $\cos^2 x - 3 \sin x - 1 = 0$

Before we can factor we first need to express this equation in terms of sine only or in terms of cosine only. Using $\cos^2 x = 1 - \sin^2 x$, we get

$\cos^2 x - 3 \sin x - 1 = 0$

$(1 - \sin^2 x) - 3 \sin x - 1 = 0$

$-\sin^2 x - 3 \sin x = 0$

$-\sin x (\sin x + 3) = 0$

$$-\sin x = 0 \qquad \text{or} \qquad \sin x + 3 = 0$$
$$x = 0,\ \pi \qquad\qquad\qquad \sin x = -3.$$

This equation has no
since $-1 \leq \sin x \leq 1$.

$x = 0 + 2k\pi = 2k\pi$, for any integer k

$x = \pi + 2k\pi = (2k + 1)\pi$, for any integer k

We can express these together as $x = k\pi$ for any integer k.

2. Solve the equation in the interval $[0, 2\pi)$.

(a) $3\tan^3 x - \tan x = 0$

$$3\tan^3 x - \tan x = 0$$
$$\tan x(3\tan^2 x - 1) = 0$$
$$\tan x\left(\sqrt{3}\tan x - 1\right)\left(\sqrt{3}\tan x + 1\right) = 0$$

$\tan x = 0 \quad \Rightarrow \quad x = 0, \pi$

$\sqrt{3}\tan x - 1 = 0 \quad \Leftrightarrow \quad \sqrt{3}\tan x = 1 \quad \Leftrightarrow$

$\tan x = \dfrac{1}{\sqrt{3}} \quad \Rightarrow \quad x = \dfrac{\pi}{6}, \dfrac{7\pi}{6}$

$\sqrt{3}\tan x + 1 = 0 \quad \Leftrightarrow \quad \sqrt{3}\tan x = -1 \quad \Leftrightarrow$

$\tan x = -\dfrac{1}{\sqrt{3}} \quad \Rightarrow \quad x = \dfrac{5\pi}{6}, \dfrac{11\pi}{6}$

So the solutions are $0, \dfrac{\pi}{6}, \dfrac{5\pi}{6}, \pi, \dfrac{7\pi}{6}, \dfrac{11\pi}{6}$.

(b) $\sec x \tan x = 4\sin x$

$\sec x \tan x = 4\sin x$ *Move all terms to left*

$\sec x \tan x - 4\sin x = 0$ *side and factor.*

It is hard to see how to factor this expression, so we first express this in terms of sine and cosine.

$$\frac{1}{\cos x} \cdot \frac{\sin x}{\cos x} - 4\sin x = 0$$

$$\sin x\left(\frac{1}{\cos^2 x} - 4\right) = \sin x\left(\sec^2 x - 4\right) = 0$$

$$\sin x(\sec x - 2)(\sec x + 2) = 0$$

So,

$\sin x = 0 \quad \Rightarrow \quad x = 0, \pi$

$\sec x - 2 = 0 \quad \Rightarrow \quad \sec x = 2 \quad \Rightarrow$

$x = \dfrac{\pi}{3}, \dfrac{5\pi}{3}$

$\sec x + 2 = 0 \quad \Rightarrow \quad \sec x = -2 \quad \Rightarrow$

$x = \dfrac{2\pi}{3}, \dfrac{4\pi}{3}$.

So the solutions are $0, \dfrac{\pi}{3}, \dfrac{2\pi}{3}, \pi, \dfrac{4\pi}{3}, \dfrac{5\pi}{3}$.

(c) Solve the equation in the interval $[0, 2\pi)$.
$$3\cos x + 1 = \cos 2x$$

$3\cos x + 1 = \cos 2x$ *Move all terms to one side.*

$\cos 2x - 3\cos x - 1 = 0$

We chose to use the double-angle identity for $\cos 2x$ that contains only cosine terms.

$$(2\cos^2 x - 1) - 3\cos 2x - 1 = 0$$

$$2\cos^2 x - 3\cos x - 2 = 0$$

$$(2\cos x + 1)(\cos x - 2) = 0$$

Set each factor equal to zero. Solve.

$2\cos x + 1 = 0$ or $\cos x - 2 = 0$

$2\cos x = -1$ $\qquad\qquad$ $\cos x = 2$

$\cos x = -\frac{1}{2}$ $\qquad$ This equation has no

$x = \frac{2\pi}{3}, \frac{4\pi}{3}$ $\qquad$ solution since

$\qquad\qquad\qquad\qquad -1 \le \cos x \le 1$.

Thus the only solutions are $x = \frac{2\pi}{3}$ and $\frac{4\pi}{3}$.

3. Sketch the graphs of f and g on the same axes and find their points of intersection.
$f(x) = 2\sin x$ and $g(x) = \tan x$

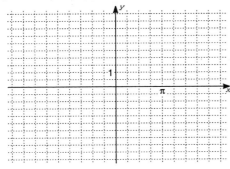

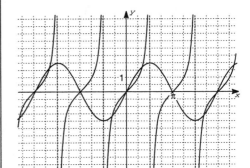

$2\sin x = \tan x$ *Solve for x in $[0, 2\pi)$.*

$2\sin x - \tan x = 0$ *Move all terms to one side.*

$2\sin x - \dfrac{\sin x}{\cos x} = 0$ *Express $\tan x$ as $\dfrac{\sin x}{\cos x}$.*

$\sin x\left(2 - \dfrac{1}{\cos x}\right) = 0$ *Factor.*

$\sin x = 0$ or $2 - \dfrac{1}{\cos x} = 0$

$x = 0, \pi$ $\qquad\qquad\qquad 2 = \dfrac{1}{\cos x}$

$\qquad\qquad\qquad\qquad \cos x = \frac{1}{2}$

$\qquad\qquad\qquad\qquad x = \frac{\pi}{3}, \frac{5\pi}{3}$

So $x = k\pi$, $\frac{\pi}{3} + 2k\pi$, or $\frac{5\pi}{3} + 2k\pi$, where k is any integer.

201

4. Solve the equation $\cos 5x = \cos 3x$.

$$\cos 5x = \cos 3x \qquad \textit{Move all terms to one}$$
$$\cos 5x - \cos 3x = 0 \qquad \textit{side. Then use the}$$
$$\textit{sum-to-product formula.}$$
$$-2\sin\frac{5x+3x}{2}\sin\frac{5x-3x}{2} = 0 \qquad \textit{Simplify.}$$
$$-2\sin 4x \sin x = 0 \qquad \textit{Divide both sides}$$
$$\sin 4x \sin x = 0 \qquad \textit{by } -2.$$

$$\sin 4x = 0 \qquad \text{or} \qquad \sin x = 0$$
$$4x = k\pi \qquad\qquad\qquad x = k\pi$$
$$x = \frac{k\pi}{4}$$

Solution: $x = \frac{k\pi}{4}$, for any integer k.

Note that $x = \frac{k\pi}{4}$ includes all solutions.

5. Use the result of Problem 7 from Section 7.2 to solve
$$-5\sin 2x + 5\cos 2x = -5\sqrt{2}$$
in the interval $[0, 2\pi)$.

In Section 7.2, Problem 7 we expressed the LHS of the equation in terms of sine only. That is,
$$-5\sin 2x + 5\cos 2x = 5\sqrt{2}\sin\left(2x + \tfrac{3}{4}\pi\right).$$

Substituting we have $5\sqrt{2}\sin\left(2x + \tfrac{3}{4}\pi\right) = -5\sqrt{2}$.

So $\sin\left(2x + \tfrac{3}{4}\pi\right) = -1$

$$2x + \tfrac{3}{4}\pi = \tfrac{3}{2}\pi + 2k\pi \qquad \textit{We add the plus } 2k\pi$$
$$2x = \tfrac{3}{2}\pi - \tfrac{3}{4}\pi + 2k\pi \qquad \textit{to make sure we}$$
$$2x = \tfrac{3}{4}\pi + 2k\pi \qquad \textit{find all } x \textit{ in the}$$
$$x = \tfrac{3}{8}\pi + k\pi \qquad \textit{interval } [0, 2\pi).$$

For $k = 0$, we have $x = \tfrac{3}{8}\pi$.

For $k = 1$, we have $x = \tfrac{3}{8}\pi + \pi = \tfrac{11}{8}\pi$.

For $k = 2$, we have $x = \tfrac{3}{8}\pi + 2\pi = \tfrac{19}{8}\pi > 2\pi$.

So the solutions are: $x = \tfrac{3}{8}\pi$ and $x = \tfrac{11}{8}\pi$.

B. In most cases, we cannot find exact solutions to equations involving both algebraic and trigonometric expressions. However, with the aid of a graphing device we can approximate the solutions by graphing each side of the equation.

6. Use a graphing device to find the solutions of each equation, correct to two decimal places.

(a) $\sin x = \frac{1}{5}x$

Since $-1 \le \sin x \le 1$, we must have $-1 \le \frac{1}{5}x \le 1$ $\Leftrightarrow$ $-5 \le x \le 5$. Setting the viewing rectangle to $[-5, 5]$ by $[-1.25, 1.25]$ we see that there are three solutions, one between -3 and -2, at $x = 0$, and one between 2 and 3. In the second viewing rectangle, $[2, 3]$ by $[0, 1]$, (by zooming in) we see that the solution is at $x \approx 2.60$. Since both functions are odd, the third solution will be at $x \approx -2.60$.

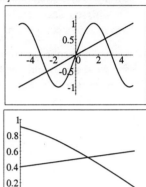

Solution: $x = 0, \pm 2.60$.

(b) $\sin^2 x = x^2 - 1$

Since $\sin^2 x \le 1$, we must have $x^2 - 1 \le 1$ $\Leftrightarrow$ $x^2 \le 2$, so $\sqrt{2} \le x \le \sqrt{2}$. Since $\sqrt{2} < 1.5$ we chose the viewing rectangle of $[-1.5, 1.5]$ by $[-1.25, 1.25]$, we see that there are two solutions, one between -1.5 and -1 and one between 1 and 1.5. In the second viewing rectangle, $[1, 1.5]$ by $[0.5, 1.25]$, by zooming in, we see that there is a solution at $x \approx 1.40$. Since both functions are even, the other solution is at $x \approx -1.40$.

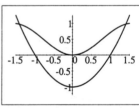

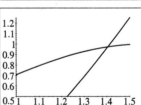

Solution: $x \approx \pm 1.40$.

Section 7.6   Trigonometric Form of Complex Numbers; DeMoivre's Theorem

Key Ideas

A. Graphing complex numbers.
B. Trigonometric form of a complex number.
C. Products and quotients of complex numbers.
D. DeMoivre's Theorem.
E. nth root of a complex number.

A. Complex numbers were introduced in Section 3.4. To graph complex numbers we need two axes, one for the real part and one for the imaginary part. These are called the **real axis** and the **imaginary axis**, respectively. The plane determined by these two axes is called the **complex plane**. To graph the complex number $a + b\,i$, we plot the ordered pair (a, b) in this plane.

1. Graph the complex numbers $z_1 = -2 - 3\,i$, $z_2 = 3 - i$, $z_1 + z_2$, and $z_1 z_2$.

We first determine $z_1 + z_2$ and $z_1 z_2$.

$$z_1 + z_2 = (-2 - 3\,i) + (3 - i)$$
$$= 1 - 4i$$

$$z_1 z_2 = (-2 - 3\,i)(3 - i)$$
$$= -6 + 2\,i - 9\,i + 3\,i^2$$
$$= -6 - 7\,i - 3 = -9 - 7\,i$$

Graphing the ordered pairs we get the following.

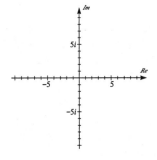

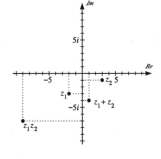

2. Graph the following set of complex numbers.
$\{z = a + b\,i \mid 0 \le a \le 5, |b| < 3\}$

Since $|b| < 3 \quad \Leftrightarrow \quad -3 < b < 3$ we get the following graph.

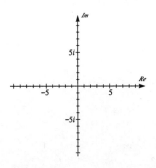

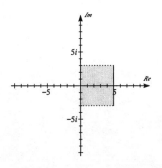

B. Let $z = a + bi$ be a complex number. Then z has the **trigonometric form**: $z = r(\cos\theta + i\sin\theta)$, where $r = |z| = \sqrt{a^2 + b^2}$ and $\tan\theta = \dfrac{b}{a}$. θ is an **argument** of z and r is called the **modulus** of z.

3. Write each complex number in trigonometric form.

    (a) $-4 + 4i$

> Since $\tan\theta = \frac{4}{-4} = -1$, an argument is $\theta = \frac{3\pi}{4}$ and
> $r = \sqrt{(-4)^2 + 4^2} = \sqrt{16 + 16} = \sqrt{32} = 4\sqrt{2}$.
> Thus $-4 + 4i = 4\sqrt{2}\left(\cos\frac{3\pi}{4} + i\sin\frac{3\pi}{4}\right)$.

    (b) $7 - i$

> To find an argument θ,
> we note that $7 - i$ is
> in quadrant IV and use
> an inverse trigonometric
> function that is defined
>
>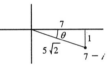
>
> on the interval $\left[-\frac{\pi}{2}, \frac{\pi}{2}\right]$. To help determine the
> angle, we sketch the point and find r.
>
> So $r = \sqrt{7^2 + (-1)^2} = \sqrt{49 + 1} = \sqrt{50} = 5\sqrt{2}$
> and $\theta = \sin^{-1}\left(\frac{1}{5\sqrt{2}}\right) = \sin^{-1}\left(\frac{\sqrt{2}}{10}\right)$.
>
> Thus $7 - i$
> $= 5\sqrt{2}\left\{\cos\left[\sin^{-1}\left(\frac{\sqrt{2}}{10}\right)\right] + i\sin\left[\sin^{-1}\left(\frac{\sqrt{2}}{10}\right)\right]\right\}$.

C. If two complex numbers z_1 and z_2 have the trigonometric form $z_1 = r_1(\cos\theta_1 + i\sin\theta_1)$ and $z_2 = r_2(\cos\theta_2 + i\sin\theta_2)$ then the product and the quotient can be expressed as

$$z_1 z_2 = r_1 r_2[\cos(\theta_1 + \theta_2) + i\sin(\theta_1 + \theta_2)]$$
$$\frac{z_1}{z_2} = \frac{r_1}{r_2}[\cos(\theta_1 - \theta_2) + i\sin(\theta_1 - \theta_2)].$$

4. Let $z_1 = 2 + 2\sqrt{3}\,i$ and let $z_2 = -i$.
   Find (a) $z_1 z_2$ and (b) $\frac{z_1}{z_2}$.

$r_1 = \sqrt{2^2 + \left(2\sqrt{3}\right)^2} = \sqrt{4 + 12} = \sqrt{16} = 4$ and

$\tan\theta_1 = \frac{2\sqrt{3}}{2} = \sqrt{3} \quad \Rightarrow \quad \theta_1 = \frac{\pi}{3}$.

$r_2 = \sqrt{0^2 + (-1)^2} = \sqrt{1} = 1$. Since $\tan\theta_2$ is not

defined and $z_2 = -i$, we must have $\theta_2 = \frac{3\pi}{2}$.

So $z_1 = 4\left(\cos\frac{\pi}{3} + i\sin\frac{\pi}{3}\right)$ and $z_2 = \cos\frac{3\pi}{2} + i\sin\frac{3\pi}{2}$.

(a)

$z_1 z_2 = (4)(1)\left[\cos\left(\frac{\pi}{3} + \frac{3\pi}{2}\right) + i\sin\left(\frac{\pi}{3} + \frac{3\pi}{2}\right)\right]$

$\qquad = 4\left(\cos\frac{11\pi}{6} + i\sin\frac{11\pi}{6}\right)$

(b)

$\frac{z_1}{z_2} = \frac{4}{1}\left[\cos\left(\frac{\pi}{3} - \frac{3\pi}{2}\right) + i\sin\left(\frac{\pi}{3} - \frac{3\pi}{2}\right)\right]$

$\qquad = 4\left(\cos\frac{-7\pi}{6} + i\sin\frac{-7\pi}{6}\right)$

$\qquad = 4\left(\cos\frac{7\pi}{6} - i\sin\frac{7\pi}{6}\right)$

5. Let $z_1 = 2 + 2i$ and let $z_2 = -\sqrt{2} - \sqrt{2}\,i$.
   Find (a) $z_1 z_2$ and (b) $\frac{z_1}{z_2}$.

$r_1 = \sqrt{2^2 + 2^2} = \sqrt{4 + 4} = \sqrt{8} = 2\sqrt{2}$ and

$\tan\theta_1 = \frac{2}{2} = 1$ with a and b both positive we have

$\theta_1 = \frac{\pi}{4}$.

$r_2 = \sqrt{\left(-\sqrt{2}\right)^2 + \left(-\sqrt{2}\right)^2} = \sqrt{2 + 2} = \sqrt{4} = 2$

and $\tan\theta_2 = \frac{-\sqrt{2}}{-\sqrt{2}} = 1$ with a and b both negative we

have $\theta_2 = \frac{5\pi}{4}$.

So $z_1 = 2\sqrt{2}\left(\cos\frac{\pi}{4} + i\sin\frac{\pi}{4}\right)$ and

$z_2 = 2\left(\cos\frac{5\pi}{4} + i\sin\frac{5\pi}{4}\right)$.

(a)

$z_1 z_2 = (2\sqrt{2})(2)\left[\cos\left(\frac{\pi}{4} + \frac{5\pi}{4}\right) + i\sin\left(\frac{\pi}{4} + \frac{5\pi}{4}\right)\right]$

$\qquad = 4\sqrt{2}\left(\cos\frac{3\pi}{2} + i\sin\frac{3\pi}{2}\right) = -4\sqrt{2}\,i$

(b)

$\frac{z_1}{z_2} = \frac{2\sqrt{2}}{2}\left[\cos\left(\frac{\pi}{4} - \frac{5\pi}{4}\right) + i\sin\left(\frac{\pi}{4} - \frac{5\pi}{4}\right)\right]$

$\qquad = \sqrt{2}\left[\cos(-\pi) + i\sin(-\pi)\right]$

$\qquad = -\sqrt{2}$

D. Repeated use of the multiplication formula gives what is known as

> **DeMoivre's Theorem**
> If $z = r(\cos\theta + i\sin\theta)$, then for any integer n, $z^n = r^n(\cos n\theta + i\sin n\theta)$.

6. Find the indicated power using DeMoivre's Theorem.

(a) $\left(4 + 4\sqrt{3}\,i\right)^5$

Find r and θ, then apply DeMoivre's Theorem.

$r = \sqrt{4^2 + \left(4\sqrt{3}\right)^2} = \sqrt{16 + 48} = \sqrt{64} = 8$ and

$\tan\theta = \frac{4\sqrt{3}}{4} = \sqrt{3}$, so we have $\theta = \frac{\pi}{3}$ and

$z = 8\left(\cos\frac{\pi}{3} + i\sin\frac{\pi}{3}\right)$. Therefore

$$\left(4 + 4\sqrt{3}\,i\right)^5 = 8^5\left[\cos\left(5\cdot\frac{\pi}{3}\right) + i\sin\left(5\cdot\frac{\pi}{3}\right)\right]$$
$$= 32768\left(\cos\frac{5\pi}{3} + i\sin\frac{5\pi}{3}\right)$$
$$= 32768\left(\frac{1}{2} - i\frac{\sqrt{3}}{2}\right)$$
$$= 16384 - 16384\sqrt{3}\,i.$$

(b) $(-2 - 2i)^6$

$r = \sqrt{(-2)^2 + (-2)^2} = \sqrt{4 + 4} = \sqrt{8} = 2\sqrt{2}$ and

$\tan\theta = 1$, so we have $\theta = \frac{5\pi}{4}$ and

$z = 2\sqrt{2}\left(\cos\frac{5\pi}{4} + i\sin\frac{5\pi}{4}\right)$. Therefore

$$(-2 - 2i)^6 = \left(2\sqrt{2}\right)^6\left[\cos\left(6\cdot\frac{5\pi}{4}\right) + i\sin\left(6\cdot\frac{5\pi}{4}\right)\right]$$
$$= 512\left(\cos\frac{15\pi}{2} + i\sin\frac{15\pi}{2}\right)$$
$$= 512(0 - i)$$
$$= -512\,i.$$

(c) $(3 - 4i)^3$

$r = \sqrt{3^2 + (-4)^2} = \sqrt{9 + 16} = \sqrt{25} = 5$. Since θ

is in the quadrant IV,

$\theta = \tan^{-1}\left(\frac{-4}{3}\right) \approx \tan^{-1}(-1.33) \approx -0.92729$ rad and

$z \approx 5(\cos(-0.92729) + i\sin(-0.92729))$. Thus

$$(3 - 4i)^3$$
$$\approx 5^3[\cos(3\cdot(-0.92729)) + i\sin(3\cdot(-0.92729))]$$
$$\approx 125(\cos(-2.7819) + i\sin(-2.7819))$$
$$\approx 125(-0.936 - 0.3520i)$$
$$\approx -117 - 44\,i.$$

Note: We could have used $\theta \approx 5.35589$, then
$3\theta \approx 16.0677$. Thus $\cos(16.0677) \approx -0.936$ and
$\sin(16.0677) \approx -0.352$. This gives the same result.

E. For n a positive integer, the n distinct nth roots of the complex number $z = r(\cos\theta + i\sin\theta)$ are given by the formula:

$$w_k = r^{1/n}\left[\cos\left(\frac{\theta + 2k\pi}{n}\right) + i\sin\left(\frac{\theta + 2k\pi}{n}\right)\right], \text{ for } k = 0, 1, 2, \ldots, n-1.$$

The main concepts are:

1. The modulus of each nth root is $r^{1/n}$.

2. The argument of the first root is $\dfrac{\theta}{n}$.

3. We repeatedly add $\dfrac{2\pi}{n}$ to get the argument of each successive root.

7. Find the indicated roots and sketch the roots in the complex plane.

(a) The cube roots of 2.

Find r and θ.

$r = \sqrt{4+0} = \sqrt{4} = 2$ and $\theta = 0$.

When $k = 0$ we have $\theta_0 = \frac{0+0}{3} = 0$ and

$w_0 = \sqrt[3]{2}[\cos(0) + i\sin(0)] = 1.260$.

When $k = 1$ we have $\theta_1 = \frac{0+2\pi}{3} = \frac{2\pi}{3}$ and

$w_1 = \sqrt[3]{2}\left[\cos\left(\frac{2\pi}{3}\right) + i\sin\left(\frac{2\pi}{3}\right)\right]$

$\approx -0.630 + 1.091\,i$.

When $k = 2$ we have $\theta_2 = \frac{0+4\pi}{3} = \frac{4\pi}{3}$ and

$w_2 = \sqrt[3]{2}\left[\cos\left(\frac{4\pi}{3}\right) + i\sin\left(\frac{4\pi}{3}\right)\right]$

$\approx -0.630 - 1.091\,i$.

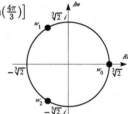

(b) The square roots of $5\sqrt{3} - 5i$.

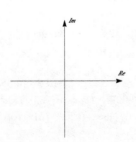

$r = \sqrt{\left(5\sqrt{3}\right)^2 + (-5)^2} = \sqrt{75 + 25} = \sqrt{100}$

$= 10$ and $\theta = \frac{11\pi}{6}$.

When $k = 0$ we have $\theta_0 = \dfrac{\frac{11\pi}{6} + 0}{2} = \frac{11\pi}{12}$ and

$w_0 = \sqrt{10}\left[\cos\left(\frac{11\pi}{12}\right) + i\sin\left(\frac{11\pi}{12}\right)\right]$

$\approx -3.055 + 0.818\,i$.

When $k = 1$ we have $\theta_1 = \dfrac{\frac{11\pi}{6} + 2\pi}{2} = \frac{23\pi}{12}$ and

$w_1 = \sqrt{10}\left[\cos\left(\frac{23\pi}{12}\right) + i\sin\left(\frac{23\pi}{12}\right)\right]$

$\approx 3.055 - 0.818\,i$.

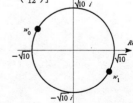

8. Solve each equation.

(a) $z^3 - i = 0$

Solve for z^3: $z^3 - i = 0 \quad \Leftrightarrow \quad z^3 = i$.

So the solutions are the cube roots of i. Since $i = \cos\frac{\pi}{2} + i\sin\frac{\pi}{2}$, we have $r = 1$ and $\theta = \frac{\pi}{2}$.

When $k = 0$, $\theta_0 = \dfrac{\frac{\pi}{2} + 0}{3} = \frac{\pi}{6}$ and

$$z_0 = \cos\left(\tfrac{\pi}{6}\right) + i\sin\left(\tfrac{\pi}{6}\right) = \tfrac{\sqrt{3}}{2} + \tfrac{1}{2}i.$$

When $k = 1$ we have $\theta_1 = \dfrac{\frac{\pi}{2} + 2\pi}{3} = \frac{5\pi}{6}$ and

$$z_1 = \cos\left(\tfrac{5\pi}{6}\right) + i\sin\left(\tfrac{5\pi}{6}\right) = -\tfrac{\sqrt{3}}{2} + \tfrac{1}{2}i.$$

When $k = 2$ we have $\theta_2 = \dfrac{\frac{\pi}{2} + 4\pi}{3} = \frac{3\pi}{2}$ and

$$z_2 = \cos\left(\tfrac{3\pi}{2}\right) + i\sin\left(\tfrac{3\pi}{2}\right) = -i.$$

(b) $z^2 + 4i = 0$

$$z^2 + 4i = 0$$
$$z^2 = -4i$$
$$z = \pm\sqrt{-4i} = \pm 2\sqrt{-i}$$

Next: find the square roots of $\sqrt{-i}$.

For $-i$, we have $r = 1$ and $\theta = \frac{3\pi}{2}$.

When $k = 0$, $\theta_0 = \dfrac{\frac{3\pi}{2} + 0}{2} = \frac{3\pi}{4}$ and

$$w_0 = \sqrt{1}\left[\cos\left(\tfrac{3\pi}{4}\right) + i\sin\left(\tfrac{3\pi}{4}\right)\right] =$$
$$= -\tfrac{\sqrt{2}}{2} + \tfrac{\sqrt{2}}{2}i.$$

When $k = 1$, $\theta_1 = \dfrac{\frac{3\pi}{2} + 2\pi}{2} = \frac{7\pi}{4}$ and

$$w_1 = \sqrt{1}\left[\cos\left(\tfrac{7\pi}{4}\right) + i\sin\left(\tfrac{7\pi}{4}\right)\right] =$$
$$= \tfrac{\sqrt{2}}{2} - \tfrac{\sqrt{2}}{2}i.$$

Since $w_1 = -w_0$, the solutions to $z^2 + 4i = 0$ are

$$z = \pm 2\sqrt{-i} = \pm 2\left(-\tfrac{\sqrt{2}}{2} + \tfrac{\sqrt{2}}{2}i\right)$$
$$= \pm\left(\sqrt{2} - \sqrt{2}i\right).$$

Section 7.7 Vectors

Key Ideas

A. Geometric description of vectors.
B. Analytic description of vectors.
C. Properties of vectors.
D. Applications to velocity and force.

A. A **vector** in the plane is a line segment with an assigned direction and length, denoted as a pair of points with an arrow above showing the direction $\overrightarrow{AB}$, or as a single bold face letter. The point A is the **initial point** and the point B is the **terminal point**. The length of the line segment is called the **magnitude**. Two vectors are **equal** if they have equal magnitudes and the same direction. The **zero vector**, **0**, has no magnitude. Geometrically the **sum** of two vectors is the vector found by placing the initial point of one vector at the terminal point of the other vector. See the figure. A constant is called

a **scalar**. If a is a scalar and **v** is a vector, then the vector $a\mathbf{v}$ is defined to have magnitude $|a|\,|\mathbf{v}|$ and to have the same direction as **v** when $a > 0$, or have direction opposite to **v** if $a < 0$. If $a = 0$, then $a\mathbf{v} = \mathbf{0}$, the zero vector. This process is called **multiplication of a vector by a scalar**. The **difference** of two vectors **u** and **v** is defined by $\mathbf{u} - \mathbf{v} = \mathbf{u} + (-\mathbf{v})$. See the figure to the right.

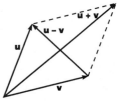

1. The vectors **u** and **v** are shown. Sketch the vectors $2\mathbf{u}$, $\mathbf{u} + \mathbf{v}$, $\mathbf{v} - \mathbf{u}$, and $\frac{1}{2}\mathbf{u} - 3\mathbf{v}$.

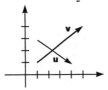

2u: This vector has twice the magnitude and the same direction as **u**.

u + v: Put the initial point of one at the terminal point of the other or make a parallelogram with **u** and **v** and take the diagonal.

v − u: Use $\mathbf{v} + (-\mathbf{u})$ or the other diagonal in the parallelogram.

$\frac{1}{2}\mathbf{u} - 3\mathbf{v}$: First find $\frac{1}{2}\mathbf{u}$ and then find $-3\mathbf{v}$. Finally, add the two vectors.

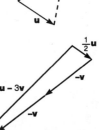

B. Analytically vectors are represented as an ordered pair of real number, $\mathbf{v} = \langle a, b \rangle$, where a is the **horizontal component** of $\mathbf{v}$ and b is the **vertical component** of $\mathbf{v}$. If a vector $\mathbf{v}$ is represented in the plane with initial point $P(x_1, y_1)$ and terminal point $Q(x_2, y_2)$, then $\mathbf{v} = \langle x_2 - x_1, y_2 - y_1 \rangle$. The zero vector is represented as $\mathbf{0} = \langle 0, 0 \rangle$. The **magnitude** or **length** of the vector $\mathbf{v} = \langle a, b \rangle$ is $|\mathbf{v}| = \sqrt{a^2 + b^2}$. The vectors $\mathbf{u} = \langle a_1, b_1 \rangle$ and $\mathbf{v} = \langle a_2, b_2 \rangle$ are equal if and only if $a_1 = a_2$ and $b_1 = b_2$. The algebraic operations on vectors are defined as:

If $\mathbf{u} = \langle a_1, b_1 \rangle$, $\mathbf{v} = \langle a_2, b_2 \rangle$, and c be any constant, then $\mathbf{u} + \mathbf{v} = \langle a_1 + a_2, b_1 + b_2 \rangle \qquad \mathbf{u} - \mathbf{v} = \langle a_1 - a_2, b_1 - b_2 \rangle \qquad c\mathbf{u} = \langle ca_1, cb_1 \rangle$

2. Let $\mathbf{u} = \langle -12, 18 \rangle$ and $\mathbf{v} = \langle 10, 8 \rangle$. Find $2\mathbf{u}$, $\mathbf{u} + \mathbf{v}$, $\mathbf{v} - \mathbf{u}$, and $\frac{1}{2}\mathbf{u} - 3\mathbf{v}$.

$$\begin{aligned} 2\mathbf{u} &= 2\langle -12, 18 \rangle \\ &= \langle 2(-12), 2(18) \rangle \\ &= \langle -24, 36 \rangle \end{aligned}$$

$$\begin{aligned} \mathbf{u} + \mathbf{v} &= \langle -12, 18 \rangle + \langle 10, 8 \rangle \\ &= \langle -12 + 10, 18 + 8 \rangle \\ &= \langle -2, 26 \rangle \end{aligned}$$

$$\begin{aligned} \mathbf{v} - \mathbf{u} &= \langle 10, 8 \rangle - \langle -12, 18 \rangle \\ &= \langle 10 + 12, 8 - 18 \rangle \\ &= \langle 22, -10 \rangle \end{aligned}$$

$$\begin{aligned} \tfrac{1}{2}\mathbf{u} - 3\mathbf{v} &= \tfrac{1}{2}\langle -12, 18 \rangle - 3\langle 10, 8 \rangle \\ &= \langle -6, 9 \rangle + \langle -30, -24 \rangle \\ &= \langle -36, -15 \rangle \end{aligned}$$

C. The properties of vector addition and multiplication of a vector by a scalar are:

Vector addition	**Multiplication of a vector by a scalar**
$\mathbf{u} + \mathbf{v} = \mathbf{v} + \mathbf{u}$	$c(\mathbf{u} + \mathbf{v}) = c\mathbf{u} + c\mathbf{v}$
$\mathbf{u} + (\mathbf{v} + \mathbf{w}) = (\mathbf{u} + \mathbf{v}) + \mathbf{w}$	$(c + d)\mathbf{u} = c\mathbf{u} + d\mathbf{u}$
$\mathbf{u} + \mathbf{0} = \mathbf{u}$	$(cd)\mathbf{u} = c(d\mathbf{u}) = d(c\mathbf{u})$
$\mathbf{u} + (-\mathbf{u}) = \mathbf{0}$	$1\mathbf{u} = \mathbf{u}$
	$0\mathbf{u} = \mathbf{0}$
Length of a vector	$c\mathbf{0} = \mathbf{0}$
$|c\mathbf{v}| = |c||\mathbf{v}|$	

A vector of length 1 is called a **unit vector**. Two special unit vectors are $\mathbf{i} = \langle 1, 0 \rangle$ and $\mathbf{j} = \langle 0, 1 \rangle$. Any vector $\mathbf{v} = \langle a, b \rangle$ can be expressed in terms of $\mathbf{i}$ and $\mathbf{j}$ by

$$\mathbf{v} = \langle a, b \rangle = a\mathbf{i} + b\mathbf{j}.$$

Let $\mathbf{v}$ be a vector with magnitude $|\mathbf{v}|$ and direction θ. Then the horizontal component of $\mathbf{v}$ is $a = |\mathbf{v}| \cos \theta$ and the vertical component of $\mathbf{v}$ is $b = |\mathbf{v}| \sin \theta$.

3. Let $\mathbf{u} = \langle -12, 5 \rangle$ and $\mathbf{v} = \langle 4, 3 \rangle$. Write $\mathbf{u}$, $\mathbf{u} + \mathbf{v}$, and $2\mathbf{u} - 3\mathbf{v}$ in terms of $\mathbf{i}$ and $\mathbf{j}$. Find the length of $\mathbf{u}$, $\mathbf{u} + \mathbf{v}$, and $2\mathbf{u} - 3\mathbf{v}$.

$\mathbf{u} = \langle -12, 5 \rangle = -12\mathbf{i} + 5\mathbf{j}$

length of $\mathbf{u} = |\mathbf{u}| = \sqrt{(-12)^2 + 5^2} = \sqrt{169} = 13$

$$\mathbf{u} + \mathbf{v} = \langle -12, 5 \rangle + \langle 4, 3 \rangle = \langle -8, 8 \rangle$$
$$= -8\mathbf{i} + 8\mathbf{j}$$

$|\mathbf{u} + \mathbf{v}| = \sqrt{(-8)^2 + 8^2} = \sqrt{128} = 8\sqrt{2}$

$$2\mathbf{u} - 3\mathbf{v} = 2\langle -12, 5 \rangle - 3\langle 4, 3 \rangle$$
$$= \langle -24, 10 \rangle + \langle -12, -9 \rangle = \langle -36, 1 \rangle$$
$$= -36\mathbf{i} + \mathbf{j}$$

$|2\mathbf{u} - 3\mathbf{v}| = \sqrt{(-36)^2 + 1^2} = \sqrt{1297}$

4. Find the direction of the vector $\mathbf{v} = -2\mathbf{i} + 2\sqrt{3}\,\mathbf{j}$. Sketch $\mathbf{v}$.

We first sketch $\mathbf{v}$.

Place the initial point at $(0, 0)$ and the terminal point at $(-2, 2\sqrt{3})$.

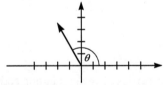

Next find the length of $\mathbf{v}$.

$$|\mathbf{v}| = \sqrt{(-2)^2 + \left(2\sqrt{3}\right)^2} = \sqrt{4 + 12} = 4$$

Since θ is in the second quadrant, we use $\cos^{-1}$ to find the direction, θ.

$$\theta = \cos^{-1}\left(\tfrac{-2}{4}\right) = \cos^{-1}\left(-\tfrac{1}{2}\right) = \tfrac{2\pi}{3}$$

5. A vector **u** has length $6\sqrt{3}$ and direction $\frac{7\pi}{6}$. Find the horizontal and vertical components and write **v** in terms of **i** and **j**.

The horizontal component of **v** is:
$$a = |\mathbf{v}| \cos \theta = 6\sqrt{3} \cos \tfrac{7\pi}{6}$$
$$= \left(6\sqrt{3}\right)\left(-\tfrac{\sqrt{3}}{2}\right) = -9.$$

The vertical component of **v** is:
$$b = |\mathbf{v}| \sin \theta = 6\sqrt{3} \sin \tfrac{7\pi}{6}$$
$$= \left(6\sqrt{3}\right)\left(-\tfrac{1}{2}\right) = -3\sqrt{3}.$$

Thus $\mathbf{v} = -9\mathbf{i} - 3\sqrt{3}\,\mathbf{j}.$

D. The **velocity** of a moving object is described by a vector whose direction is the direction of motion and whose magnitude is the speed. **Force** is also a quantity represented by a vector. The magnitude of force is pounds (newtons, in the metric system) and the direction is the direction the force is applied in.

6. A plane flies true north at 550 mph against a head wind blowing at S45°E at 75 mph. Find the true speed and direction of the plane.

Let **p** represent the velocity of the plane and Let **w** represent the velocity of the wind.
Since the plane is flying due north at 550 mph,
$\mathbf{p} = \langle 0,\ 550 \rangle$.
Since the wind is S45°E, we apply the formula for the components of a vector,
$$\mathbf{w} = \langle -75 \cos(-45°),\ 75 \sin(-45°) \rangle$$
$$= \langle 37.5\sqrt{2},\ -37.5\sqrt{2} \rangle.$$

The result of these two velocities is the vector
$$\mathbf{p} + \mathbf{w} = \langle 0,\ 550 \rangle + \langle 37.5\sqrt{2},\ -37.5\sqrt{2} \rangle$$
$$= \langle 37.5\sqrt{2},\ 550 - 37.5\sqrt{2} \rangle$$
$$\approx \langle 53.033,\ 496.967 \rangle.$$

$$|\mathbf{p} + \mathbf{w}| \approx \sqrt{2812.5 + 246976.19}$$
$$\approx \sqrt{249788.69} \approx 499.8 \text{ mph}$$

We use $\tan^{-1}$ to find the direction, θ.
$$\theta \approx \tan^{-1}\left(\tfrac{496.967}{53.033}\right) \approx \tan^{-1}(9.371) \approx 83.9°$$

Therefore the direction is N83.9°E.

Note: Degrees is the standard for this application.

7. A beam weighing 600 pounds is supported by two ropes as shown in the figure at right. Find the magnitudes of the tensions $|\mathbf{T_1}|$ and $|\mathbf{T_2}|$.

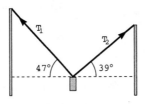

Add the force $\mathbf{T_3}$ for the weight of the beam. Since the beam is in a secured position $\mathbf{T_1} + \mathbf{T_2} + \mathbf{T_3} = 0$.

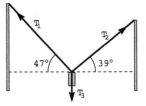

$\mathbf{T_1} = \langle a \cos 133°, \ a \sin 133° \rangle = \langle -0.68a, 0.73a \rangle$

$\mathbf{T_2} = \langle b \cos 39°, \ b \sin 39° \rangle = \langle 0.78b, \ 0.63b \rangle$

$\mathbf{T_3} = \langle 0, \ -600 \rangle$

Here a and b are the magnitudes of the tensions. Set the components equal to zero and solve.

$-0.68a + 0.78b = 0$

$0.73a + 0.63b - 600 = 0$

Solve the first equation for a in terms of b.

$-0.68a + 0.78b = 0 \quad \Leftrightarrow \quad a = 1.15b$

Substitute for a in the second equation.

$0.73(1.15b) + 0.63b = 600 \quad \Leftrightarrow$

$0.84b + 0.63b = 1.47b = 600 \quad \Leftrightarrow$

$b = 408.2$ pounds

$a = 1.15b = 1.15(408.2) = 469.4$ pounds.

Section 7.8 The Dot Product

Key Ideas
A. The dot product of vectors.
B. Component and projections.
C. Work.

A. If $\mathbf{u} = \langle a_1, b_1 \rangle$ and $\mathbf{v} = \langle a_2, b_2 \rangle$ are vectors, then their **dot product**, denoted by $\mathbf{u} \cdot \mathbf{v}$, is defined by:

$$\mathbf{u} \cdot \mathbf{v} = a_1 a_2 + b_1 b_2.$$

The dot product has the following properties:

$\mathbf{u} \cdot \mathbf{v} = \mathbf{v} \cdot \mathbf{u}$	$(a\mathbf{u}) \cdot \mathbf{v} = a(\mathbf{u} \cdot \mathbf{v}) = \mathbf{u} \cdot (a\mathbf{v})$
$(\mathbf{u} + \mathbf{v}) \cdot \mathbf{w} = \mathbf{u} \cdot \mathbf{w} + \mathbf{v} \cdot \mathbf{w}$	$|\mathbf{u}|^2 = \mathbf{u} \cdot \mathbf{u}$

Let $\mathbf{u}$ and $\mathbf{v}$ be vectors with initial points at the origin. Then the smaller of the angles formed by these representations of $\mathbf{u}$ and $\mathbf{v}$ is called the **angle between $\mathbf{u}$ and $\mathbf{v}$**. If θ is the angle between two nonzero vectors $\mathbf{u}$ and $\mathbf{v}$, then $\mathbf{u} \cdot \mathbf{v} = |\mathbf{u}||\mathbf{v}|\cos\theta$. This can also be expressed as $\cos\theta = \dfrac{\mathbf{u} \cdot \mathbf{v}}{|\mathbf{u}||\mathbf{v}|}$. This allows us to use the dot product to test whether two vectors are perpendicular. Two vectors $\mathbf{u}$ and $\mathbf{v}$ are perpendicular if and only if $\mathbf{u} \cdot \mathbf{v} = 0$.

In Problems 1 and 2, find (a) $\mathbf{u} \cdot \mathbf{v}$ and (b) the angle between $\mathbf{u}$ and $\mathbf{v}$ to the nearest degree.

1. $\mathbf{u} = \langle -2, 5 \rangle$, $\mathbf{v} = \langle 4, 3 \rangle$

Find the pieces and apply the theorem.

$\mathbf{u} \cdot \mathbf{v} = (-2)(4) + (5)(3) = -8 + 15 = 7$

$|\mathbf{u}| = \sqrt{(-2)^2 + 5^2} = \sqrt{29}$

$|\mathbf{v}| = \sqrt{4^2 + 3^2} = \sqrt{25} = 5$

$\cos\theta = \dfrac{\mathbf{u} \cdot \mathbf{v}}{|\mathbf{u}||\mathbf{v}|} = \dfrac{7}{5\sqrt{29}} \approx 0.260$

$\theta \approx \cos^{-1}(0.260) \approx 75°$

2. $\mathbf{u} = \langle 4, -3 \rangle$, $\mathbf{v} = \langle 6, -8 \rangle$

Find the pieces and apply the theorem.

$\mathbf{u} \cdot \mathbf{v} = (4)(6) + (-3)(-8) = 24 + 24 = 48$

$|\mathbf{u}| = \sqrt{4^2 + (-3)^2} = \sqrt{25} = 5$

$|\mathbf{v}| = \sqrt{6^2 + (-8)^2} = \sqrt{100} = 10$

$\cos\theta = \dfrac{\mathbf{u} \cdot \mathbf{v}}{|\mathbf{u}||\mathbf{v}|} = \dfrac{48}{5 \cdot 10} = 0.96$

$\theta \approx \cos^{-1}(0.96) \approx 16°$

3. Determine whether the given vectors are orthogonal.

   (a) $\mathbf{u} = \langle -2, 4 \rangle$, $\mathbf{v} = \langle 7, 3 \rangle$

        Translation: Is $\mathbf{u} \cdot \mathbf{v} = 0$?

$$\mathbf{u} \cdot \mathbf{v} = (-2)(7) + (4)(3) = -14 + 12$$
$$= -2 \neq 0$$

        The vectors $\mathbf{u}$ and $\mathbf{v}$ are not orthogonal.

   (b) $\mathbf{u} = \langle -4, 14 \rangle$, $\mathbf{v} = \langle -7, -2 \rangle$

$$\mathbf{u} \cdot \mathbf{v} = (-4)(-7) + (14)(-2) = 28 + (-28)$$
$$= 0$$

        The vectors $\mathbf{u}$ and $\mathbf{v}$ are orthogonal.

   (c) $\mathbf{u} = 9\mathbf{i} - 7\mathbf{j}$, $\mathbf{v} = 21\mathbf{i} + 27\mathbf{j}$

$$\mathbf{u} \cdot \mathbf{v} = (9)(21) + (-7)(27) = 189 + (-189)$$
$$= 0$$

        The vectors $\mathbf{u}$ and $\mathbf{v}$ are orthogonal.

B. The **component of u along v** (or the **component of u in the direction of v**) is defined to be $|\mathbf{u}| \cos \theta = \dfrac{\mathbf{u} \cdot \mathbf{v}}{|\mathbf{v}|}$. The **projection of u onto v**, denoted by $\text{proj}_{\mathbf{v}}\, \mathbf{u}$ is the vector whose *direction* is the same as $\mathbf{v}$ and whose *length* is the component of $\mathbf{u}$ along $\mathbf{v}$. The projection of $\mathbf{u}$ onto $\mathbf{v}$ is given by:

$$\text{proj}_{\mathbf{v}}\, \mathbf{u} = (\text{comopnent of } \mathbf{u} \text{ along } \mathbf{v})(\text{unit vector in direction of } \mathbf{v})$$

$$= \left(\frac{\mathbf{u} \cdot \mathbf{v}}{|\mathbf{v}|} \right) \frac{\mathbf{v}}{|\mathbf{v}|} = \left(\frac{\mathbf{u} \cdot \mathbf{v}}{|\mathbf{v}|^2} \right) \mathbf{v}$$

We often need to **resolve** a vector $\mathbf{u}$ into $\mathbf{u}_1$ and $\mathbf{u}_2$ where $\mathbf{u}_1$ is parallel to $\mathbf{v}$ and $\mathbf{u}_2$ is orthogonal to $\mathbf{v}$. So $\mathbf{u}_1 = \text{proj}_{\mathbf{v}}\, \mathbf{u}$ and $\mathbf{u}_2 = \mathbf{u} - \text{proj}_{\mathbf{v}}\mathbf{u}$.

4. Let $\mathbf{u} = \langle -1, 5 \rangle$ and $\mathbf{v} = \langle 4, 3 \rangle$. Find the component of $\mathbf{u}$ along $\mathbf{v}$. Then find $\text{proj}_{\mathbf{v}}\mathbf{u}$. Sketch $\mathbf{u}$, $\mathbf{v}$, and $\text{proj}_{\mathbf{v}}\mathbf{u}$.

        The component of $\mathbf{u}$ along $\mathbf{v}$ is

$$\frac{\mathbf{u} \cdot \mathbf{v}}{|\mathbf{v}|} = \frac{(-1)(4) + (5)(3)}{\sqrt{4^2 + 3^2}}$$

$$= \frac{-4 + 15}{5} = \frac{11}{5}$$

        The component of $\mathbf{u}$ along $\mathbf{v}$ is

$$\text{proj}_{\mathbf{v}}\mathbf{u} = \left(\frac{\mathbf{u} \cdot \mathbf{v}}{|\mathbf{v}|^2} \right) \mathbf{v} = \frac{11}{5^2} \langle 4, 3 \rangle$$

$$= \left\langle \tfrac{44}{25}, \tfrac{33}{25} \right\rangle$$

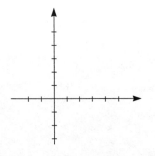

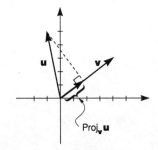

C. Work, W, is force $\times$ distance. The work, W, done by a force $\mathbf{F}$ in moving along a vector $\mathbf{D}$ is:
$$W = \mathbf{F} \cdot \mathbf{D}.$$

5. Find the work done by the force $\mathbf{F} = 2\mathbf{i} + 3\mathbf{j}$ in moving an object from $P(4, 7)$ to $Q(5, 10)$.

Find $\mathbf{D}$ and apply the formula.

$\mathbf{D} = (5 - 4)\mathbf{i} + (10 - 7)\mathbf{j} = \mathbf{i} + 3\mathbf{j}$

Therefore the work done is:

$W = \mathbf{F} \cdot \mathbf{D} = (2)(1) + (3)(3) = 2 + 9 = 11$

Since we are given no units, we leave the answer without units.

6. A man pulls a a wagon up a 15° incline by exerting a force of 40 lb on the handle. If the handle make an angle of 45° with the incline, find the work done in moving the wagon 20 feet up this incline.

   Hint: Sketch.

The sketch of the problem is on the right. From this we see that the handle makes an angle of $45 + 15 = 60°$ angle with the horizontal. So,

$\mathbf{F} = 40 \cos 60°\mathbf{i} + 40 \sin 60°\mathbf{j}$
$\quad = 20\mathbf{i} + 20\sqrt{3}\,\mathbf{j}$

$\mathbf{D} = 20 \cos 15°\mathbf{i} + 20 \sin 15°\mathbf{j}$
$\quad = 19.32\mathbf{i} + 5.18\,\mathbf{j}$

$W = \mathbf{F} \cdot \mathbf{D} = (20)(19.32) + (20\sqrt{3})(5.18)$
$\qquad = 565.84 \text{ foot-lb.}$

Chapter 8
Systems of Equations and Inequalities

Section 8.1   Systems of Equations

Key Ideas

A. Substitution method.

B. Elimination method.

C. Using graphing devices to solve systems of equations.

A. A **system of equations** is a set of equations with common variables. A **solution** to a system of equations is a point that simultaneously makes each equation of the system a true statement. Use the **substitution method** when an equation can be solved for one variable in terms of the other variables. The steps of the substitution method are

> **Step 1.** **Solve one equation** for one variable in terms of the other variables.
>
> **Step 2.** **Substitute** this value into the *other equation*. (It is very important that you substitute this value into a *different* equation.). Solve for the second variable.
>
> **Step 3.** **Back-substitute**--substitute the value you found in Step 2 back into the expression found in Step 1 to solve for the remaining variable.

1. Find all real solutions (x, y) of the system of equations.

$$\begin{cases} x^2 - xy - y^2 = 1 \\ 2x + y = 1 \end{cases}$$

Solve the second equation for y.

$\underline{y = 1 - 2x}$

Substitute this value into the first equation.

$$x^2 - x(1 - 2x) - (1 - 2x)^2 = 1$$
$$x^2 - x + 2x^2 - (1 - 4x + 4x^2) = 1$$
$$x^2 - x + 2x^2 - 1 + 4x - 4x^2 = 1$$
$$-x^2 + 3x - 1 = 1$$
$$-x^2 + 3x - 2 = 0$$
$$x^2 - 3x + 2 = 0$$
$$(x - 2)(x - 1) = 0$$
$\underline{x = 2 \text{ or } x = 1}$

Back-substitute $x = 2$ and $x = 1$.

When $x = 2$, then $y = 1 - 2(2) = -3$.

When $x = 1$, then $y = 1 - 2(1) = -1$.

So the possible solutions are $(2, -3)$ and $(1, -1)$.

Check: $(2, -3)$.

$(2)^2 - (2)(-3) - (-3)^2 = 4 + 6 - 9 = 1$

$2(2) + (-3) = 4 - 3 = 1$ This checks.

Check: $(1, -1)$.

$(1)^2 - (1)(-1) - (-1)^2 = 1 + 1 - 1 = 1$

$2(1) + (-1) = 2 - 1 = 1$ This checks.

2. Find all real solutions (x, y) of the system of equations.
$$\begin{cases} 3x + y = -1 \\ -2x + 3y = 8 \end{cases}$$

Solve $3x + y = -1$ for y (since it is easier);
$y = -3x - 1$.
Substitute for y in the second equation.
$$-2x + 3(-3x - 1) = 8$$
$$-2x - 9x - 3 = 8$$
$$-11x = 11$$
$$x = -1$$

Substituting this value of x back into $y = -3x - 1$
yields $y = -3(-1) - 1 = 3 - 1 = 2$.

So the possible solution is $(-1, 2)$.
Check: $(-1, 2)$.
$3(-1) + (2) = -3 + 2 = -1$
$-2(-1) + 3(2) = 2 + 6 = 8$ This checks.

B. Use the **elimination method** when a variable can be eliminated by the addition of a multiple of one equation to a multiple of another equation. The steps of this method are:

> Step 1. **Pick the variable to eliminate. Adjust the coefficients** of this variable by multiplying *each* equation by the appropriate *nonzero* numbers so that the coefficients of this variable add to zero.
>
> Step 2. **Add the two equations** to eliminate the variable you chose in Step 1. Solve this new equation for the other variable.
>
> Step 3. **Back-substitute**--substitute the value you found in Step 2 back into the expression found in Step 1 to solve for the remaining variable.

3. Find all real solutions (x, y) of the system of equations.
$$\begin{cases} 3x + 2y = -5 \\ x + 4y = 5 \end{cases}$$

$3x + 2y = -5$
$x + 4y = 5$
Eliminate the x variable. Multiply the second equation by -3 and add it to the first.

$$\begin{array}{ll} 3x + 2y = -5 & 3x + 2y = -5 \\ -3(x + 4y = 5) & -3x - 12y = -15 \\ \hline & -10y = -20 \\ & y = 2 \end{array}$$

Substitute for y and solve for x, so we get
$$x + 4(2) = 5 \quad \Leftrightarrow \quad x + 8 = 5 \quad \Leftrightarrow \quad x = -3.$$

OR

Eliminate the y variable. Multiply the first equation by -2 and add it to the second.

$$\begin{array}{ll} -2(3x + 2y = -5) & -6x - 4y = 10 \\ x + 4y = 5 & x + 4y = 5 \\ \hline & -5x = 15 \\ & x = -3 \end{array}$$

Substitute for x and solve for y, so we get
$$-3 + 4y = 5 \quad \Leftrightarrow \quad 4y = 8 \quad \Leftrightarrow \quad y = 2.$$

By either method the solution is $(-3, 2)$.

4. Find all real solutions (x, y) of the system of equations.
$$\begin{cases} x^2 + y^2 + 2y = 10 \\ x^2 + 4y = 7 \end{cases}$$

The x^2 term can be eliminated by subtracting the second equation from the first equation.
$$\begin{array}{r} x^2 + y^2 + 2y = 10 \\ -x^2 \qquad -4y = -7 \\ \hline y^2 - 2y = 3 \end{array}$$
$$y^2 - 2y - 3 = 0 \quad \Leftrightarrow \quad (y-3)(y+1) = 0.$$
So $y = 3$ or $y = -1$.

Substituting $y = 3$ into the second equation gives
$$x^2 + 4(3) = 7 \quad \Leftrightarrow \quad x^2 + 12 = 7 \quad \Leftrightarrow$$
$$x^2 = -5, \text{ which has no real number solution.}$$

Substituting $y = -1$ into the second equation gives
$$x^2 + 4(-1) = 7 \quad \Leftrightarrow \quad x^2 - 4 = 7 \quad \Leftrightarrow$$
$$x^2 = 11 \quad \Leftrightarrow \quad x = \pm\sqrt{11}.$$
The solutions are $(-\sqrt{11}, -1)$ and $(\sqrt{11}, -1)$.

Be sure to check that each is a solution to this system of equations.

C. Graphing devices are sometimes useful in solving systems of equations in two variables. However, with most graphing devices, an equation must first be expressed in terms of one or more functions of the form $y = f(x)$, before the equation can be graphed. Not all equations can be expressed in this way, so not all systems can be solved this way.

5. Find all real solutions to the system of equations, correct to two decimal places.
$$\begin{cases} x^2y + 3xy + 4y - 4 = 0 \\ x^2 + 4x - y = 6 \end{cases}$$

$x^2y + 3xy + 4y - 4 = 0$ *Express y as a function*
$x^2y + 3xy + 4y = 4$ *of x. Group terms with*
$y(x^2 + 3x + 4) = 4$ *y to one side. Factor.*
$$y = \frac{4}{x^2 + 3x + 4}$$
And
$x^2 + 4x - y = 6$ *Express y as a function of x.*
$-y = -x^2 - 4x + 6$
$y = x^2 + 4x - 6.$

$$[-8, 5] \text{ by } [-2, 3]$$

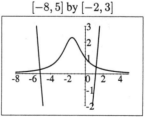

The solutions are $(-5.20, 0.26)$ and $(1.23, 0.43)$.

6. Find all real solutions to the system of equations, correct to two decimal places.
$$\begin{cases} e^x y = x^2 \\ (x+2)^2 + (y-2)^2 = 9 \end{cases}$$

Express each equation as a function of x.

$e^x y = x^2 \quad \Leftrightarrow \quad y = \dfrac{x^2}{e^x}$

And

$(x+2)^2 + (y-2)^2 = 9$

$(y-2)^2 = 9 - (x+2)^2$ *Take square root of*

$y - 2 = \pm\sqrt{9 - (x+2)^2}$ *both sides.*

$y = 2 \pm \sqrt{9 - (x+2)^2}.$

To solve, we graph $y = 2 - \sqrt{9 - (x+2)^2}$ and $y = 2 + \sqrt{9 - (x+2)^2}$, and find where they intersect the equation $y = \dfrac{x^2}{e^x}$.

$[-3, 2]$ by $[-2, 5]$

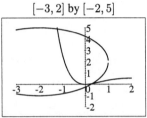

The solutions are $(-1.21, 4.90)$ and $(0.29, 0.06)$.

7. Find all real solutions to the system of equations, correct to three decimal places.
$$\begin{cases} xy = x^2 - 3x + 2 \\ x = 3y \end{cases}$$

Express each equation as a function of x.

$yx = x^2 - 3x + 2$ *Divide by x to write*

$y = \dfrac{x^2 - 3x + 2}{x}$ *y as a function of x.*

and

$x = 3y \quad \Leftrightarrow \quad y = \tfrac{1}{3}x.$

To solve, we graph the equations $y = \dfrac{x^2 - 3x + 2}{x}$ and $y = \tfrac{1}{3}x$, and find the points of intersection.

$[-10, 10]$ by $[-10, 10]$

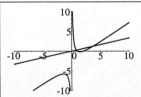

The solutions are $(0.814, 0.271)$ and $(3.686, 1.229)$.

Section 8.2   Systems of Linear Equations in Two Variables

Key Ideas

A. Solutions to systems of equations.

B. Applied linear systems.

A. A system of two linear equations in two unknowns can have *infinitely many* solutions, *exactly one* solution, or *no* solutions. A system that has no solution is said to be **inconsistent,** while a system with infinitely many solutions is called **dependent.**

In problems 1-4, solve the system. If the system has infinitely many solutions, express them in terms of x.

1. $\begin{cases} 3x + 2y = 3 \\ x - 2y = 3 \end{cases}$

To eliminate the y variable, we add the two equations.

$$\begin{array}{l} \begin{cases} 3x + 2y = 3 \\ \underline{x - 2y = 3} \end{cases} \\ 4x = 6 \quad \Leftrightarrow \quad x = \frac{6}{4} = \frac{3}{2}. \end{array}$$

Back-substituting into the second equation gives

$$\left(\tfrac{3}{2}\right) - 2y = 3 \quad \Leftrightarrow \quad -2y = \tfrac{3}{2} \quad \Leftrightarrow \quad y = -\tfrac{3}{4}.$$

Thus the solution is $\left(\frac{3}{2}, -\frac{3}{4}\right)$.

2. $\begin{cases} 2x - 3y = 9 \\ -4x + 6y = 4 \end{cases}$

To eliminate the x variable, we multiply the first equation by 2 and add it to the second equation.

$$\begin{cases} 2x - 3y = 9 \\ -4x + 6y = 4 \end{cases} \quad \Leftrightarrow \quad \begin{array}{l} \begin{cases} 4x - 6y = 18 \\ \underline{-4x + 6y = 4} \end{cases} \\ 0 = 22 \end{array}$$

Since this last statement is never true, this system has no solution. Further examination of this system shows that these lines are parallel.

3. $\begin{cases} \frac{3}{4}x + y = 1 \\ 3x + 2y = 1 \end{cases}$

To eliminate the x variable, we multiply the first equation by -4 and add it to the second equation.

$\begin{cases} \frac{3}{4}x + y = 1 \\ 3x + 2y = 1 \end{cases} \Leftrightarrow \begin{cases} -3x - 4y = -4 \\ \underline{3x + 2y = 1} \\ -2y = -3 \end{cases}$

So $y = \frac{3}{2}$.

Back-substituting into the second equation gives

$3x + 2\left(\frac{3}{2}\right) = 1 \quad \Leftrightarrow \quad 3x + 3 = 1 \quad \Leftrightarrow$

$3x = -2 \quad \Leftrightarrow \quad x = -\frac{2}{3}.$

So, the solution to the system is $\left(-\frac{2}{3}, \frac{3}{2}\right)$.

4. $\begin{cases} \frac{3}{4}x - \frac{1}{2}y = 1 \\ -3x + 2y = -4 \end{cases}$

To eliminate the x variable, we multiply the first equation by 4 and add it to the second equation.

$\begin{cases} \frac{3}{4}x - \frac{1}{2}y = 1 \\ -3x + 2y = -4 \end{cases} \Leftrightarrow \begin{cases} 3x - 2y = 4 \\ \underline{-3x + 2y = -4} \\ 0 = 0 \end{cases}$

So the two equations in the original system are two different ways of expressing the same equation. So, we solve the second equation for y.

$-3x + 2y = -4 \quad \Leftrightarrow \quad 2y = 3x - 4 \quad \Leftrightarrow$

$y = \frac{3}{2}x - 2$

So, the solutions to the system are $\left(x, \frac{3}{2}x - 2\right)$, where x is any number.

B. Systems of linear equations frequently appear in science applications and models in other areas. Use these guidelines when modeling with systems of equations.

1. **Identify the variables.** Identify the quantities the problem asks you to find. These are usually determined by carefully reading the question posed. Assign letters to these quantities in the problem and define what each letter represents. This helps you and others to understand what value you are looking for.

2. **Express all unknown quantities in terms of the variables.** Read the problem again and express all the quantities mentioned in the problem in terms of the variables you defined in Step 1.

3. **Set up a system of equations.** Find the crucial facts in the problem that give the relationships between the expressions you found in Step 2. Set up a system of equations that expresses these relationships.

4. **Solve the system and interpret the results.** Solve the system found in Step 3, check your solutions, and state your final answer as a sentence that answers the question posed.

5. The sum of two numbers is 43 and their difference is 11. Find the two numbers.

Let $x =$ the larger number.

Let $y =$ the other number.

Then "the sum of two numbers is 43" translates to $x + y = 43$. And "their difference is 11" translates to $x - y = 11$. So we get the system
$$\begin{cases} x + y = 43 \\ x - y = 11. \end{cases}$$

We add to solve.
$$\begin{array}{r} x + y = 43 \\ x - y = 11 \\ \hline 2x = 54 \end{array}$$

So $x = 27$. Substituting into the first equation yields
$$27 + y = 43 \quad \Leftrightarrow \quad y = 16.$$
So the numbers are 27 and 16.

6. A movie theater charges $4.50 for children and $8.50 for adults. On a certain day, 625 people saw a movie and the theater collected $3632.50 in receipts. How many children and how many adults were admitted?

Let $x =$ the number of children admitted.

Let $y =$ the number of adults admitted.

Then "625 people saw a movie" translates to $x + y = 625$. And "the theater collected $3632.50 in receipts" translates to $4.50x + 8.50y = 3632.50$.

So we get the system $\begin{cases} x + y = 625 \\ 4.50x + 8.50y = 3632.50 \end{cases}$.

To solve, multiply the first equation by -4.5 and add it to the second equation.
$$\begin{array}{r} -4.5x - 4.5y = -2812.50 \\ 4.5x + 8.5y = 3632.50 \\ \hline 4y = 820 \quad \Leftrightarrow \quad y = 205 \end{array}$$

Substituting into the first equation yields
$$x + 205 = 625 \quad \Leftrightarrow \quad x = 420.$$
So the theater admitted 420 children and 205 adults.

7. A city plans to plant 50 new trees in a subdivision. If birch trees cost $40 each and oak trees cost $55 each, how many of each kind of trees can they plant for $2300?

Let $x =$ the number of birch trees.

Let $y =$ the number of oak trees.

Then "plan to plant 50 new trees" translates to $x + y = 50$. And "birch trees cost $40 each and oak trees cost $55 each" translates to the cost of the trees as $40x + 55y = 2300$. So we get the system $\begin{cases} x + y = 50 \\ 40x + 55y = 2300 \end{cases}$.

Multiply the first equation by -40 and add, to get

$$\begin{array}{r} -40x - 40y = -2000 \\ 40x + 55y = 2300 \\ \hline 15y = 300 \end{array} \quad \Leftrightarrow \quad y = 20.$$

Substituting into the first equation yields

$$x + 20 = 50 \quad \Leftrightarrow \quad x = 30.$$

So the city should plant 30 birch trees and 20 oak trees in the subdivision.

226

Section 8.3   Systems of Linear Equations in Several Variables

Key Ideas

A. Linear equations and triangular form.
B. Equivalent systems and Guassian elimination.
C. Inconsistent systems and dependent systems.
D. Curve fitting.

A. A **linear equation in n variables** is an equation of the form $a_1x_1 + a_2x_2 + a_3x_3 + \cdots + a_nx_n = c$, where $a_1, a_2, a_3, \ldots, a_n$, and c are constants and $x_1, x_2, x_3, \ldots$, and x_n are variables. When $n \leq 4$, the letters x, y, z, and w, are usually used in place of the variables x_1, x_2, x_3, and x_4. Each term in a linear equation is either a constant or a constant multiple of *one* of the variables raised to the first power.

A system is in **triangular form** when the x variable does not appear in the second equation, the x and y variables do not appear in the third equation, etc. It is easy to solve a system that is in triangular form using back-substitution. For a system of three equations, the idea is eliminate all but one variable from one equation, then all but two variables from another equation.

1. State whether the equation is linear.

   (a) $\sqrt{21}\, x + 3z = 5$

   Yes, $\sqrt{21}$, 3, and 5 are all constants, so this is a linear equation.

   (b) $x + 2\sqrt{y} - 3z = 23$

   No, in the second term, $2\sqrt{y}$, the variable y is not raised to the first power, so this is not a linear equation.

   (c) $\dfrac{5}{x} + 2y - 13z + 7w = 0$

   No, in the first term, $\dfrac{5}{x}$, the variable x is not raised to the first power, so this equation is not linear.

2. Solve the following system using back-substitution.
$$\begin{cases} x + 3y + 3z = -1 \\ \quad\;\; y - z = -7 \\ \quad\qquad z = 2 \end{cases}$$

   This system is in triangular form. Solving we have $z = 2$. Back-substituting into the second equation gives $y - (2) = -7 \;\Leftrightarrow\; y = -5$. Back-substituting these values into the first equation gives
   $$x + 3(-5) + 3(2) = -1 \quad\Leftrightarrow\quad x = 8.$$
   Thus the solution is $(8, -5, 2)$.

B. Two systems of linear equations are **equivalent** if they have the same solution set. The only allowable algebraic operations for transforming one system into an equivalent system are:

   1. Add a multiple of one equation to another equation.
   2. Multiply an equation by a *nonzero* constant.
   3. Rearrange the order of the equations.

Gaussian elimination is a systematic process to transform a system of linear equations using elimination into a triangular system.

1. Use two equations to eliminate one of the variables, then use a different pair of equations to eliminate the same variable.
2. Use these two new equations to eliminate another variable.
3. Back-substitute to find the solution.

Remember to check the solution found in all three original equations.

3. Solve the system using Guassian elimination.

$$\begin{cases} x + y + 3z = 1 \\ 3x + 2y + 3z = 1 \\ 2x + y + 2z = -3 \end{cases}$$

We use the first equation to eliminate the x variable in the second and third equations.

$$\begin{cases} x + y + 3z = 1 \\ 3x + 2y + 3z = 1 \\ 2x + y + 2z = -3 \end{cases}$$

$$\begin{array}{ll} -3(x + y + 3z = 1) & \Leftrightarrow \quad -3x - 3y - 9z = -3 \\ 3x + 2y + 3z = 1 & \qquad 3x + 2y + 3z = 1 \\ \hline & \qquad \qquad -y - 6z = -2 \end{array}$$

$$\begin{array}{ll} -2(x + y + 3z = 1) & \Leftrightarrow \quad -2x - 2y - 6z = -2 \\ 2x + y + 2z = -3 & \qquad 2x + y + 2z = -3 \\ \hline & \qquad \qquad -y - 4z = -5 \end{array}$$

This gives the new system

$$\begin{cases} x + y + 3z = 1 \\ -y - 6z = -2 \\ -y - 4z = -5 \end{cases} \Leftrightarrow \begin{cases} x + y + 3z = 1 \\ y + 6z = 2 \\ y + 4z = 5 \end{cases}.$$

Subtract the two new equation to get

$$\begin{array}{l} y + 6z = 2 \\ -y - 4z = -5 \\ \hline 2z = -3 \quad \Leftrightarrow \quad z = -\tfrac{3}{2}. \end{array}$$

This gives the triangular system
$$\begin{cases} x + y + 3z = 1 \\ y + 6z = 2 \\ z = -\tfrac{3}{2} \end{cases}.$$

Back-substituting for z in the second equation gives

$$y + 6\left(-\tfrac{3}{2}\right) = 2 \quad \Leftrightarrow \quad y - 9 = 2 \quad \Leftrightarrow \quad y = 11.$$

Back-substituting for y and z in the first equation gives $x + (11) + 3\left(-\tfrac{3}{2}\right) = 1 \quad \Leftrightarrow \quad x + 11 - \tfrac{9}{2} = 1$

$\Leftrightarrow \quad x + \tfrac{13}{2} = 1 \quad \Leftrightarrow \quad x = -\tfrac{11}{2}.$

So the possible solution is $\left(-\tfrac{11}{2}, 11, -\tfrac{3}{2}\right)$.

Check.
$$-\tfrac{11}{2} + (11) + 3\left(-\tfrac{3}{2}\right) = -\tfrac{11}{2} + \tfrac{22}{2} - \tfrac{9}{2} = \tfrac{2}{2} = 1 \checkmark$$

$$3\left(-\tfrac{11}{2}\right) + 2(11) + 3\left(-\tfrac{3}{2}\right) = -\tfrac{33}{2} + \tfrac{44}{2} - \tfrac{9}{2} = 1 \checkmark$$

$$2\left(-\tfrac{11}{2}\right) + (11) + 2\left(-\tfrac{3}{2}\right) = -11 + 11 - 3 = -3 \checkmark$$

Thus the solution is $\left(-\tfrac{11}{2}, 11, -\tfrac{3}{2}\right)$.

4. Solve the system using Guassian elimination.

$$\begin{cases} x + 3y - z = 4 \\ 2x + 5y + z = -1 \\ 3x - y + 2z = -3 \end{cases}$$

We use the first equation to eliminate the x variable in the second and third equations.

$$\begin{cases} x + 3y - z = 4 \\ 2x + 5y + z = -1 \\ 3x - y + 2z = -3 \end{cases}$$

$$\begin{array}{l} -2(x + 3y - z = 4) \\ 2x + 5y + z = -1 \end{array} \quad \Leftrightarrow \quad \begin{array}{r} -2x - 6y + 2z = -8 \\ 2x + 5y + \ z = -1 \\ \hline -y + 3z = -9 \end{array}$$

$$\begin{array}{l} -3(x + 3y - z = 4) \\ 3x - y + 2z = -3 \end{array} \quad \Leftrightarrow \quad \begin{array}{r} -3x - 9y + 3z = -12 \\ 3x - \ y + 2z = -3 \\ \hline -10y + 5z = -15 \end{array}$$

This gives the new system

$$\begin{cases} x + 3y - z = 4 \\ -y + 3z = -9 \\ -10y + 5z = -15 \end{cases} \quad \Leftrightarrow \quad \begin{cases} x + 3y - z = 4 \\ -y + 3z = -9. \\ 2y - z = 3 \end{cases}$$

(Divide the third equation by -5, to simplify it.) Now eliminate the y variable in the new third equation, using the new second equation.

$$\begin{array}{l} 2(-y + 3z = -9) \\ 2y - z = 3 \end{array} \quad \Leftrightarrow \quad \begin{array}{r} -2y + 6z = -18 \\ 2y - \ z = 3 \\ \hline 5z = -15 \end{array}$$

So $z = -3$.

This gives the triangular system $\begin{cases} x + 3y - z = 4 \\ -y + 3z = -9. \\ z = -3 \end{cases}$

Back-substituting for z in the second equation gives

$$-y + 3(-3) = -9 \quad \Leftrightarrow \quad -y - 9 = -9 \quad \Leftrightarrow$$
$$y = 0.$$

Back-substituting for y and z in the first equation gives $x + 3(0) - (-3) = 4 \quad \Leftrightarrow \quad x + 3 = 4 \quad \Leftrightarrow$
$$x = 1.$$

So the possible solution is $(1, 0, -3)$.

Check.

$(1) + 3(0) - (-3) = 1 + 0 + 3 = 4$ ✓

$2(1) + 5(0) + (-3) = 2 + 0 - 3 = -1$ ✓

$3(1) - (0) + 2(-3) = 3 - 0 - 6 = -3$ ✓

Thus the solution is $(1, 0, -3)$.

C. A system that has no solution is called an **inconsistent system**. A system is inconsistent when we try to use Gaussian elimination to get a triangular system and we end up with an equation that is always false. The false equation always has the form $0 = N$, where N is nonzero number. A system that has infinitely many solutions is called a **dependent system**. The complete solution to such a system will have one or more variables that are arbitrary, and the other variables will depend on the arbitrary one(s). If we use Gaussian elimination to try to change a system to triangular form, discard equations of the form $0 = 0$, and if the resulting system has more variables than equations, then the system is dependent.

For problem 5 and 6, find the complete solution to each system of equations, or show that none exists.

5. $\begin{cases} x \quad\;\; -2z + \;\; w = 2 \\ x + y + 5z + \;\; w = 7 \\ 2x + y + 3z + 2w = 9 \\ 3x + y + \;\; z + 3w = 11 \end{cases}$

We use the first equation to eliminate the x variable in the second, third, and fourth equations.

$\begin{cases} x \quad\;\; -2z + \;\; w = 2 \\ x + y + 5z + \;\; w = 7 \\ 2x + y + 3z + 2w = 9 \\ 3x + y + \;\; z + 3w = 11 \end{cases}$

$\begin{array}{l} -x \quad\quad\; + 2z - w = -2 \\ \underline{x + y + 5z + w = 7} \\ \quad\quad\; y + 7z \quad\quad = 5 \end{array}$

$\begin{array}{l} -2x \quad\quad\; + 4z - 2w = -4 \\ \underline{2x + y + 3z + 2w = 9} \\ \quad\quad\; y + 7z \quad\quad = 5 \end{array}$

$\begin{array}{l} -3x \quad\quad\; + 6z - 3w = -6 \\ \underline{3x + y + \;\; z + 3w = 11} \\ \quad\quad\; y + 7z \quad\quad = 5 \end{array}$

This gives the new system

$\begin{cases} x \;\; -2z + w = 2 \\ y + 7z \quad\quad = 5 \\ y + 7z \quad\quad = 5 \\ y + 7z \quad\quad = 5 \end{cases}$ Since the last three equations are the same, this system equivalent to the system

$\begin{cases} x \;\; -2z + w = 2 \\ y + 7z \quad\quad = 5 \end{cases}$.

This system is dependent since this it has only 2 equations in 4 variables. We solve the first equation for x in terms of z and w, and then we solve the second equation for y in terms of z.

$x - 2z + w = 2 \quad \Leftrightarrow \quad x = 2 + 2z - w$

$y + 7z = 5 \quad \Leftrightarrow \quad y = 5 - 7z$

Letting $z = t$ and $w = s$ we get the solutions $(2 + 2t - s, 5 - 7t, t, s)$ where t and s are any real numbers.

6. $\begin{cases} x - 7y + 5z = 2 \\ 2x + 3y - z = 5 \\ 4x - 11y + 9z = 10 \end{cases}$

We use the first equation to eliminate the x variable in the second and third equations.

$\begin{cases} x - 7y + 5z = 2 \\ 2x + 3y - z = 5 \\ 4x - 11y + 9z = 10 \end{cases}$

$\begin{array}{l} -2(x - 7y + 5z = 2) \\ 2x + 3y - z = 5 \end{array} \Leftrightarrow \begin{array}{l} -2x + 14y - 10z = -4 \\ \underline{2x + 3y - z = 5} \\ 17y - 11z = 1 \end{array}$

$\begin{array}{l} -4(x - 7y + 5z = 2) \\ 4x - 11y + 9z = 10 \end{array} \Leftrightarrow \begin{array}{l} -4x + 28y - 20z = -8 \\ \underline{4x - 11y + 9z = 10} \\ 17y - 11z = 2 \end{array}$

This gives the new system

$\begin{cases} x - 7y + 5z = 2 \\ 17y - 11z = 1 \\ 17y - 11z = 2 \end{cases}$

Subtract the two new equation to get

$\begin{array}{l} -(17y - 11z = 1) \\ 17y - 11z = 2 \end{array} \Leftrightarrow \begin{array}{l} -17y + 11z = -1 \\ \underline{17y - 11z = 2} \\ 0 = 1. \end{array}$

Since the last equation, $0 = 1$, is always false, there is no solution to this system.

D. Two points in the coordinate plane determine a unique line (first degree polynomial) and three non-collinear points determine a unique quadratic (second degree) polynomial. Of course, the x coordinates must be distinct, otherwise the points can not be those of a function.

7. Find an equation of a function through the points $(3, 7)$ and $(-8, 2)$.

Since we are given two points, we are looking for a first degree equation (a line) of the form $y = Ax + B$. Substituting the for x and y we get:
$(3, 7): \ 3A + B = 7$ and $(-8, 2): \ -8A + B = 2$.

This gives the system $\begin{cases} 3A + B = 7 \\ -8A + B = 2 \end{cases}$

We use the first equation to eliminate the variable B in the second equation.

$\begin{array}{l} -(3A + B = 7) \\ -8A + B = 2 \end{array} \Leftrightarrow \begin{array}{l} -3A - B = -7 \\ \underline{-8A + B = 2} \\ -11A = -5 \end{array}$

So $-11A = -5 \ \Leftrightarrow \ A = \frac{5}{11}$.

Back-substituting we have

$3\left(\frac{5}{11}\right) + B = 7 \ \Leftrightarrow \ B = \frac{77}{11} - \frac{15}{11} = \frac{62}{11}$.

Thus the equation of the curve is $y = \frac{5}{11}x + \frac{62}{11}$.

$\frac{5}{11}(3) + \frac{62}{11} = \frac{15}{11} + \frac{62}{11} = \frac{77}{11} = 7 \ \checkmark$

$\frac{5}{11}(-8) + \frac{62}{11} = \frac{-40}{11} + \frac{62}{11} = \frac{22}{11} = 2 \ \checkmark$

231

8. Find the equation of the curve through the points $(-1, -2)$, $(2, -5)$, and $(3, 2)$.

Since we are given three points, we are looking for a second degree equation of the form $y = Ax^2 + Bx + C$. Substituting the for x and y we get:

$(-1, -2)$: $A - B + C = -2$,
$(2, -5)$: $4A + 2B + C = -5$ and
$(3, 2)$: $9A + 3B + C = 2$.

This gives the system

$$\begin{cases} A - B + C = -2 \\ 4A + 2B + C = -5. \\ 9A + 3B + C = 2 \end{cases}$$

We use the first equation to eliminate the variable A in the second and third equations.

$$\begin{array}{ll} -4(A - B + C = -2) & \Leftrightarrow \\ 4A + 2B + C = -5 & \end{array} \quad \begin{array}{l} -4A + 4B - 4C = 8 \\ \underline{4A + 2B + C = -5} \\ 6B - 3C = 3 \end{array}$$

$$\begin{array}{ll} -9(A - B + C = -2) & \Leftrightarrow \\ 9A + 3B + C = 2 & \end{array} \quad \begin{array}{l} -9A + 9B - 9C = 18 \\ \underline{9A + 3B + C = 2} \\ 12B - 8C = 20 \end{array}$$

This gives the new system

$$\begin{cases} A - B + C = -2 \\ 6B - 3C = 3 \\ 12B - 8C = 20 \end{cases} \Leftrightarrow \begin{cases} A - B + C = -2 \\ 2B - C = 1 \\ 3B - 2C = 5 \end{cases}.$$

(Divide the second equation by 3 and the third equation by 4, to simplify each.)
Now eliminate the variable B in the new third equation, using the new second equation.

$$\begin{array}{ll} -3(2B - C = 1) & \Leftrightarrow \\ 2(3B - 2C = 5) & \end{array} \quad \begin{array}{l} -6B + 3C = -3 \\ \underline{6B - 4C = 10} \\ -C = 7 \end{array}$$

So $C = -7$.

This gives the triangular system $\begin{cases} A - B + C = -2 \\ 2B - C = 1 \\ C = -7 \end{cases}$.

Back-substituting for C in the second equation gives
$2B - (-7) = 1 \quad \Leftrightarrow \quad 2B = -6 \quad \Leftrightarrow \quad B = -3$.
Back-substituting for B and C in the first equation gives $A - (-3) + (-7) = -2 \quad \Leftrightarrow \quad A = 2$.

Thus the equation of the curve is $y = 2x^2 - 3x - 7$.

Check.

$2(-1)^2 - 3(-1) - 7 = 2 + 3 - 7 = -2 \ \checkmark$

$2(2)^2 - 3(2) - 7 = 8 - 6 - 7 = -5 \ \checkmark$

$2(3)^2 - 3(3) - 7 = 18 - 9 - 7 = 2 \ \checkmark$

Section 8.4   Systems of Linear Equations: Matrices

Key Ideas

A. Matrix.
B. Elementary row operations.
C. Echelon Form and Reduced Echelon Form.
D. Inconsistent systems and dependent systems.

A. An $m \times n$ **matrix** (shown on the right) is a rectangular array of numbers with m rows and n columns. The numbers are called **entries** of the matrix. The entry a_{ij} indicates that it is in the ith row jth column. Instead of writing the equations of a system in full, we may write only the coefficients and constants in a matrix call an **augmented matrix**. Each equation corresponds to a row of the matrix and each variable to a column of the matrix. The constants also contribute a column.

$$\begin{bmatrix} a_{11} & a_{12} & a_{13} & \cdots & a_{1n} \\ a_{21} & a_{22} & a_{23} & \cdots & a_{2n} \\ a_{31} & a_{32} & a_{33} & \cdots & a_{3n} \\ \vdots & \vdots & \vdots & \vdots & \vdots \\ a_{m1} & a_{m2} & a_{m3} & \cdots & a_{mn} \end{bmatrix}$$

1. Write the augmented matrix of the system of equations.

$$\begin{cases} x - 7y + 3z = 5 \\ 2x + 3y - 2z = 11 \\ -x + 5y + 4z = -2 \end{cases}$$

Since the system has three equations in three variables, the augmented matrix we seek is a 3×4 matrix.

$$\begin{bmatrix} 1 & -7 & 3 & 5 \\ 2 & 3 & -2 & 11 \\ -1 & 5 & 4 & -2 \end{bmatrix}$$

2. Write the system of equations that corresponds to the augmented matrix below and solve.

$$\begin{bmatrix} 1 & 6 & 0 & 8 \\ 0 & 1 & -5 & -2 \\ 0 & 0 & 1 & 3 \end{bmatrix}$$

This is a 3×4 matrix, so we seek a three equation three unknown system. We get

$$\begin{cases} x + 6y & = 8 \\ y - 5z & = -2 \\ z & = 3 \end{cases}.$$

This is a triangular system, using back-substitution we have $y - 5(3) = -2 \iff y = 13$ and $x + 6(13) = 8 \iff x = -70$.
Thus the solution is $(-70, 13, 3)$.

B. One matrix is transformed into an equivalent matrix by these **elementary row operations**:

1. Add a multiple of one row to another row.
2. Multiply a row by a nonzero constant.
3. Interchange two rows.

There is no *one way* to solve a system of equations using matrices, so it is helpful to yourself (and others who read your work) to leave a trail of the row operations you perform. We use the following notation to do this.

$R_i + kR_j \to R_i$	Replace the ith row by adding k times row j to it.
kR_i	Multiply the ith row by k.
$R_i \leftrightarrow R_j$	Interchange the ith and jth rows.

3. Write the system in matrix form and solve.

$$\begin{cases} x + 2y + 3z = 1 \\ 2x - y - 3z = 1 \\ x + 3y + 4z = -2 \end{cases}$$

In matrix form, we have $\begin{bmatrix} 1 & 2 & 3 & 1 \\ 2 & -1 & -3 & 1 \\ 1 & 3 & 4 & -2 \end{bmatrix}$

$\xrightarrow[R_3 - R_1 \to R_3]{R_2 - 2R_1 \to R_2} \begin{bmatrix} 1 & 2 & 3 & 1 \\ 0 & -5 & -9 & -1 \\ 0 & 1 & 1 & -3 \end{bmatrix}$

$\xrightarrow{R_2 \leftrightarrow R_3} \begin{bmatrix} 1 & 2 & 3 & 1 \\ 0 & 1 & 1 & -3 \\ 0 & -5 & -9 & -1 \end{bmatrix}$

$\xrightarrow{R_3 + 5R_2 \to R_3} \begin{bmatrix} 1 & 2 & 3 & 1 \\ 0 & 1 & 1 & -3 \\ 0 & 0 & -4 & -16 \end{bmatrix}$

$\xrightarrow{-\frac{1}{4}R_3} \begin{bmatrix} 1 & 2 & 3 & 1 \\ 0 & 1 & 1 & -3 \\ 0 & 0 & 1 & 4 \end{bmatrix}.$

So $\underline{z = 4}$.
Back-substituting into $y + z = -3$, we have
$y + (4) = -3 \quad \Leftrightarrow \quad \underline{y = -7}.$
Back-substituting into $x + 2y + 3z = 1$, we have
$x + 2(-7) + 3(4) = 1 \quad \Leftrightarrow \quad x - 14 + 12 = 1$
$\Leftrightarrow \quad \underline{x = 3}.$

So the solution is $(3, -7, 4)$.

Check the solution to make sure it satisfies all three equations in the original system.

4. Write the system in matrix form and solve.

$$\begin{cases} x + 2y + 5z = 6 \\ -x + 2y - 5z - 4w = -2 \\ x - 2y + w = 1 \\ 2x + 10z + w = 2 \end{cases}$$

$$\left[\begin{array}{ccccc} 1 & 2 & 5 & 0 & 6 \\ -1 & 2 & -5 & -4 & -2 \\ 1 & -2 & 0 & 1 & 1 \\ 2 & 0 & 10 & 1 & 2 \end{array}\right]$$
Don't forget to insert 0 for the coefficients of missing terms.

$$\begin{array}{c} R_2+R_1 \to R_2 \\ \overline{R_3-R_1 \to R_3} \\ R_4-2R_1 \to R_4 \end{array} \left[\begin{array}{ccccc} 1 & 2 & 5 & 0 & 6 \\ 0 & 4 & 0 & -4 & 4 \\ 0 & -4 & -5 & 1 & -5 \\ 0 & -4 & 0 & 1 & -10 \end{array}\right]$$

It is often easier to use an entry when it is 1 to eliminate the entries in the column below it. In this case, we would like to create a 1 in the second row, second column.

$$\begin{array}{c} \frac{1}{4}R_2 \\ \longrightarrow \end{array} \left[\begin{array}{ccccc} 1 & 2 & 5 & 0 & 6 \\ 0 & 1 & 0 & -1 & 1 \\ 0 & -4 & -5 & 1 & -5 \\ 0 & -4 & 0 & 1 & -10 \end{array}\right]$$

$$\begin{array}{c} R_3+4R_2 \to R_3 \\ \overline{R_4+4R_2 \to R_4} \end{array} \left[\begin{array}{ccccc} 1 & 2 & 5 & 0 & 6 \\ 0 & 1 & 0 & -1 & 1 \\ 0 & 0 & -5 & -3 & -1 \\ 0 & 0 & 0 & -3 & -6 \end{array}\right]$$

$$\begin{array}{c} -\frac{1}{3}R_4 \\ \longrightarrow \end{array} \left[\begin{array}{ccccc} 1 & 2 & 5 & 0 & 6 \\ 0 & 1 & 0 & -1 & 1 \\ 0 & 0 & -5 & -3 & -1 \\ 0 & 0 & 0 & 1 & 2 \end{array}\right]$$

So $\underline{w = 2}$.

Back-substituting into $-5z - 3w = -1$, we have
$-5z - 3(2) = -1 \quad \Leftrightarrow \quad \underline{z = -1}$.

Back-substituting into $y - w = 1$, we have
$y - (2) = 1 \quad \Leftrightarrow \quad \underline{y = 3}$.

Back-substituting into $x + 2y + 5z = 6$, we have
$x + 2(3) + 5(-1) = 6 \Leftrightarrow \quad \underline{x = 5}$.

So the solution is $(5, 3, -1, 2)$.

Check the solution to make sure it satisfies all four equations in the original system.

C. In general, to solve a system of equations using a matrix, we use elementary row operations to arrive at a matrix in a certain form. This form is called **echelon form**.

> A matrix is in echelon form if it satisfies the following conditions.
>
> 1. The first nonzero number in each row (reading from left to right) is 1. This is called the **leading entry**.
>
> 2. The leading entry in each row is to the right of the of the leading entry in the row immediately above it.
>
> 3. All rows consisting entirely of zeros are at the bottom of the matrix.
>
> A matrix in echelon form is in **reduced echelon form** if it also satisfies the following condition.
>
> 4. Every entry above and below each leading entry is a 0.

The systemic technique of using elementary row operation to arrive at a matrix in echelon form is called **Gaussian elimination**, and the process that results in a matrix in reduced echelon form is called **Gauss-Jordan elimination**.

5. Determine whether the matrix is in echelon form or reduced echelon form. Then write the system of equations that correspond to the matrix.

(a) $\begin{bmatrix} 1 & 0 & 3 & 1 \\ 0 & 1 & 2 & 3 \\ 0 & 0 & 1 & 4 \end{bmatrix}$

This matrix is in echelon form. But the matrix is NOT in reduced echelon form. The entries in the third column above the leading 1 of the third row are not zero.

This matrix corresponds to the system
$$\begin{cases} x + 3z = 1 \\ y + 2z = 3 \\ z = 4 \end{cases}.$$

(b) $\begin{bmatrix} 1 & 4 & 0 & 0 & 7 \\ 0 & 0 & 1 & 0 & -2 \\ 0 & 0 & 0 & 1 & 6 \end{bmatrix}$

This matrix is in reduced echelon form. The 4 in the first row, second column does not violate any of the conditions, since the second column does not contain the leading entry of a row.

This matrix corresponds to the system
$$\begin{cases} x + 4y = 7 \\ z = -2 \\ w = 6 \end{cases}.$$

6. Find the unique solution using Gauss-Jordan elimination.
$$\begin{cases} x + 3y = 2 \\ 2x + 5y = 7 \end{cases}$$

$$\begin{cases} x + 3y = 2 \\ 2x + 5y = 7 \end{cases} \Leftrightarrow \begin{bmatrix} 1 & 3 & 2 \\ 2 & 5 & 7 \end{bmatrix}$$

$$\xrightarrow{R_2 - 2R_1 \to R_2} \begin{bmatrix} 1 & 3 & 2 \\ 0 & -1 & 3 \end{bmatrix} \xrightarrow{-R_2}$$

$$\begin{bmatrix} 1 & 3 & 2 \\ 0 & 1 & -3 \end{bmatrix} \xrightarrow{R_1 - 3R_2 \to R_1} \begin{bmatrix} 1 & 0 & 11 \\ 0 & 1 & -3 \end{bmatrix}$$

So $x = 11$ and $y = -3$. The solution of the system is $(11, -3)$.

7. Find the unique solution using Gauss-Jordan elimination.
$$\begin{cases} x - 3y + 3z = -2 \\ 2x + 3y - z = 1 \\ 3x - 10y + 6z = 1 \end{cases}$$

$$\begin{cases} x - 3y + 3z = -2 \\ 2x + 3y - z = 1 \\ 3x - 10y + 6z = 1 \end{cases} \Leftrightarrow \begin{bmatrix} 1 & -3 & 3 & -2 \\ 2 & 3 & -1 & 1 \\ 3 & -10 & 6 & 1 \end{bmatrix}$$

$$\xrightarrow[R_3 - 3R_1 \to R_3]{R_2 - 2R_1 \to R_2} \begin{bmatrix} 1 & -3 & 3 & -2 \\ 0 & 9 & -7 & 5 \\ 0 & -1 & -3 & 7 \end{bmatrix}$$

$$\xrightarrow{R_2 \leftrightarrow -R_3} \begin{bmatrix} 1 & -3 & 3 & -2 \\ 0 & 1 & 3 & -7 \\ 0 & 9 & -7 & 5 \end{bmatrix}$$

$$\xrightarrow{R_3 - 9R_2 \to R_3} \begin{bmatrix} 1 & -3 & 3 & -2 \\ 0 & 1 & 3 & -7 \\ 0 & 0 & -34 & 68 \end{bmatrix}$$

$$\xrightarrow{-\frac{1}{34}R_3} \begin{bmatrix} 1 & -3 & 3 & -2 \\ 0 & 1 & 3 & -7 \\ 0 & 0 & 1 & -2 \end{bmatrix}$$

$$\xrightarrow[R_2 - 3R_3 \to R_2]{R_1 - 3R_3 \to R_1} \begin{bmatrix} 1 & -3 & 0 & 4 \\ 0 & 1 & 0 & -1 \\ 0 & 0 & 1 & -2 \end{bmatrix}$$

$$\xrightarrow{R_1 + 3R_2 \to R_1} \begin{bmatrix} 1 & 0 & 0 & 1 \\ 0 & 1 & 0 & -1 \\ 0 & 0 & 1 & -2 \end{bmatrix}$$

So the solution is $(1, -1, -2)$.

D. A system that has no solution is called an **inconsistent system**. A system is inconsistent when we try to use Gaussian elimination to get a triangular system and we end up with an equation that is always false. The false equation always has the form $0 = N$, where N is nonzero number. A system that has infinitely many solutions is called a **dependent system**. The complete solution to such a system will have one or more variables that are arbitrary, and the other variables will depend on the arbitrary one(s).

For problems 8-11, find the complete solution to each system of equations, or show that none exists.

8. $\begin{cases} x - 5y + 3z = 3 \\ 2x + 3y - z = 5 \\ 4x - 7y + 5z = 8 \end{cases}$

$$\begin{bmatrix} 1 & -5 & 3 & 3 \\ 2 & 3 & -1 & 5 \\ 4 & -7 & 5 & 8 \end{bmatrix}$$

$$\xrightarrow[R_3-4R_1 \to R_3]{R_2-2R_1 \to R_2} \begin{bmatrix} 1 & -5 & 3 & 3 \\ 0 & 13 & -7 & -1 \\ 0 & 13 & -7 & -4 \end{bmatrix}$$

$$\xrightarrow[R_3-R_2 \to R_3]{} \begin{bmatrix} 1 & -5 & 3 & 3 \\ 0 & 13 & -7 & -1 \\ 0 & 0 & 0 & -3 \end{bmatrix}$$

Since the last row corresponds to the equation $0 = -3$, which is always false, there is no solution.

9. $\begin{cases} x - 2z + w = 2 \\ x + y + 5z + w = 7 \\ 2x + y + 3z + 2w = 9 \\ 3x + y + z + 3w = 13 \end{cases}$

$$\begin{bmatrix} 1 & 0 & -2 & 1 & 2 \\ 1 & 1 & 5 & 1 & 7 \\ 2 & 1 & 3 & 2 & 9 \\ 3 & 1 & 1 & 3 & 13 \end{bmatrix}$$

$$\xrightarrow[\substack{R_3-2R_1 \to R_3 \\ R_4-3R_1 \to R_4}]{R_2-R_1 \to R_2} \begin{bmatrix} 1 & 0 & -2 & 1 & 6 \\ 0 & 1 & 7 & 0 & 5 \\ 0 & 1 & 7 & 0 & 5 \\ 0 & 1 & 7 & 0 & 7 \end{bmatrix}$$

$$\xrightarrow[R_4-R_2 \to R_4]{R_3-R_2 \to R_3} \begin{bmatrix} 1 & 0 & -2 & 1 & 6 \\ 0 & 1 & 7 & 0 & 5 \\ 0 & 0 & 0 & 0 & 0 \\ 0 & 0 & 0 & 0 & 2 \end{bmatrix}$$

Since the last row corresponds to the equation $0 = 2$, which is always false, there is no solution.

238

$$10. \quad \begin{cases} x - 2y - 2z + 2w = -3 \\ x + 2y + 2z + 4w = 5 \\ x - 6y - 6z + \quad = -11 \\ 2x - 2y - 2z + 5w = -2 \end{cases}$$

$$\begin{bmatrix} 1 & -2 & -2 & 2 & -3 \\ 1 & 2 & 2 & 4 & 5 \\ 1 & -6 & -6 & 0 & -11 \\ 2 & -2 & -2 & 5 & -2 \end{bmatrix}$$

$$\begin{array}{c} R_2-R_1 \rightarrow R_2 \\ \hline R_3-R_1 \rightarrow R_3 \\ R_4-2R_1 \rightarrow R_4 \end{array} \begin{bmatrix} 1 & -2 & -2 & 2 & -3 \\ 0 & 4 & 4 & 2 & 8 \\ 0 & -4 & -4 & -2 & -8 \\ 0 & 2 & 2 & 1 & 4 \end{bmatrix}$$

$$\xrightarrow{R_2 \leftrightarrow R_4} \begin{bmatrix} 1 & -2 & -2 & 2 & -3 \\ 0 & 2 & 2 & 1 & 4 \\ 0 & -4 & -4 & -2 & -8 \\ 0 & 4 & 4 & 2 & 8 \end{bmatrix}$$

$$\begin{array}{c} R_3+2R_2 \rightarrow R_3 \\ \hline R_4-2R_2 \rightarrow R_4 \end{array} \begin{bmatrix} 1 & -2 & -2 & 2 & -3 \\ 0 & 2 & 2 & 1 & 4 \\ 0 & 0 & 0 & 0 & 0 \\ 0 & 0 & 0 & 0 & 0 \end{bmatrix}$$

$$\xrightarrow{\frac{1}{2}R_2} \begin{bmatrix} 1 & -2 & -2 & 2 & -3 \\ 0 & 1 & 1 & \frac{1}{2} & 2 \\ 0 & 0 & 0 & 0 & 0 \\ 0 & 0 & 0 & 0 & 0 \end{bmatrix}$$

$$\xrightarrow{R_1+2R_2 \rightarrow R_1} \begin{bmatrix} 1 & 0 & 0 & 3 & 1 \\ 0 & 1 & 1 & \frac{1}{2} & 2 \\ 0 & 0 & 0 & 0 & 0 \\ 0 & 0 & 0 & 0 & 0 \end{bmatrix}$$

This system is dependent since this matrix has 2 equations in 4 variables. We solve the first equation for x in terms of w, and then we solve the second equation for y in terms of z and w.

$x + 3w = 1 \quad \Leftrightarrow \quad \underline{x = 1 - 3w}$

$y + z + \frac{1}{2}w = 2 \quad \Leftrightarrow \quad \underline{y = 2 - z - \frac{1}{2}w}$

Solution:

$(1 - 3t, 2 - s - \frac{1}{2}t, s, t)$ where s and t are any real numbers.

11. $\begin{cases} x + 2y \quad - w = 0 \\ x + y + z + w = 8 \\ 3x + 4y - z + w = -5 \\ 2x + 3y - z \quad = -6 \end{cases}$

$$\begin{bmatrix} 1 & 2 & 0 & -1 & 0 \\ 1 & 1 & 1 & 1 & 8 \\ 3 & 4 & -1 & 1 & -5 \\ 2 & 3 & -1 & 0 & -6 \end{bmatrix}$$

$\xrightarrow[\substack{R_2 - R_1 \to R_2 \\ R_3 - 3R_1 \to R_3 \\ R_4 - 2R_1 \to R_4}]{} \begin{bmatrix} 1 & 2 & 0 & -1 & 0 \\ 0 & -1 & 1 & 2 & 8 \\ 0 & -2 & -1 & 4 & -5 \\ 0 & -1 & -1 & 2 & -6 \end{bmatrix}$

$\xrightarrow{-R_2} \begin{bmatrix} 1 & 2 & 0 & -1 & 0 \\ 0 & 1 & -1 & -2 & -8 \\ 0 & -2 & -1 & 4 & -5 \\ 0 & -1 & -1 & 2 & -6 \end{bmatrix}$

$\xrightarrow[\substack{R_3 + 2R_2 \to R_3 \\ R_4 + R_2 \to R_4}]{} \begin{bmatrix} 1 & 2 & 0 & -1 & 0 \\ 0 & 1 & -1 & -2 & -8 \\ 0 & 0 & -3 & 0 & -21 \\ 0 & 0 & -2 & 0 & -14 \end{bmatrix}$

Again, since it is easier to work with a 1, we create a 1 in row 3, column 3 position.

$\xrightarrow{-\frac{1}{3}R_3} \begin{bmatrix} 1 & 2 & 0 & -1 & 0 \\ 0 & 1 & -1 & -2 & -8 \\ 0 & 0 & 1 & 0 & 7 \\ 0 & 0 & -2 & 0 & -14 \end{bmatrix}$

$\xrightarrow{R_4 + 2R_3 \to R_4} \begin{bmatrix} 1 & 2 & 0 & -1 & 0 \\ 0 & 1 & -1 & -2 & -8 \\ 0 & 0 & 1 & 0 & 7 \\ 0 & 0 & 0 & 0 & 0 \end{bmatrix}$

$\xrightarrow{R_2 + R_3 \to R_2} \begin{bmatrix} 1 & 2 & 0 & -1 & 0 \\ 0 & 1 & 0 & -2 & -1 \\ 0 & 0 & 1 & 0 & 7 \\ 0 & 0 & 0 & 0 & 0 \end{bmatrix}$

$\xrightarrow{R_1 - 2R_2 \to R_1} \begin{bmatrix} 1 & 0 & 0 & 3 & 2 \\ 0 & 1 & 0 & -2 & -1 \\ 0 & 0 & 1 & 0 & 7 \\ 0 & 0 & 0 & 0 & 0 \end{bmatrix}$

$\underline{z = 7}$

$y - 2w = -1 \quad \Leftrightarrow \quad \underline{y = -1 + 2w}$

$x + 3w = 2 \quad \Leftrightarrow \quad \underline{x = 2 - 3w}$

Solution:

$(2 - 3t, -1 + 2t, 7, t)$ where t is any real number.

Section 8.5　The Algebra of Matrices

Key Ideas

A.　Matrix terminology, addition/subtraction, and scalar multiplication.

B.　Inner product and matrix multiplication.

A.　The **dimension** of a matrix is the number of rows by the number of columns, expressed as $R \times C$ and read as 'R by C'. The numbers in a matrix are called **entries**, and are referred to by their row and column position. In the matrix A, the term in the ith-row jth-column position is call the (i, j) entry, and denoted by a_{ij}. Two matrices are **equal** if and only if they have the same dimensions and the corresponding entries are equal.

> If A and B are matrices of the same dimension and if c is any real number, then
>
> 1.　　The **sum** $A + B$ is a matrix of the same dimension as A and B whose (i, j) entry is $a_{ij} + b_{ij}$.
>
> 2.　　The **difference** $A - B$ is a matrix of the same dimension as A and B whose (i, j) entry is $a_{ij} - b_{ij}$.
>
> 3.　　The **scalar product** cA is a matrix of the same dimension as A whose (i, j) entry is ca_{ij}.

In problems 1-4, carry out the indicated matrix algebraic operation, or explain why it cannot be performed.

1.　$\begin{bmatrix} 2 & -5 \\ 0 & 7 \end{bmatrix} + \begin{bmatrix} 8 & -5 \\ -5 & 4 \end{bmatrix}$

Adding the corresponding entries, we get

$$\begin{bmatrix} 2 & -5 \\ 0 & 7 \end{bmatrix} + \begin{bmatrix} 8 & -5 \\ -5 & 4 \end{bmatrix} = \begin{bmatrix} 2+8 & -5-5 \\ 0-5 & 7+4 \end{bmatrix}$$

$$= \begin{bmatrix} 10 & -10 \\ -5 & 11 \end{bmatrix}.$$

2.　$\begin{bmatrix} 2 & 1 & 8 \\ 11 & -3 & 0 \\ 7 & 0 & 9 \\ -2 & 1 & 8 \end{bmatrix} - \begin{bmatrix} 8 & -2 & -1 \\ 0 & -5 & 4 \\ 1 & 3 & -2 \\ 1 & 0 & 3 \end{bmatrix}$

Subtract the corresponding entries, we get

$$\begin{bmatrix} 2 & 1 & 8 \\ 11 & -3 & 0 \\ 7 & 0 & 9 \\ -2 & 1 & 8 \end{bmatrix} - \begin{bmatrix} 8 & -2 & -1 \\ 0 & -5 & 4 \\ 1 & 3 & -2 \\ 1 & 0 & 3 \end{bmatrix}$$

$$= \begin{bmatrix} 2-8 & 1-(-2) & 8-(-1) \\ 11-0 & -3-(-5) & 0-4 \\ 7-1 & 0-3 & 9-(-2) \\ -2-1 & 1-0 & 8-3 \end{bmatrix}$$

$$= \begin{bmatrix} -6 & 3 & 9 \\ 11 & 2 & -4 \\ 6 & -3 & 11 \\ -3 & 1 & 5 \end{bmatrix}.$$

3. $2\begin{bmatrix} 4 & 0 & -2 & 3 & 4 \\ -3 & 2 & -6 & 0 & -5 \\ 6 & 2 & -5 & 4 & 1 \\ 3 & -2 & 1 & 0 & 3 \end{bmatrix}$

Multiply each entry by 2, to get

$2\begin{bmatrix} 4 & 0 & -2 & 3 & 4 \\ -3 & 2 & -6 & 0 & -5 \\ 6 & 2 & -5 & 4 & 1 \\ 3 & -2 & 1 & 0 & 3 \end{bmatrix}$

$= \begin{bmatrix} 2\cdot 4 & 2\cdot 0 & 2\cdot -2 & 2\cdot 3 & 2\cdot 4 \\ 2\cdot -3 & 2\cdot 2 & 2\cdot -6 & 2\cdot 0 & 2\cdot -5 \\ 2\cdot 6 & 2\cdot 2 & 2\cdot -5 & 2\cdot 4 & 2\cdot 1 \\ 2\cdot 3 & 2\cdot -2 & 2\cdot 1 & 2\cdot 0 & 2\cdot 3 \end{bmatrix}$

$= \begin{bmatrix} 8 & 0 & -4 & 6 & 8 \\ -6 & 4 & -12 & 0 & -10 \\ 12 & 4 & -10 & 8 & 2 \\ 6 & -4 & 2 & 0 & 6 \end{bmatrix}.$

4. $2\begin{bmatrix} 4 & 0 & 7 \\ 3 & 4 & -11 \\ 2 & -6 & 0 \end{bmatrix} - \begin{bmatrix} 6 & 2 \\ -5 & 4 \\ 1 & 3 \\ -2 & 1 \\ 0 & -9 \end{bmatrix}$

Although the scalar product, $2\begin{bmatrix} 4 & 0 & 7 \\ 3 & 4 & -11 \\ 2 & -6 & 0 \end{bmatrix}$
is possible, the result is a 3×3 matrix and that cannot be added to a 5×2 matrix.

B. If A is a **row matrix** (dimension $1 \times n$) and B is a **column matrix** (dimension $n \times 1$) then the **inner product**, AB is the sum of the products of the corresponding entries. That is,

$$[a_1 \quad a_2 \quad \cdots \quad a_n] \cdot \begin{bmatrix} b_1 \\ b_2 \\ \vdots \\ b_n \end{bmatrix} = a_1 b_1 + a_2 b_2 + \cdots + a_n b_n.$$

Suppose that A is an $m \times n$ matrix and B is an $n \times k$ matrix; then the **product** $C = AB$ of two matrices is an $m \times k$ matrix, where the (i, j) entry, c_{ij}, is the inner product of the ith row of A and the jth column of B. Always check the dimension of the matrices first to make sure that the product is possible and to find the dimension of the resulting matrix.

Let A, B, C, and D be matrices for which the following products are defined, then

$(AB)C = A(BC)$ Associate Property

$A(B + C) = AB + AC$ $(B + C)D = BD + CD$ Distributive Property

The matrix product is not commutative, that is, $AB \neq BA$.

Find each matrix product or explain why it cannot be performed.

5. $\begin{bmatrix} -3 & 1 \\ -4 & 5 \\ -1 & 2 \end{bmatrix} \cdot \begin{bmatrix} -7 & 2 \\ 3 & 1 \end{bmatrix}$

This is the product of a 3×2 matrix times a 2×2 matrix. The product is possible and the result is a 3×2 matrix.

$$\begin{bmatrix} -3 & 1 \\ -4 & 5 \\ -1 & 2 \end{bmatrix} \cdot \begin{bmatrix} -7 & 2 \\ 3 & 1 \end{bmatrix}$$

$$= \begin{bmatrix} -3(-7) + 1(3) & -3(2) + 1(1) \\ -4(-7) + 5(3) & -4(2) + 5(1) \\ -1(-7) + 2(3) & -1(2) + 2(1) \end{bmatrix}$$

$$= \begin{bmatrix} 21 + 3 & -6 + 1 \\ 28 + 15 & -8 + 5 \\ 7 + 6 & -2 + 2 \end{bmatrix}$$

$$= \begin{bmatrix} 24 & -5 \\ 43 & -3 \\ 13 & 0 \end{bmatrix}$$

6. $\begin{bmatrix} 2 & 1 & 0 & 4 & 2 \\ 0 & 0 & -3 & -1 & 2 \end{bmatrix} \cdot \begin{bmatrix} -7 & 2 \\ 2 & 5 \\ 0 & -2 \\ 4 & 3 \end{bmatrix}$

This is the product of a 2×5 matrix times a 4×2 matrix. Since the 5 and 4 are not equal this product is not possible.

7. $\begin{bmatrix} 0 & 9 & -3 & 2 \\ 1 & 0 & 3 & -4 \\ 5 & 0 & -1 & 2 \end{bmatrix} \cdot \begin{bmatrix} 1 & 0 \\ 2 & 1 \\ 3 & 0 \\ 0 & 5 \end{bmatrix}$

This is the product of a 3×4 matrix times a 4×2 matrix. The product is possible and the result is a 3×2 matrix.

$$\begin{bmatrix} 0 & 9 & -3 & 2 \\ 1 & 0 & 3 & -4 \\ 5 & 0 & -1 & 2 \end{bmatrix} \cdot \begin{bmatrix} 1 & 0 \\ 2 & 1 \\ 3 & 0 \\ 0 & 5 \end{bmatrix}$$

$$= \begin{bmatrix} 0+18-9+0 & 0+9-0+10 \\ 1+0+9-0 & 0+0+0-20 \\ 5+0-3+0 & 0+0-0+10 \end{bmatrix}$$

$$= \begin{bmatrix} 9 & 19 \\ 10 & -20 \\ 2 & 10 \end{bmatrix}$$

8. $\begin{bmatrix} 4 & 0 & 7 \\ 3 & 4 & -11 \\ 2 & -6 & 0 \end{bmatrix} \cdot \begin{bmatrix} 6 & 2 & -5 \\ 4 & 1 & 3 \\ -2 & 1 & 3 \end{bmatrix}$

This is the product of a 3×3 matrix times a 3×3 matrix. The product is possible and the result is a 3×3 matrix.

$$\begin{bmatrix} 4 & 0 & 7 \\ 3 & 4 & -11 \\ 2 & -6 & 0 \end{bmatrix} \cdot \begin{bmatrix} 6 & 2 & -5 \\ 4 & 1 & 3 \\ -2 & 1 & 3 \end{bmatrix}$$

$$= \begin{bmatrix} 24+0-14 & 8+0+7 & -20+0+21 \\ 18+16+22 & 6+4-11 & -15+12-33 \\ 12-24+0 & 4-6+0 & -10-18+0 \end{bmatrix}$$

$$= \begin{bmatrix} 10 & 15 & 1 \\ 56 & -1 & -36 \\ -12 & -2 & -28 \end{bmatrix}$$

9. Let $A = \begin{bmatrix} 4 & 3 & 3 \\ -1 & 0 & -1 \\ -4 & -4 & -3 \end{bmatrix}$. Find A^2 and A^3.

Since A is a 3×3 matrix, both A^2 and A^3 are possible, and the result is a 3×3 matrix.

$$A^2 = \begin{bmatrix} 4 & 3 & 3 \\ -1 & 0 & -1 \\ -4 & -4 & -3 \end{bmatrix} \cdot \begin{bmatrix} 4 & 3 & 3 \\ -1 & 0 & -1 \\ -4 & -4 & -3 \end{bmatrix}$$

$$= \begin{bmatrix} 16-3-12 & 12+0-12 & 12-3-9 \\ -4+0+4 & -3+0+4 & -3+0+3 \\ -16+4+12 & -12+0+12 & -12+4+9 \end{bmatrix}$$

$$= \begin{bmatrix} 1 & 0 & 0 \\ 0 & 1 & 0 \\ 0 & 0 & 1 \end{bmatrix}$$

$A^3 = A \cdot A^2$

$$= \begin{bmatrix} 4 & 3 & 3 \\ -1 & 0 & -1 \\ -4 & -4 & -3 \end{bmatrix} \cdot \begin{bmatrix} 1 & 0 & 0 \\ 0 & 1 & 0 \\ 0 & 0 & 1 \end{bmatrix}$$

$$= \begin{bmatrix} 4 & 3 & 3 \\ -1 & 0 & -1 \\ -4 & -4 & -3 \end{bmatrix}$$

Section 8.6 Inverses of Matrices and Matrix Equations

Key Ideas

A. The identity matrix and the inverses of 2 by 2 matrices.

B. Determinants and inverses of larger matrices.

C. Matrix equations.

A. The main diagonal of a square $n \times n$ matrix is the set of entries $a_{11}, a_{22}, a_{33}, \ldots, a_{nn}$. The **identity matrix, I_n,** is the $n \times n$ matrix where each entry of the main diagonal is 1, and all other entries are 0. Identity matrices behave like the number 1 does in the sense that $A \cdot I_n = A$ and $I_n \cdot B = B$, whenever these products are defined. If A and B are $n \times n$ matrices where $A \cdot B = B \cdot A = I_n$, then we say that B is the **inverse** of A, and we write $B = A^{-1}$. The concept of the inverse of a matrix is analogous to that of a reciprocal of a real number; however, *not all square matrices have inverses*. When A is a 2×2 matrix, the inverse of A can be calculated (if it exists) with the following formula. If

$$A = \begin{bmatrix} a & b \\ c & d \end{bmatrix}, \text{ then } A^{-1} = \frac{1}{ad - bc} \begin{bmatrix} d & -b \\ -c & a \end{bmatrix}, \text{ provided } ad - bc \neq 0. \text{ The quantity, } ad - bc \text{ is}$$

called the **determinant** of the matrix A. Matrix A has an inverse if, and only if, the determinant of A does not equal 0.

1. Are the following pairs of matrices inverses of each other?

(a) $\begin{bmatrix} 4 & 7 \\ 3 & 3 \end{bmatrix}$ and $\begin{bmatrix} -\frac{1}{3} & \frac{7}{9} \\ \frac{1}{3} & -\frac{4}{9} \end{bmatrix}$.

| This question asks: Does $AB = I_n$?
Check by finding the product.

Does $\begin{bmatrix} 4 & 7 \\ 3 & 3 \end{bmatrix} \cdot \begin{bmatrix} -\frac{1}{3} & \frac{7}{9} \\ \frac{1}{3} & -\frac{4}{9} \end{bmatrix} = I_2$?

$\begin{bmatrix} 4 & 7 \\ 3 & 3 \end{bmatrix} \cdot \begin{bmatrix} -\frac{1}{3} & \frac{7}{9} \\ \frac{1}{3} & -\frac{4}{9} \end{bmatrix} = \begin{bmatrix} 1 & 0 \\ 0 & 1 \end{bmatrix}$

So, these two matrices are inverses of each other.

(b) $\begin{bmatrix} 1 & 1 & -1 \\ 1 & 2 & -1 \\ -1 & -1 & 2 \end{bmatrix}$ and $\begin{bmatrix} 3 & -1 & 1 \\ -1 & 1 & 0 \\ 1 & 0 & 1 \end{bmatrix}$.

Does $\begin{bmatrix} 1 & 1 & -1 \\ 1 & 2 & -1 \\ -1 & -1 & 2 \end{bmatrix} \cdot \begin{bmatrix} 3 & -1 & 1 \\ -1 & 1 & 0 \\ 1 & 0 & 1 \end{bmatrix} = I_3$?

$\begin{bmatrix} 1 & 1 & -1 \\ 1 & 2 & -1 \\ -1 & -1 & 2 \end{bmatrix} \cdot \begin{bmatrix} 3 & -1 & 1 \\ -1 & 1 & 0 \\ 1 & 0 & 1 \end{bmatrix}$

$= \begin{bmatrix} 1 & 0 & 0 \\ 0 & 1 & 0 \\ 0 & 0 & 1 \end{bmatrix}$

So, these two matrices are inverses of each other.

2. Find the inverse of each 2×2 matrix, if it exists.

(a) $\begin{bmatrix} 4 & 7 \\ 3 & 5 \end{bmatrix}$.

Use the formula: if $A = \begin{bmatrix} a & b \\ c & d \end{bmatrix}$, then

$$A^{-1} = \frac{1}{ad - bc} \begin{bmatrix} d & -b \\ -c & a \end{bmatrix}.$$

We first find $ad - bc = 4 \cdot 5 - 7 \cdot 3 = -1$. So A^{-1} exists. Thus,

$$A^{-1} = \frac{1}{-1} \begin{bmatrix} 5 & -7 \\ -3 & 4 \end{bmatrix} = \begin{bmatrix} -5 & 7 \\ 3 & -4 \end{bmatrix}.$$

(b) $\begin{bmatrix} 9 & 12 \\ 6 & 8 \end{bmatrix}$.

Using the same formula, as in Problem 2(a), we first find $ad - bc = 9 \cdot 8 - 12 \cdot 6 = 0$. So this matrix has no inverse.

(c) $\begin{bmatrix} 2 & 3 \\ -3 & 5 \end{bmatrix}$

Using the same formula, as in Problem 2(a), we first find $ad - bc = 2 \cdot 5 - 3 \cdot (-3) = 19$. So

$$A^{-1} = \frac{1}{19} \begin{bmatrix} 5 & -3 \\ 3 & 2 \end{bmatrix} = \begin{bmatrix} \frac{5}{19} & -\frac{3}{19} \\ \frac{3}{19} & \frac{2}{19} \end{bmatrix}.$$

B. One technique to calculate the inverse of an $n \times n$ matrix A is to augment A with I_n, the identity matrix. Then use row operations to change the left side to the identity matrix. That is, we start with $[A \mid I_n]$ and use row operations to get the matrix $[I_n \mid A^{-1}]$.

3. Find the inverse of each matrix.

(a) $A = \begin{bmatrix} 1 & 0 & 2 \\ 6 & 1 & 16 \\ 2 & 0 & 3 \end{bmatrix}$

Augment A with I_3.

$$\begin{bmatrix} 1 & 0 & 2 & | & 1 & 0 & 0 \\ 6 & 1 & 16 & | & 0 & 1 & 0 \\ 2 & 0 & 3 & | & 0 & 0 & 1 \end{bmatrix}$$

$\xrightarrow[R_3-2R_1 \to R_3]{R_2-6R_1 \to R_2}$ $\begin{bmatrix} 1 & 0 & 2 & | & 1 & 0 & 0 \\ 0 & 1 & 4 & | & -6 & 1 & 0 \\ 0 & 0 & -1 & | & -2 & 0 & 1 \end{bmatrix}$

$\xrightarrow{-R_3}$ $\begin{bmatrix} 1 & 0 & 2 & | & 1 & 0 & 0 \\ 0 & 1 & 4 & | & -6 & 1 & 0 \\ 0 & 0 & 1 & | & 2 & 0 & -1 \end{bmatrix}$

$\xrightarrow[R_2-4R_3 \to R_2]{R_1-2R_3 \to R_1}$ $\begin{bmatrix} 1 & 0 & 0 & | & -3 & 0 & 2 \\ 0 & 1 & 0 & | & -14 & 1 & 4 \\ 0 & 0 & 1 & | & 2 & 0 & -1 \end{bmatrix}$

So $A^{-1} = \begin{bmatrix} -3 & 0 & 2 \\ -14 & 1 & 4 \\ 2 & 0 & -1 \end{bmatrix}$.

Check:

$$\begin{bmatrix} 1 & 0 & 2 \\ 6 & 1 & 16 \\ 2 & 0 & 3 \end{bmatrix} \cdot \begin{bmatrix} -3 & 0 & 2 \\ -14 & 1 & 4 \\ 2 & 0 & -1 \end{bmatrix} = \begin{bmatrix} 1 & 0 & 0 \\ 0 & 1 & 0 \\ 0 & 0 & 1 \end{bmatrix}$$

(b) $B = \begin{bmatrix} 1 & 2 & 1 \\ 1 & 1 & -1 \\ 3 & 4 & -1 \end{bmatrix}$

Augment B with I_3.

$$\begin{bmatrix} 1 & 2 & 1 & | & 1 & 0 & 0 \\ 1 & 1 & -1 & | & 0 & 1 & 0 \\ 3 & 4 & -1 & | & 0 & 0 & 1 \end{bmatrix}$$

$\xrightarrow[R_3-3R_1 \to R_3]{R_2-R_1 \to R_2}$ $\begin{bmatrix} 1 & 2 & 1 & | & 1 & 0 & 0 \\ 0 & -1 & -2 & | & -1 & 1 & 0 \\ 0 & -2 & -4 & | & -3 & 0 & 1 \end{bmatrix}$

$\xrightarrow{-R_2}$ $\begin{bmatrix} 1 & 2 & 1 & | & 1 & 0 & 0 \\ 0 & 1 & 2 & | & 1 & -1 & 0 \\ 0 & -2 & -4 & | & -3 & 0 & 1 \end{bmatrix}$

$\xrightarrow{R_3+2R_2 \to R_3}$ $\begin{bmatrix} 1 & 2 & 1 & | & 1 & 0 & 0 \\ 0 & 1 & 2 & | & 1 & -1 & 0 \\ 0 & 0 & 0 & | & -1 & -2 & 1 \end{bmatrix}$

Since the last row of the left half of the matrix is all 0, matrix B has no inverse.

$\begin{bmatrix} 1 & 2 & 1 \\ 0 & 1 & 2 \\ 0 & 0 & 0 \end{bmatrix}$

(c) $C = \begin{bmatrix} 7 & -3 & -3 \\ -1 & 1 & 0 \\ -1 & 0 & 1 \end{bmatrix}$

Augment C with I_3.

$$\begin{bmatrix} 7 & -3 & -3 & | & 1 & 0 & 0 \\ -1 & 1 & 0 & | & 0 & 1 & 0 \\ -1 & 0 & 1 & | & 0 & 0 & 1 \end{bmatrix}$$

$\xrightarrow{R_1 \leftrightarrow -R_2}$
$\begin{bmatrix} 1 & -1 & 0 & | & 0 & -1 & 0 \\ 7 & -3 & -3 & | & 1 & 0 & 0 \\ -1 & 0 & 1 & | & 0 & 0 & 1 \end{bmatrix}$

$\xrightarrow[R_3 + R_1 \to R_3]{R_2 - 7R_1 \to R_2}$
$\begin{bmatrix} 1 & -1 & 0 & | & 0 & -1 & 0 \\ 0 & 4 & -3 & | & 1 & 7 & 0 \\ 0 & -1 & 1 & | & 0 & -1 & 1 \end{bmatrix}$

$\xrightarrow{R_2 \leftrightarrow -R_3}$
$\begin{bmatrix} 1 & -1 & 0 & | & 0 & -1 & 0 \\ 0 & 1 & -1 & | & 0 & 1 & -1 \\ 0 & 4 & -3 & | & 1 & 7 & 0 \end{bmatrix}$

$\xrightarrow{R_3 - 4R_2 \to R_3}$
$\begin{bmatrix} 1 & -1 & 0 & | & 0 & -1 & 0 \\ 0 & 1 & -1 & | & 0 & 1 & -1 \\ 0 & 0 & 1 & | & 1 & 3 & 4 \end{bmatrix}$

$\xrightarrow{R_2 + R_3 \to R_2}$
$\begin{bmatrix} 1 & -1 & 0 & | & 0 & -1 & 0 \\ 0 & 1 & 0 & | & 1 & 4 & 3 \\ 0 & 0 & 1 & | & 1 & 3 & 4 \end{bmatrix}$

$\xrightarrow{R_1 + R_2 \to R_1}$
$\begin{bmatrix} 1 & 0 & 0 & | & 1 & 3 & 3 \\ 0 & 1 & 0 & | & 1 & 4 & 3 \\ 0 & 0 & 1 & | & 1 & 3 & 4 \end{bmatrix}$

So $C^{-1} = \begin{bmatrix} 1 & 3 & 3 \\ 1 & 4 & 3 \\ 1 & 3 & 4 \end{bmatrix}$.

Check:

$$\begin{bmatrix} 7 & -3 & -3 \\ -1 & 1 & 0 \\ -1 & 0 & 1 \end{bmatrix} \cdot \begin{bmatrix} 1 & 3 & 3 \\ 1 & 4 & 3 \\ 1 & 3 & 4 \end{bmatrix} = \begin{bmatrix} 1 & 0 & 0 \\ 0 & 1 & 0 \\ 0 & 0 & 1 \end{bmatrix}$$

C. A system of linear equations can be written as a **matrix equation**, $A \cdot X = B$, where A is the **coefficient matrix**, X is the **matrix of unknowns**, and B is the **constant matrix**. The solution to the matrix equation is found by multiplying both sides by A^{-1}, provided the inverse exists. Then

$$A \cdot X = B$$
$$A^{-1}(A \cdot X) = A^{-1}B$$
$$(A^{-1}A) \cdot X = A^{-1}B$$
$$I_n \cdot X = A^{-1}B$$
$$X = A^{-1}B.$$

Remember, the matrix product is associative but it is not commutative, so the side you use to multiply by A^{-1} is important.

In problems 4-9, write the system of equations as a matrix equation. Then use the inverses you found in Problems 1, 2, and 3 to solve.

$A \cdot X = B$ can be solved by the matrix product $X = A^{-1}B$, provided A^{-1} exists.

4. $\begin{cases} 4x + 7y = 7 \\ 3x + 3y = -3 \end{cases}$

The system can be represented by the matrix equation

$$\begin{bmatrix} 4 & 7 \\ 3 & 3 \end{bmatrix} \cdot \begin{bmatrix} x \\ y \end{bmatrix} = \begin{bmatrix} 7 \\ -3 \end{bmatrix}.$$

The inverse of $\begin{bmatrix} 4 & 7 \\ 3 & 3 \end{bmatrix}$ was found in Problem 1(a),

it is $\begin{bmatrix} -\frac{1}{3} & \frac{7}{9} \\ \frac{1}{3} & -\frac{4}{9} \end{bmatrix}$. Since $X = A^{-1}B$, we have

$$\begin{bmatrix} x \\ y \end{bmatrix} = \begin{bmatrix} -\frac{1}{3} & \frac{7}{9} \\ \frac{1}{3} & -\frac{4}{9} \end{bmatrix} \cdot \begin{bmatrix} 7 \\ -3 \end{bmatrix} = \begin{bmatrix} -\frac{1}{3}(7) + \frac{7}{9}(-3) \\ \frac{1}{3}(7) - \frac{4}{9}(-3) \end{bmatrix}$$

$$= \begin{bmatrix} -\frac{14}{3} \\ \frac{11}{3} \end{bmatrix}.$$

So $x = -\frac{14}{3}$ and $y = \frac{11}{3}$.

Check:
$4\left(-\frac{14}{3}\right) + 7\left(\frac{11}{3}\right) = -\frac{56}{3} + \frac{77}{3} = \frac{21}{3} = 7$ ✓
$3\left(-\frac{14}{3}\right) + 3\left(\frac{11}{3}\right) = -14 + 11 = -3$ ✓

5. $\begin{cases} 4x + 7y = 2 \\ 3x + 5y = 1 \end{cases}$

The system can be represented by the matrix equation

$$\begin{bmatrix} 4 & 7 \\ 3 & 5 \end{bmatrix} \cdot \begin{bmatrix} x \\ y \end{bmatrix} = \begin{bmatrix} 2 \\ 1 \end{bmatrix}.$$

The inverse of $\begin{bmatrix} -5 & 7 \\ 3 & -4 \end{bmatrix}$ was found in

Problem 2(a).

$$X = A^{-1}B$$

$$\begin{bmatrix} x \\ y \end{bmatrix} = \begin{bmatrix} -5 & 7 \\ 3 & -4 \end{bmatrix} \cdot \begin{bmatrix} 2 \\ 1 \end{bmatrix} = \begin{bmatrix} -3 \\ 2 \end{bmatrix}$$

So $x = -3$ and $y = 2$.

Check:
$4(-3) + 7(2) = -12 + 14 = 2$ ✓
$3(-3) + 5(2) = -9 + 10 = 1$ ✓

6. $\begin{cases} 2x + 3y = -95 \\ -3x + 5y = 76 \end{cases}$

The system can be represented by the matrix equation

$$\begin{bmatrix} 2 & 3 \\ -3 & 5 \end{bmatrix} \cdot \begin{bmatrix} x \\ y \end{bmatrix} = \begin{bmatrix} -95 \\ 76 \end{bmatrix}.$$

The inverse of $\begin{bmatrix} \frac{5}{19} & -\frac{3}{19} \\ \frac{3}{19} & \frac{2}{19} \end{bmatrix}$ was found in Problem 2(c).

$$X = A^{-1}B$$

$$\begin{bmatrix} x \\ y \end{bmatrix} = \begin{bmatrix} \frac{5}{19} & -\frac{3}{19} \\ \frac{3}{19} & \frac{2}{19} \end{bmatrix} \cdot \begin{bmatrix} -95 \\ 76 \end{bmatrix} = \begin{bmatrix} -37 \\ -7 \end{bmatrix}$$

So $x = -37$ and $y = -7$.

Check:

$2(-37) + 3(-7) = -74 - 21 = -95 \ \checkmark$

$-3(-37) + 5(-7) = 111 - 35 = 76 \ \checkmark$

7. $\begin{cases} x + y - z = -3 \\ x + 2y - z = 6 \\ -x - y + 2z = 4 \end{cases}$

The system can be represented by the matrix equation

$$\begin{bmatrix} 1 & 1 & -1 \\ 1 & 2 & -1 \\ -1 & -1 & 2 \end{bmatrix} \cdot \begin{bmatrix} x \\ y \\ z \end{bmatrix} = \begin{bmatrix} -3 \\ 6 \\ 4 \end{bmatrix}.$$

The inverse of $\begin{bmatrix} 3 & -1 & 1 \\ -1 & 1 & 0 \\ 1 & 0 & 1 \end{bmatrix}$ was given in Problem 1(b).

$$X = A^{-1}B$$

$$\begin{bmatrix} x \\ y \\ z \end{bmatrix} = \begin{bmatrix} 3 & -1 & 1 \\ -1 & 1 & 0 \\ 1 & 0 & 1 \end{bmatrix} \cdot \begin{bmatrix} -3 \\ 6 \\ 4 \end{bmatrix} = \begin{bmatrix} -11 \\ 9 \\ 1 \end{bmatrix}$$

So $x = -11$, $y = 9$, and $z = 1$.

Check:

$(-11) + (9) - (1) = -11 + 9 - 1 = -3 \ \checkmark$

$(-11) + 2(9) - (1) = -11 + 18 - 1 = 6 \ \checkmark$

$-(-11) - (9) + 2(1) = 11 - 9 + 2 = 4 \ \checkmark$

8. $\begin{cases} x + 2z = 5 \\ 6x + y + 16z = 3 \\ 2x + 3z = -5 \end{cases}$

The system can be represented by the matrix equation

$$\begin{bmatrix} 1 & 0 & 2 \\ 6 & 1 & 16 \\ 2 & 0 & 3 \end{bmatrix} \cdot \begin{bmatrix} x \\ y \\ z \end{bmatrix} = \begin{bmatrix} 5 \\ 3 \\ -5 \end{bmatrix}.$$

The inverse of $\begin{bmatrix} -3 & 0 & 2 \\ -14 & 1 & 4 \\ 2 & 0 & -1 \end{bmatrix}$ was found in

Problem 3(a).

$$X = A^{-1}B$$

$$\begin{bmatrix} x \\ y \\ z \end{bmatrix} = \begin{bmatrix} -3 & 0 & 2 \\ -14 & 1 & 4 \\ 2 & 0 & -1 \end{bmatrix} \cdot \begin{bmatrix} 5 \\ 3 \\ -5 \end{bmatrix} = \begin{bmatrix} -25 \\ -87 \\ 15 \end{bmatrix}$$

So $x = -25$, $y = -87$, and $z = 15$.

Check:

$(-25) + 2(15) = -25 - 30 = 5$ ✓

$6(-25) + (-87) + 16(15) = -150 - 87 + 240$
$$= 3 \text{ ✓}$$

$2(-25) + 3(15) = -50 + 45 = -5$ ✓

9. $\begin{cases} 7x - 3y - 3z = 4 \\ -x + y = -2 \\ -x + z = 1 \end{cases}$

The system can be represented by the matrix equation

$$\begin{bmatrix} 7 & -3 & -3 \\ -1 & 1 & 0 \\ -1 & 0 & 1 \end{bmatrix} \cdot \begin{bmatrix} x \\ y \\ z \end{bmatrix} = \begin{bmatrix} 4 \\ -2 \\ 1 \end{bmatrix}.$$

The inverse of $\begin{bmatrix} 1 & 3 & 3 \\ 1 & 4 & 3 \\ 1 & 3 & 4 \end{bmatrix}$ was found in

Problem 3(c).

$$X = A^{-1}B$$

$$\begin{bmatrix} x \\ y \\ z \end{bmatrix} = \begin{bmatrix} 1 & 3 & 3 \\ 1 & 4 & 3 \\ 1 & 3 & 4 \end{bmatrix} \cdot \begin{bmatrix} 4 \\ -2 \\ 1 \end{bmatrix} = \begin{bmatrix} 1 \\ -1 \\ 2 \end{bmatrix}$$

So $x = 1$, $y = -1$, and $z = 2$.

Check:

$7(1) - 3(-1) - 3(2) = 7 + 3 - 6 = 4$ ✓

$-(1) + (-1) = -1 - 1 = -2$ ✓

$-(1) + (2) = -1 + 2 = 1$ ✓

Section 8.7   Determinants and Cramer's Rule

Key Ideas

A. Determinants, minors, and cofactors.
B. Invertibility, or "When does matrix A have an inverse?"
C. Cramer's Rule.
D. Area of a triangle.

A. The **determinant** of the square matrix A is a value denoted by the symbol $|A|$. For a 1×1 matrix, the determinant is its only entry. When A is the 2×2 matrix $\begin{bmatrix} a & b \\ c & d \end{bmatrix}$ the determinant of A is

$$|A| = \begin{vmatrix} a & b \\ c & d \end{vmatrix} = ad - bc.$$ To define the determinant for a general $n \times n$ matrix A, we need some additional terminology. The **minor**, M_{ij}, of the element a_{ij}, is the determinant of the matrix obtained by deleting the ith row and jth column of A. The **cofactor** A_{ij} associated with the element a_{ij} is $A_{ij} = (-1)^{i+j} M_{ij}$. The determinant of A is obtained by multiplying each element of the first row by its cofactor, and then adding the results. In symbols, this is

$$|A| = \begin{vmatrix} a_{11} & a_{12} & \cdots & a_{1n} \\ a_{21} & a_{22} & \cdots & a_{2n} \\ \vdots & \vdots & \ddots & \vdots \\ a_{n1} & a_{n2} & \cdots & a_{nn} \end{vmatrix} = a_{11}A_{11} + a_{12}A_{12} + \cdots + a_{1n}A_{1n}.$$

This is called **expanding the determinant by the first row**.

Hints:

1. The cofactor A_{ij} is just the minor M_{ij} multiplied by either 1 or -1, depending on whether $i + j$ is even or odd.

2. The determinant can be found by expanding by any row or column in the matrix; the result will be the same. Usually the row or column with the most zeros is used.

---

1. Find the determinant of each matrix.

(a) $\begin{bmatrix} 5 & 3 \\ 7 & -2 \end{bmatrix}$

$$\begin{vmatrix} 5 & 3 \\ 7 & -2 \end{vmatrix} = 5(-2) - 3(7) = -10 - 21 = -31$$

(b) $\begin{bmatrix} 0 & 1 & -2 \\ 4 & 0 & 3 \\ -1 & 2 & 1 \end{bmatrix}$

$$\begin{vmatrix} 0 & 1 & -2 \\ 4 & 0 & 3 \\ -1 & 2 & 1 \end{vmatrix} = (-1)^{1+1}(0)\begin{vmatrix} 0 & 3 \\ 2 & 1 \end{vmatrix} +$$

$$(-1)^{1+2}(1)\begin{vmatrix} 4 & 3 \\ -1 & 1 \end{vmatrix} + (-1)^{1+3}(-2)\begin{vmatrix} 4 & 0 \\ -1 & 2 \end{vmatrix}$$

$$= 0 - (4 + 3) - 2(8 - 0)$$

$$= -7 - 16 = -23$$

(c) $\begin{bmatrix} 0 & 1 & 0 & 2 \\ 1 & 0 & 1 & 0 \\ 4 & 0 & 3 & 2 \\ -3 & 1 & 2 & 1 \end{bmatrix}$

Expanding by the first row,

$$\begin{vmatrix} 0 & 1 & 0 & 2 \\ 1 & 0 & 1 & 0 \\ 4 & 0 & 3 & 2 \\ -3 & 1 & 2 & 1 \end{vmatrix} = 0 - (1)\begin{vmatrix} 1 & 1 & 0 \\ 4 & 3 & 2 \\ -3 & 2 & 1 \end{vmatrix} + 0$$

$$- (2)\begin{vmatrix} 1 & 0 & 1 \\ 4 & 0 & 3 \\ -3 & 1 & 2 \end{vmatrix}$$

$$= -1 \cdot (-11) - 2 \cdot (1) = 9,$$

since

$$\begin{vmatrix} 1 & 1 & 0 \\ 4 & 3 & 2 \\ -3 & 2 & 1 \end{vmatrix} = 1\begin{vmatrix} 3 & 2 \\ 2 & 1 \end{vmatrix} - 1\begin{vmatrix} 4 & 2 \\ -3 & 1 \end{vmatrix}$$

$$= (3 - 4) - (4 + 6) = -11,$$

and since

$$\begin{vmatrix} 1 & 0 & 1 \\ 4 & 0 & 3 \\ -3 & 1 & 2 \end{vmatrix} = 1\begin{vmatrix} 0 & 3 \\ 1 & 2 \end{vmatrix} - 0 + 1\begin{vmatrix} 4 & 0 \\ -3 & 1 \end{vmatrix}$$

$$= (0 - 3) + (4 - 0) = 1.$$

B. A square matrix with a nonzero determinant is called **invertible**. A square matrix is invertible if, and only if, A^{-1} exists. So, A^{-1} exists if, and only if, $|A| \neq 0$. Also, if the matrix B is obtained from A by adding a multiple of one row to another, or a multiple of one column to another, then $|A| = |B|$.

2. Find the determinant of $A = \begin{bmatrix} 2 & 1 & 3 \\ 1 & 0 & -4 \\ -2 & 0 & 10 \end{bmatrix}$ and the determinant of the matrix obtained by replacing row three with $R_3 + 2R_2$.

Expanding by the second column (because it contains two 0), this yields

$$\begin{vmatrix} 2 & 1 & 3 \\ 1 & 0 & -4 \\ -2 & 0 & 10 \end{vmatrix} = (-1)\begin{vmatrix} 1 & -4 \\ -2 & 10 \end{vmatrix}$$

$$= -1 \cdot (10 - 8) = -2.$$

$$\begin{bmatrix} 2 & 1 & 3 \\ 1 & 0 & -4 \\ -2 & 0 & 10 \end{bmatrix} \xrightarrow{R_3 + 2R_2 \to R_3} \begin{bmatrix} 2 & 1 & 3 \\ 1 & 0 & -4 \\ 0 & 0 & 2 \end{bmatrix}$$

Expanding by the third row yields

$$\begin{vmatrix} 2 & 1 & 3 \\ 1 & 0 & -4 \\ 0 & 0 & 2 \end{vmatrix} = (-1)^{3+3}(2)\begin{vmatrix} 2 & 1 \\ 1 & 0 \end{vmatrix}$$

$$= 2 \cdot (0 - 1) = -2.$$

C. The solution of linear equations can sometimes be expressed using determinants. An n variable, n equation linear system written in matrix equation form is

$$\begin{bmatrix} a_{11} & a_{12} & \cdots & a_{1n} \\ a_{21} & a_{22} & \cdots & a_{2n} \\ \vdots & \vdots & \ddots & \vdots \\ a_{n1} & a_{n2} & \cdots & a_{nn} \end{bmatrix} \cdot \begin{bmatrix} x_1 \\ x_2 \\ \vdots \\ x_n \end{bmatrix} = \begin{bmatrix} b_1 \\ b_2 \\ \vdots \\ b_n \end{bmatrix}.$$

Let D be the coefficient matrix and let D_{xi} be the matrix obtained from D by replacing the ith column of D by the numbers $b_1, b_2, \ldots, b_n$. Cramer's Rule states that if $|D| \neq 0$, then

$$x_1 = \frac{|D_{x_1}|}{|D|}, \ x_2 = \frac{|D_{x_2}|}{|D|}, \ \ldots, \ x_n = \frac{|D_{x_n}|}{|D|}.$$

3. Use Cramer's to solve the matrix equation

$$\begin{bmatrix} 1 & 0 & 2 \\ 2 & 7 & 6 \\ 2 & -1 & 3 \end{bmatrix} \cdot \begin{bmatrix} x \\ y \\ z \end{bmatrix} = \begin{bmatrix} 0 \\ 2 \\ 3 \end{bmatrix}.$$

We find the matrices D_x, D_y, and D_z.

$$D = \begin{bmatrix} 1 & 0 & 2 \\ 2 & 7 & 6 \\ 2 & -1 & 3 \end{bmatrix} \quad D_x = \begin{bmatrix} 0 & 0 & 2 \\ 2 & 7 & 6 \\ 3 & -1 & 3 \end{bmatrix}$$

$$D_y = \begin{bmatrix} 1 & 0 & 2 \\ 2 & 2 & 6 \\ 2 & 3 & 3 \end{bmatrix} \quad D_z = \begin{bmatrix} 1 & 0 & 0 \\ 2 & 7 & 2 \\ 2 & -1 & 3 \end{bmatrix}$$

To find $|D|$ we expand by row 1 to get

$$\begin{vmatrix} 1 & 0 & 2 \\ 2 & 7 & 6 \\ 2 & -1 & 3 \end{vmatrix} = 1 \begin{vmatrix} 7 & 6 \\ -1 & 3 \end{vmatrix} - 0 + 2 \begin{vmatrix} 2 & 7 \\ 2 & -1 \end{vmatrix}$$

$$= (21 + 6) + 2(-2 - 14) = -5.$$

To find $|D_x|$ we expand by row 1 to get

$$\begin{vmatrix} 0 & 0 & 2 \\ 2 & 7 & 6 \\ 3 & -1 & 3 \end{vmatrix} = 0 - 0 + 2 \begin{vmatrix} 2 & 7 \\ 3 & -1 \end{vmatrix}$$

$$= 2(-2 - 21) = -46.$$

To find $|D_y|$ we expand by row 1 to get

$$\begin{vmatrix} 1 & 0 & 2 \\ 2 & 2 & 6 \\ 2 & 3 & 3 \end{vmatrix} = 1 \begin{vmatrix} 2 & 6 \\ 3 & 3 \end{vmatrix} - 0 + 2 \begin{vmatrix} 2 & 2 \\ 2 & 3 \end{vmatrix}$$

$$= (6 - 18) + 2(6 - 4) = -8.$$

To find $|D_z|$ we expand by row 1 to get

$$\begin{vmatrix} 1 & 0 & 0 \\ 2 & 7 & 2 \\ 2 & -1 & 3 \end{vmatrix} = 1 \begin{vmatrix} 7 & 2 \\ -1 & 3 \end{vmatrix} - 0 + 0$$

$$= 21 + 2 = 23.$$

So , $x = \dfrac{|D_x|}{|D|} = \dfrac{-46}{-5} = \dfrac{46}{5}$,

$y = \dfrac{|D_y|}{|D|} = \dfrac{-8}{-5} = \dfrac{8}{5}$, and

$z = \dfrac{|D_z|}{|D|} = \dfrac{23}{-5} = -\dfrac{23}{5}$.

D. If a triangle in the coordinate plane has vertices (a_1, b_1), (a_2, b_2), and (a_3, b_3), then its area is

$$\text{area} = \pm \frac{1}{2} \begin{vmatrix} a_1 & b_1 & 1 \\ a_2 & b_2 & 1 \\ a_3 & b_3 & 1 \end{vmatrix}.$$

4. The vertices of a triangle are $(-3, -2)$, $(-4, 7)$, and $(5, 2)$. Find the area of the triangle.

Using the formula above we have

$$\text{area} = \pm \frac{1}{2} \begin{vmatrix} -3 & -2 & 1 \\ -4 & 7 & 1 \\ 5 & 2 & 1 \end{vmatrix}.$$

Expanding by the third column we have

$$= \pm \frac{1}{2} \left[1 \begin{vmatrix} -4 & 7 \\ 5 & 2 \end{vmatrix} - 1 \begin{vmatrix} -3 & -2 \\ 5 & 2 \end{vmatrix} \right.$$

$$\left. + 1 \begin{vmatrix} -3 & -2 \\ -4 & 7 \end{vmatrix} \right]$$

$$= \pm \frac{1}{2} [(-8 - 35) - (-6 + 10) + (-21 - 8)]$$

$$= \pm \frac{1}{2} [-43 - 4 - 29]$$

$$= \pm \frac{1}{2} [-76] = 38.$$

Section 8.8   Systems of Inequalities

Key Ideas

A. Graphing inequalities in two variables.
B. Systems of inequalities.

A. The graph of an inequality, in general, consists of a region in the plane whose boundary is the graph of the equation obtained by replacing the inequality sign by an equal sign. The points on the boundary are included when the inequality sign is $\geq$ or $\leq$, and this is indicated by a solid boundary. When the inequality sign is $>$ or $<$, the points on the boundary are excluded, and this is indicated by a broken curve. To determine which region of the plane gives the solution set of the inequality, pick a point *not on the curve*, called a **test point**, and test to see if it satisfies the inequality. If it does, then every point in the same region will satisfy the inequality. If it does not, then every point in the same region will not.

Graph the following inequalities.

1. $x + y^2 \geq 4$

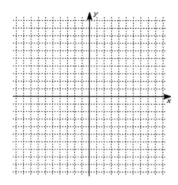

$x + y^2 \geq 4 \quad \Leftrightarrow \quad x \geq -y^2 + 4$ is a parabola, with a vertex at $(4, 0)$ opening to the left. Using the point $(0, 0)$ as a test point, we get

Test pt.	Test
$(0,0)$	$0 + (0)^2 \overset{?}{\geq} 4$ No.

Since $(0, 0)$ does not satisfy the inequality and $(0, 0)$

is inside the region, the solution lies outside of the parabola. Because $\leq$ is used, we graph the parabola with a solid line.

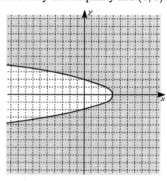

2. $4x^2 + 9y^2 < 36$

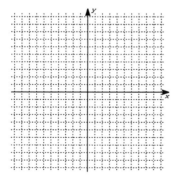

$4x^2 + 9y^2 < 36$ is an ellipse.
Using the point $(0, 0)$ as a test point, we have

Test pt.	Test
$(0,0)$	$4(0)^2 + 9(0)^2 \overset{?}{<} 36$ Yes.

Since $(0, 0)$ satisfies the inequality and $(0, 0)$ is inside the region, the solution lies inside the ellipse. Because $<$ is used, we graph the ellipse with a broken line.

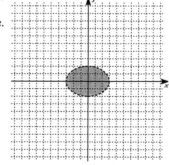

3. $xy > 4$

$xy > 4 \iff y = \dfrac{4}{x}$. We use the test points $(-5, -5)$, $(0, 0)$, and $(5, 5)$.

Test pt.	Test	
$(-5, -5)$	$(-5)(-5) \overset{?}{>} 4$	Yes.
$(0, 0)$	$(0)(0) \overset{?}{>} 4$	No.
$(5, 5)$	$(5)(5) \overset{?}{>} 4$	Yes.

Here we took more test points than we needed to, we did this to test all regions.

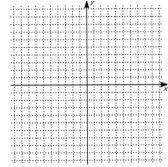

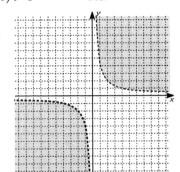

B. The solution to a system of inequalities consists of the intersection of the graphs of each inequality in the system. The **vertices** of a solution set are points where two (or more) inequalities intersect. Vertices may or may not be part of the solution, but they show where the corners of the solution set lie. An inequality is **linear** if the equality is the equation of a straight line, that is, if it is one of the forms

$$ax + by \geq c \qquad ax + by \leq c \qquad ax + by > c \qquad ax + by < c.$$

4. Graph the solution set of the system of inequalities.
$$\begin{cases} x + y < 3 \\ 3x + 2y \geq 5 \end{cases}$$

Graph $x + y = 3$ with a broken line and graph $3x + 2y = 5$ with a solid line. The vertex is obtained by solving the system of equations $\begin{cases} x + y = 3 \\ 3x + 2y = 5 \end{cases}$.
We solve the system by substitution. Solving the first equation for y gives $y = 3 - x$, and substitution into the second equation gives $3x + 2(3 - x) = 5$ $\iff$ $3x + 6 - 2x = 5$ $\iff$ $x = -1$. Thus $y = 3 - (-1) = 4$, and the vertex is at $(-1, 4)$. We use the test point $(0, 0)$.

Test pt.	Test	
$(0, 0)$	$(0) + (0) \overset{?}{<} 3$	Yes.
$(0, 0)$	$3(0) + 2(0) \overset{?}{\geq} 5$	No.

So $(0, 0)$ satisfies the $x + y < 3$, but does not satisfy $3x + 2y \geq 5$.

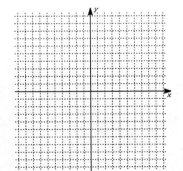

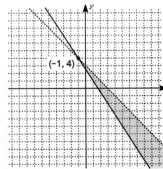

(-1, 4)

5. Graph the solution set of the system of inequalities.
$$\begin{cases} x^2 + 4x + y \le -6 \\ x + y > -6 \end{cases}$$

Graph $x^2 + 4x + y = -6$ with a solid line and graph $x + y = -6$ with a broken line. We solve the system by subtracting the second equation from the first. So

$$\begin{aligned} x^2 + 4x + y &= -6 \\ -x - y &= 6 \\ \hline x^2 + 3x &= 0 \end{aligned} \quad \Leftrightarrow \quad x(x+3) = 0 \quad \Leftrightarrow$$

$x = 0$ and $x = -3$. If $x = 0$, then $y = -6$, and if $x = -3$, then $y = -3$. So the vertices are at $(0, -6)$ and $(-3, -3)$. We use $(0, 0)$ as a test point.

Test pt.	Test	
$(0,0)$	$(0)^2 + 4(0) + (0) \overset{?}{\le} -6$	No.
$(0,0)$	$(0) + (0) \overset{?}{>} -6$	Yes.

The test point $(0, 0)$ does not satisfy the inequality $x^2 + 4x + y \le -6$, but does satisfy $x + y > -6$.

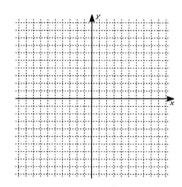

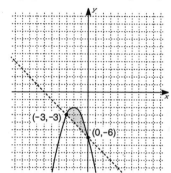

6. Graph the solution set of the system of inequalities.
$$\begin{cases} 3x + y > 9 \\ 2x + y \ge 8 \\ -x + 3y \ge 3 \end{cases}$$

Graph $2x + y = 8$ and $-x + 3y = 3$ with a solid line and $3x + y = 9$ with a broken line. We solve the system. We use $(0, 0)$ as a test point.

Test pt.	Test	
$(0,0)$	$3(0) + (0) \overset{?}{>} 9$	No.
$(0,0)$	$2(0) + (0) \overset{?}{\ge} 8$	No.
$(0,0)$	$-(0) + 3(0) \overset{?}{\ge} 3$	No.

So $(0, 0)$ does not satisfy any of the inequalities. Thus the solution to each inequality does not contain the origin. Next, we find the vertices of the region. The equations $3x + y = 9$ and $2x + y = 8$ intersect at $(1, 6)$. The equations $2x + y = 8$ and $-x + 3y = 3$ intersect at $(3, 2)$.

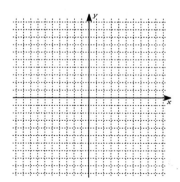

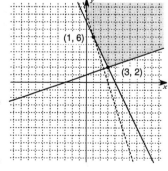

Section 8.9   Partial Fractions

Key Ideas

A. Concepts of partial fractions, four types of denominator factors.

B. The steps of partial fraction decomposition.

A. Let r be the proper rational function $r(x) = \dfrac{P(x)}{Q(x)}$, where the degree of $P(x)$ is less than the degree of $Q(x)$. The goal is to express the proper rational function $r(x)$ as a sum of simpler functions called **partial fractions**. Every polynomial with real coefficients can be factored completely into linear factors and irreducible quadratic factors, that is, into factors of the form $ax + b$ and of the form $ax^2 + bx + c$, where a, b, and c are real numbers. There are four possible types of factors.

1. Each <u>linear factor</u> that occurs *exactly once* in the denominator leads to a term of the form $\dfrac{A}{ax + b}$, where $A \neq 0$.

2. Each <u>irreducible quadratic</u> factor that occurs *exactly once* in the denominator leads to a term of the form $\dfrac{Ax + B}{ax^2 + bx + c}$, where A and B are not both zero.

3. When the linear factor is repeated k times in the denominator, then the corresponding partial fraction decomposition will contain the terms $\dfrac{A_1}{ax + b} + \dfrac{A_2}{(ax + b)^2} + \cdots + \dfrac{A_k}{(ax + b)^k}$.

4. When the irreducible quadratic factor is repeated m times in the denominator, then the corresponding partial fraction decomposition will contain the terms
$$\dfrac{A_1 x + B_1}{ax^2 + bx + c} + \dfrac{A_2 x + B_2}{(ax^2 + bx + c)^2} + \cdots + \dfrac{A_m x + B_m}{(ax^2 + bx + c)^m}.$$
The numbers k and m in types 3 and 4 above are called the **multiplicity** of the factor.

B. The technique of partial fraction decomposition can be broken down into the following steps.

> 1. If the rational function is not proper, then perform long division before applying partial fraction decomposition to the (proper) remainder.
>
> 2. <u>Factor</u> the denominator into linear and irreducible quadratic factors. Express the rational function in its partial fraction decomposition with unknown constants as discussed in part A.
>
> 3. <u>Multiply</u> both sides by the common denominator. Simplify the result by grouping like terms.
>
> 4. <u>Equate</u> corresponding coefficients. That is, the coefficients of the x^k term on either side of the equation must be equal. As a result, this leads to a system of linear equations based on these coefficients. <u>Solve</u> the resulting system of linear equations.

Find the partial fraction decomposition of the given rational function.

1. $\dfrac{-2x + 15}{x^2 - x - 12}$

Since $x^2 - x - 12 = (x - 4)(x + 3)$, we have

$\dfrac{-2x + 15}{x^2 - x - 12} = \dfrac{A}{x - 4} + \dfrac{B}{x + 3}$. Multiply both

sides by $(x + 3)(x - 4)$ gives

$-2x + 15 = A(x + 3) + B(x - 4)$
$\qquad = Ax + 3A + Bx - 4B$
$\qquad = (A + B)x + (3A - 4B)$

$\begin{cases} -2 = A + B \\ 15 = 3A - 4B \end{cases}$. To solve, we multiply the first

equation by 4 and add. This gives

$-8 = 4A + 4B$
$\underline{15 = 3A - 4B}$
$\quad 7 = 7A \qquad \Leftrightarrow \quad A = 1$. Substituting, we have

$-2 = (1) + B \quad \Leftrightarrow \quad B = -3$.

Thus $\dfrac{-2x + 15}{x^2 - x - 12} = \dfrac{1}{x - 4} + \dfrac{-3}{x + 3}$.

2. $\dfrac{3x^3 + 2x^2 - 3x - 3}{x^4 + x^3}$

Since $x^4 + x^3 = x^3(x + 1)$, the factor x has multiplicity 3. Thus, we have

$\dfrac{3x^3 + 2x^2 - 3x - 3}{x^4 + x^3} = \dfrac{A}{x} + \dfrac{B}{x^2} + \dfrac{C}{x^3} + \dfrac{D}{x + 1}$.

Multiplying both sides by $x^3(x + 1)$ gives

$3x^3 + 2x^2 - 3x - 3$
$\qquad = Ax^2(x + 1) + Bx(x + 1) + C(x + 1) + Dx^3$.

Simplifying, we have

$\qquad = Ax^3 + Ax^2 + Bx^2 + Bx + Cx + C + Dx^3$
$\qquad = (A + D)x^3 + (A + B)x^2 + (B + C)x + C$

$\begin{cases} 3 = A + D \\ 2 = A + B \\ -3 = B + C \\ -3 = C \end{cases}$. From the last equation, $C = -3$, then

back-substituting, we get $B = 0$, $A = 2$, and $D = 1$.

Thus $\dfrac{3x^3 + 2x^2 - 3x - 3}{x^4 + x^3} = \dfrac{2}{x} + \dfrac{-3}{x^3} + \dfrac{1}{x + 1}$.

3. $\dfrac{x^5}{x^4 - 5x^2 + 4}$

Since $\dfrac{x^5}{x^4 - 5x^2 + 4}$ is not a proper fraction, we must first do long division to get a proper fraction.

$$\frac{x^5}{x^4 - 5x^2 + 4} = x + \frac{5x^3 - 4x}{x^4 - 5x^2 + 4}$$

$$x^4 - 5x^2 + 4 = (x^2 - 1)(x^2 - 4)$$
$$= (x - 1)(x + 1)(x - 2)(x + 2)$$

So

$$\frac{5x^3 - 4x}{x^4 - 5x^2 + 4} = \frac{A}{x - 1} + \frac{B}{x + 1} + \frac{C}{x - 2} + \frac{D}{x + 2}.$$

Multiplying both sides by $x^4 - 5x^2 + 4$ gives

$$5x^3 - 4x = A(x + 1)(x^2 - 4) + B(x - 1)(x^2 - 4)$$
$$+ C(x + 2)(x^2 - 1) + D(x - 2)(x^2 - 1).$$

Instead of writing a system of 4 equations in 4 unknowns, we use substitution to find the values A, B, C, and D.

If $x = 1$, then the equations becomes
$$5(1)^3 - 4(1) = A(2)(-3) + 0B + 0C + 0D$$
$$\Leftrightarrow \quad 5 - 4 = -6A \quad \Leftrightarrow \quad A = -\tfrac{1}{6}.$$

If $x = -1$, then the equations becomes
$$5(-1)^3 - 4(-1) = 0A + B(-2)(-3) + 0C + 0D$$
$$\Leftrightarrow \quad -5 + 4 = 6B \quad \Leftrightarrow \quad B = -\tfrac{1}{6}.$$

If $x = 2$, then the equations becomes
$$5(2)^3 - 4(2) = 0A + 0B + C(4)(3) + 0D$$
$$\Leftrightarrow \quad 40 - 8 = 12C \quad \Leftrightarrow \quad C = \tfrac{32}{12} = \tfrac{8}{3}.$$

If $x = -2$, then the equations becomes
$$5(-2)^3 - 4(-2) = 0A + 0B + 0C + D(-4)(3)$$
$$\Leftrightarrow \quad -40 + 8 = -12D \quad \Leftrightarrow \quad D = \tfrac{32}{12} = \tfrac{8}{3}.$$

Thus $\dfrac{x^5}{x^4 - 5x^2 + 4}$

$$= x + \frac{-\tfrac{1}{6}}{x - 1} + \frac{-\tfrac{1}{6}}{x + 1} + \frac{\tfrac{8}{3}}{x - 2} + \frac{\tfrac{8}{3}}{x + 2}.$$

4. $\dfrac{3x^3 - x^2 + 15x - 1}{x^4 + 8x^2 + 15}$

Factor the denominator.

$$x^4 + 8x^2 + 15 = (x^2 + 3)(x^2 + 5)$$

Since both factors are irreducible quadratics, we must have

$$\frac{3x^3 - x^2 + 15x - 1}{x^4 + 8x^2 + 15} = \frac{Ax + B}{x^2 + 3} + \frac{Cx + D}{x^2 + 5}.$$

Multiply both sides by $x^4 + 8x^2 + 15$ gives

$$3x^3 - x^2 + 15x - 1$$
$$= (Ax + B)(x^2 + 5) + (Cx + D)(x^2 + 3)$$
$$= Ax^3 + Bx^2 + 5Ax + 5B + Cx^3 + Dx^2$$
$$\qquad\qquad + 3Cx + 3D$$
$$= (A + C)x^3 + (B + D)x^2 + (5A + 3C)x$$
$$\qquad\qquad + (5B + 3D).$$

Set the coefficients equal and solve.

x^3 terms: $3 = A + C$
x^2 terms $-1 = B + D$
x terms: $15 = 5A + 3C$
constants: $-1 = 5B + 3D$

Subtracting 3 times the first equation from the third equation gives

$$\begin{array}{r} 15 = 5A + 3C \\ -9 = -3A - 3C \\ \hline 6 = 2A \end{array}$$

$\Leftrightarrow$ $A = 3$. Substituting into the first equation gives $3 = (3) + C$ $\Leftrightarrow$ $C = 0$.

Subtracting 3 times the second equation from the fourth equation gives

$$\begin{array}{r} -1 = 5B + 3D \\ 3 = -3B - 3D \\ \hline 2 = 2B \end{array}$$

$\Leftrightarrow$ $B = 1$. Substituting into the second equation gives $-1 = (1) + D$ $\Leftrightarrow$ $D = -2$.

Thus

$$\frac{3x^3 - x^2 + 15x - 1}{x^4 + 8x^2 + 15} = \frac{3x + 1}{x^2 + 3} + \frac{-2}{x^2 + 5}.$$

5. $\dfrac{3x^3 + 13x^2 + 18x + 14}{(x^2 + 3x + 4)(x - 1)(x + 2)}$

The denominator is already factored and $x^2 + 3x + 4$ is an irreducible quadratic. So we have

$$\frac{3x^3 + 13x^2 + 18x + 14}{(x^2 + 3x + 4)(x - 1)(x + 2)}$$
$$= \frac{Ax + B}{x^2 + 3x + 4} + \frac{C}{x - 1} + \frac{D}{x + 2}.$$

Multiplying both sides by $(x^2 + 3x + 4)(x - 1)(x + 2)$ gives

$$3x^3 + 13x^2 + 18x + 14$$
$$= (Ax + B)(x - 1)(x + 2)$$
$$+ C(x^2 + 3x + 4)(x + 2)$$
$$+ D(x^2 + 3x + 4)(x - 1).$$

We use substitution to find the values of A, B, C, and D.

If $x = 1$, then the LHS of the equation gives $3(1)^3 + 13(1)^2 + 18(1) + 14 = 48$, and the RHS of the equation gives $(A + B)(0)(3) + C(8)(3) + D(8)(0) = 24C$.
So $48 = 24C \iff C = 2$.

If $x = -2$, then the LHS of the equation gives $3(-2)^3 + 13(-2)^2 + 18(-2) + 14 = 6$, and the RHS of the equation gives $(-2A + B)(-3)(0) + C(2)(0) + D(2)(-3) = -6D$.
So $6 = -6D \iff D = -1$.

If $x = 0$, then the LHS of the equation gives $3(0)^3 + 13(0)^2 + 18(0) + 14 = 14$, and the RHS of the equation gives $(0 + B)(-1)(2) + C(4)(2) + D(4)(-1)$
$$= -2B + 8C - 4D.$$
So $14 = -2B + 8C - 4D$. Substituting the values for C and D, we have $14 = -2B + 8(2) - 4(-1)$
$\iff \quad 14 = -2B + 20 \iff B = 3$.

If $x = -1$, then the LHS of the equation gives $3(-1)^3 + 13(-1)^2 + 18(-1) + 14 = 6$, and the RHS of the equation gives $(-A + B)(-2)(1) + C(2)(1) + D(2)(-2)$
$$= 2A - 2B + 2C - 4D.$$
So $6 = 2A - 2B + 2C - 4D$. Substituting the values for B, C, and D, we have
$6 = 2A - 2(3) + 2(2) - 4(-1) \iff 6 = 2A + 2$
$\iff \quad A = 2$.

So $\dfrac{3x^3 + 13x^2 + 18x + 14}{(x^2 + 3x + 4)(x - 1)(x + 2)}$
$$= \frac{2x + 3}{x^2 + 3x + 4} + \frac{2}{x - 1} + \frac{-1}{x + 2}.$$

Chapter 9
Topics in Analytic Geometry

Section 9.1   Parabolas

Key Ideas

A. Parabola definitions.

B. Relationship between parts.

A. Graphs of equations of the form $y = ax^2 + bx + c$ are **parabolas**. A parabola is the set of points in a plane *equidistant* from a fixed point (**focus**) and a fixed line ℓ (**directrix**). The **vertex** is the point where the parabola changes direction. The parabola is symmetric about its **axis**.

B. Parabolas with focus at $F(0, p)$ and directrix $y = -p$ are given by the equation $x^2 = 4py$, their vertex is at $(0, 0)$ and they are symmetric about the y-axis. Parabolas with focus at $F(p, 0)$ and directrix $x = -p$ are given by the equation $y^2 = 4px$, their vertex is at (0, 0) and they are symmetric about the x-axis. The **latus rectum** is the line segment that runs through the focus perpendicular to the axis. Its length is called the **focal diameter** and is $|4p|$.

In problems 1 and 2, find the focus, directrix, and focal diameter for each parabola. Sketch the graph.

1. $-4x^2 = y$

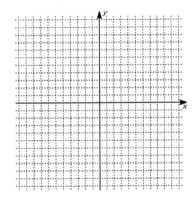

First solve for the squared term, x^2.

$$-4x^2 = y \quad \Leftrightarrow \quad x^2 = -\tfrac{1}{4}y$$

So $4p = -\tfrac{1}{4} \quad \Leftrightarrow \quad p = -\tfrac{1}{16}$.

Focus: $\left(0, -\tfrac{1}{16}\right)$.

Directrix: $y = -\left(-\tfrac{1}{16}\right)$

$\qquad = \tfrac{1}{16}$.

Focal diameter:

$|4p| = \left|4\left(-\tfrac{1}{16}\right)\right| = \tfrac{1}{4}$.

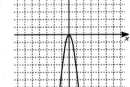

2. $20x - y^2 = 0$

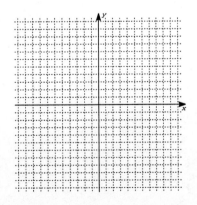

First solve for the squared term, y^2.

$$20x - y^2 = 0 \quad \Leftrightarrow \quad y^2 = 20x$$

So $4p = 20 \quad \Leftrightarrow \quad p = 5$.

Focus: $(5, 0)$.

Directrix: $x = -5$.

Focal diameter:

$|4p| = |4 \cdot 5| = 20$.

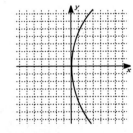

3. Find an equation for each parabola whose graph is shown.

(a)

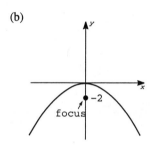

Since the directrix of this parabola is at $x = \frac{1}{2}$, we have $p = -\frac{1}{2}$.
Since this parabola opens horizontally, we use the form: $y^2 = 4px$.
Applying this form we get
$y^2 = 4\left(-\frac{1}{2}\right)x$
$y^2 = -2x$.

(b)

Since the focus of this parabola is at $(0, -2)$, we have $p = -2$.
Since this parabola opens vertically, we use the form: $x^2 = 4py$.
Applying this form we get
$x^2 = 4(-2)y$
$x^2 = -8y$.

4. Find an equation for the parabola that has its vertex at the origin and satisfies the given condition(s).

(a) Focus $F(0, -5)$.

Since its vertex is the origin and the focus is below the x-axis, this parabola opens downward and we seek a parabola of the form: $x^2 = 4py$.
Since $p = -5$, we have
$x^2 = 4(-5)y$
$x^2 = -20y$.

(b) Directrix $x = -7$.

Since the directrix is the vertical line $x = -7$, this parabola is of the form $y^2 = 4px$.
The directrix of this type of parabola is at $x = -p$,
so $-p = -7 \quad \Leftrightarrow \quad p = 7$.
Thus the equation is
$y^2 = 4(7)x \quad \Leftrightarrow \quad y^2 = 28x$.

Section 9.2   Ellipses

Key Ideas

A. Ellipses and their parts.

A. An **ellipse** is the set of points in the plane the sum of whose distance from two fixed points (**foci**) is a constant.

Ellipse	Horizontal Major Axis	Vertical Major Axis
Equation	$\dfrac{x^2}{a^2} + \dfrac{y^2}{b^2} = 1$	$\dfrac{x^2}{b^2} + \dfrac{y^2}{a^2} = 1$
Characterized	$a > b$	$a > b$
Vertices	$(\pm a, 0)$	$(0, \pm a)$
Major axes	$(-a, 0)$ to $(a, 0)$	$(0, -a)$ to $(0, a)$
Length of major axis	$2a$	$2a$
Minor axes	$(0, -b)$ to $(0, b)$	$(-b, 0)$ to $(b, 0)$
Length of minor axis	$2b$	$2b$
Foci	$(\pm c, 0)$	$(0, \pm c)$
	$a^2 = b^2 + c^2$	$a^2 = b^2 + c^2$
Eccentricity	$\dfrac{c}{a}$	$\dfrac{c}{a}$

In problems 1 and 2, find the vertices, foci, and eccentricity of the ellipse. Determine the lengths of the major and minor axes. Sketch the graph.

1. $\dfrac{x^2}{49} + \dfrac{y^2}{64} = 1$

First, express the ellipse in the form $\dfrac{x^2}{7^2} + \dfrac{y^2}{8^2} = 1$.

Since $8 > 7$, the major axis of this ellipse is vertical.

Vertices: $(0, -8)$ and $(0, 8)$.

Length of the major axis: $2(8) = 16$.

Length of the minor axis: $2(7) = 14$.

To find the foci and eccentricity we need to solve the equation $a^2 = b^2 + c^2$ for c.

$8^2 = 7^2 + c^2 \quad \Leftrightarrow \quad 64 = 49 + c^2 \quad \Leftrightarrow \quad c^2 = 15$

$\Rightarrow \quad c = \sqrt{15}$

Foci: $\left(0, -\sqrt{15}\right)$ and $\left(0, \sqrt{15}\right)$.

Eccentricity:

$\dfrac{\sqrt{15}}{8}$.

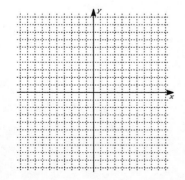

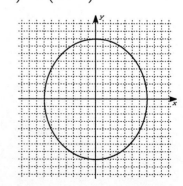

2. $4x^2 + 25y^2 = 1$

We first need to express the equation in the form

$\dfrac{x^2}{a^2} + \dfrac{y^2}{b^2} = 1$. Since $w = \dfrac{1}{\left(\frac{1}{w}\right)}$, we can express

$4x^2 + 25y^2 = \dfrac{x^2}{\left(\frac{1}{4}\right)} + \dfrac{y^2}{\left(\frac{1}{25}\right)}$. So the equation of the

ellipse can be expressed as

$$\dfrac{x^2}{\left(\frac{1}{2}\right)^2} + \dfrac{y^2}{\left(\frac{1}{5}\right)^2} = 1.$$

So $a = \frac{1}{2}$ and $b = \frac{1}{5}$. Since $\frac{1}{2} > \frac{1}{5}$, the major axis of this ellipse is horizontal.

Vertices: $\left(-\frac{1}{2}, 0\right)$ and $\left(\frac{1}{2}, 0\right)$.

Length of the major axis: $2\left(\frac{1}{2}\right) = 1$.

Length of the minor axis: $2\left(\frac{1}{5}\right) = \frac{2}{5}$.

To find the foci and eccentricity we need to solve the equation $a^2 = b^2 + c^2$ for c.

$\left(\frac{1}{2}\right)^2 = \left(\frac{1}{5}\right)^2 + c^2 \quad \Leftrightarrow \quad \frac{1}{4} = \frac{1}{25} + c^2 \quad \Leftrightarrow$

$25 = 4 + 100c^2 \quad \Leftrightarrow \quad 100c^2 = 21 \quad \Leftrightarrow$

$c^2 = \frac{21}{100} \quad \Rightarrow \quad c = \sqrt{\frac{21}{100}} = \frac{\sqrt{21}}{10}$

Foci: $\left(-\frac{\sqrt{21}}{10}, 0\right)$ and $\left(\frac{\sqrt{21}}{10}, 0\right)$.

Eccentricity: $\dfrac{\frac{\sqrt{21}}{10}}{\frac{1}{2}} = \dfrac{\sqrt{21}}{5}$.

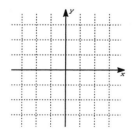

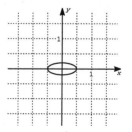

3. Find an equation for an ellipse that satisfies the given conditions.

(a) Foci $(\pm 5, 0)$, vertices $(\pm 13, 0)$.

Since the foci are on the x-axis, we use the form $\dfrac{x^2}{a^2} + \dfrac{y^2}{b^2} = 1$, we are given $a = 13$ and $c = 5$. Next we find b^2. Since $a^2 = b^2 + c^2$, substituting we have $13^2 = b^2 + 5^2 \quad \Leftrightarrow \quad 169 = b^2 + 25 \quad \Leftrightarrow$ $b^2 = 144$. Thus the equation of the ellipse is

$$\frac{x^2}{169} + \frac{y^2}{144} = 1.$$

(b) Length of major axis is 8, length of minor axis is 5, foci are on the x-axis.

Since the foci are on the x-axis, the major axis is horizontal. We are given $2a = 8$ and $2b = 5$, so $a = 4$ and $b = \frac{5}{2}$. Thus the equation of the ellipse is

$$\frac{x^2}{4^2} + \frac{y^2}{\left(\frac{5}{2}\right)^2} = 1 \quad \Leftrightarrow \quad \frac{x^2}{16} + \frac{4y^2}{25} = 1.$$

(c) Length of major axis is 34, foci $(0, \pm 8)$.

Since the foci are on the y-axis, the major axis is vertical. We are given $c = 8$ and $2a = 34$, so $a = 17$. Substituting, we have $17^2 = b^2 + 8^2 \quad \Leftrightarrow$ $289 = b^2 + 64 \quad \Leftrightarrow \quad b^2 = 225$. Thus the equation of the ellipse is

$$\frac{x^2}{225} + \frac{y^2}{289} = 1.$$

Section 9.3   Hyperbolas

Key Ideas

A. Hyperbolas and their parts.

A. A **hyperbola** is the set of points in the plane, the difference of whose distance from two fixed points (**foci**) is a constant. A hyperbola consists of two parts called **branches**. Each branch approaches the **asymptotes** $y = \pm\left(\dfrac{b}{a}\right)x$. There are two types of standard hyperbolas, those with a horizontal transverse axis and those with a vertical transverse axis. A convenient way to sketch a hyperbola is to draw a box $2a$ by $2b$ centered at the origin. The diagonals of the box are the asymptotes and the points where the sides intersect the x or y axes are the vertices.

Hyperbola	Horizontal Transverse Axis	Vertical Transverse Axis
Equation	$\dfrac{x^2}{a^2} - \dfrac{y^2}{b^2} = 1$	$\dfrac{y^2}{b^2} - \dfrac{x^2}{a^2} = 1$
Vertices	$(\pm a, 0)$	$(0, \pm b)$
Asymptotes	$y = \pm\left(\dfrac{b}{a}\right)x$	$y = \pm\left(\dfrac{b}{a}\right)x$
Foci	$(\pm c, 0)$	$(0, \pm c)$
	$c^2 = a^2 + b^2$	$c^2 = a^2 + b^2$

In problems 1 and 2 find the vertices, foci, and asymptotes of the hyperbola. Then sketch the graph.

1. $\dfrac{x^2}{64} - \dfrac{y^2}{36} = 1$

Since x^2 is the positive term in the equation $\dfrac{x^2}{64} - \dfrac{y^2}{36} = 1$, this hyperbola has horizontal transverse axis (that is, it opens horizontally).

Vertices: $(-8, 0)$ and $(8, 0)$.

To find the foci, we solve the following equation for c: $c^2 = 64 + 36 = 100 \quad \Rightarrow \quad c = 10$.

Foci: $(-10, 0)$ and $(10, 0)$.

Asymptotes: $y = \pm\frac{6}{8}x = \pm\frac{3}{4}x$.

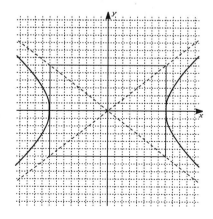

2. $4x^2 + 36 = 9y^2$

Before we can do anything, we must first express this equation in a standard form. First isolate the constant term.

$$4x^2 + 36 = 9y^2 \quad \Leftrightarrow \quad 36 = 9y^2 - 4x^2$$

Next divide by 36 to make the constant 1 and find a and b.

$$1 = \frac{y^2}{4} - \frac{x^2}{9} \quad \Leftrightarrow \quad 1 = \frac{y^2}{2^2} - \frac{x^2}{3^2}$$

This hyperbola opens vertically.

Vertices: $(0, -2)$ and $(0, 2)$.

To find the foci, we need to solve the following equation for c: $c^2 = 4 + 9 = 13 \quad \Rightarrow \quad c = \sqrt{13}$.

Foci: $(0, -\sqrt{13})$ and $(0, \sqrt{13})$.

Asymptotes: $y = \pm \frac{2}{3}x$.

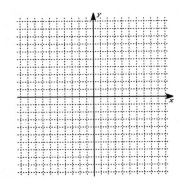

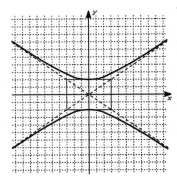

3. Find an equation for the hyperbola that satisfies the given conditions.

   (a) Foci $(\pm 15, 0)$, vertices $(\pm 9, 0)$.

This hyperbola has a horizontal transverse axis, and we are given $a = 9$ and $c = 15$. Next we find b. So

$$15^2 = 9^2 + b^2 \quad \Leftrightarrow \quad 225 = 81 + b^2 \quad \Leftrightarrow$$

$$b^2 = 144 \quad \Rightarrow \quad b = 12.$$

Thus the equation of the hyperbola is

$$\frac{x^2}{9^2} - \frac{y^2}{12^2} = 1 \quad \Leftrightarrow \quad \frac{x^2}{81} - \frac{y^2}{144} = 1.$$

   (b) Vertices $(0, \pm 10)$, asymptotes $y = \pm \frac{5}{3}x$.

This hyperbola has a vertical transverse axis, and we are given $b = 10$ and $\frac{b}{a} = \frac{5}{3}$. So $\frac{10}{a} = \frac{5}{3} \quad \Leftrightarrow$

$a = 6$.

Thus the equation of the hyperbola is

$$\frac{y^2}{10^2} - \frac{x^2}{6^2} = 1 \quad \Leftrightarrow \quad \frac{y^2}{100} - \frac{x^2}{36} = 1.$$

Section 9.4 Shifted Conics

Key Ideas

A. Equations of shifted conics, general equation of a conic.

A. When a conic is shifted h units in the x direction and k units in the y direction, then we can find the equation of the conic by replacing x with $x - h$ and y with $y - k$. The vertex of the parabola is now (h, k) and the center of the ellipse and hyperbola is (h, k). The standard forms of the equation of these conics are summarized in the table below.

Conic	Horizontal	Vertical
Parabola vertex at (h, k).	$(y - k)^2 = 4p(x - h)$	$(x - h)^2 = 4p(y - k)$
Ellipse centered at (h, k) with $a \geq b$.	$\dfrac{(x - h)^2}{a^2} + \dfrac{(y - k)^2}{b^2} = 1$	$\dfrac{(x - h)^2}{b^2} + \dfrac{(y - k)^2}{a^2} = 1$
Hyperbola centered at (h, k).	$\dfrac{(x - h)^2}{a^2} - \dfrac{(y - k)^2}{b^2} = 1$	$\dfrac{(y - k)^2}{b^2} - \dfrac{(x - h)^2}{a^2} = 1$

The general equation of a conic section is $Ax^2 + Cy^2 + Dx + Ey + F = 0$, where A and C are not both zero, representing a conic or a degenerate conic. In the nondegenerate cases, the graph is

1. a parabola if A or C is zero;

2. an ellipse if A and C have the same sign (or a circle if $A = C$);

3. a hyperbola if A and C have opposite signs.

The *completing the square* technique is used to determine the center or vertex of the conic and aid in graphing. A **degenerate conic** is a conic where the graph of the equation turns out to be just a pair of lines, a single point, or the equation may have no graph at all.

1. Find the center, foci, and vertices of the ellipse, and determine the lengths of the major and minor axes. Then sketch the graph.
$$\frac{(x - 3)^2}{9} + \frac{(y + 2)^2}{16} = 1$$

First express in standard form.
$$\frac{(x - 3)^2}{3^2} + \frac{(y + 2)^2}{4^2} = 1$$
Center: $(3, -2)$ (since $y + 2 = y - (-2)$).
Since $a = 3$ and $b = 4$, the major axis is vertical.
Vertices: $(h, k \pm b)$ which are $(3, -6)$ and $(3, 2)$.
Length of the major axis: $2b = 8$.
Length of the minor axis: $2a = 6$.
Foci: $(h, k \pm c)$; to find c we solve
$$4^2 = 3^2 + c^2 \quad \Leftrightarrow \quad 16 = 9 + c^2 \quad \Leftrightarrow \quad c^2 = 5$$
$$\Rightarrow \quad c = \sqrt{5}.$$
So foci are at
$(3, -2 - \sqrt{5})$
and
$(3, -2 + \sqrt{5})$.

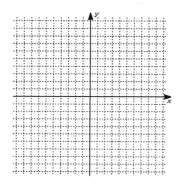

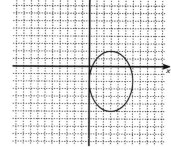

2. Find the vertex, focus, and directrix of the parabola. Then sketch the graph.

$$(x + 2)^2 = 2y + 3$$

First express in standard form.

$$(x + 2)^2 = 2\left(y + \tfrac{3}{2}\right)$$

$$[x - (-2)]^2 = 2\left[y - \left(-\tfrac{3}{2}\right)\right]$$

Thus $h = -2$ and $k = -\tfrac{3}{2}$.

Vertex: $\left(-2, -\tfrac{3}{2}\right)$.

Since $4p = 2$, we have $p = \tfrac{1}{2}$.

Focus: $(h, k + p)$, which is $(-2, -1)$.

Directrix: $y = k - p = -\tfrac{3}{2} - \tfrac{1}{2} = -2$.

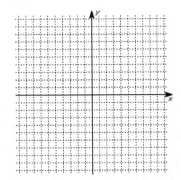

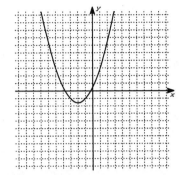

3. Find the center, foci, vertices, and asymptotes of the hyperbola. Then sketch the graph.

$$\frac{(x + 4)^2}{4} - \frac{(y - 1)^2}{9} = 1$$

First express in standard form.

$$\frac{(x + 4)^2}{2^2} - \frac{(y - 1)^2}{3^2} = 1$$

This hyperbola has a horizontal transverse axis, that is, it opens horizontally.

Center: $(-4, 1)$.

Vertices: $(h \pm a, k)$ which are $(-6, 1)$ and $(-2, 1)$.

To find the foci, we first need to find c.

$$c^2 = 2^2 + 3^2 = 4 + 9 = 13 \quad \Rightarrow \quad c = \sqrt{13}.$$

Foci: $(h \pm c, k)$, which are $(-4 - \sqrt{13}, 1)$ and $(-4 + \sqrt{13}, 1)$.

Asymptotes: $y - k = \pm\frac{b}{a}(x - h)$, so they are $y - 1 = \pm\frac{3}{2}(x + 4)$.

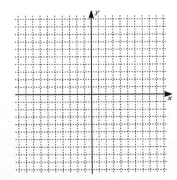

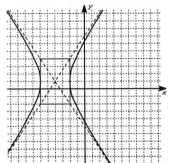

4. Complete the square to determine whether the equation represents an ellipse, a parabola, a hyperbola, or a degenerate conic.

(a) $x^2 + 4y^2 - 8x + 16y + 16 = 0$

Since the x^2 and y^2 terms have the same sign, this is the equation of an ellipse. The following steps are used in completing the square.

1. Group the x terms and y terms and move the constant term to the other side.

$$(x^2 - 8x) + (4y^2 + 16y) = -16$$

2. Factor out the coefficients of the x^2 and y^2 terms.

$$(x^2 - 8x) + 4(y^2 + 4y) = -16$$

3. Complete the square in x and in y.

$$(x^2 - 8x + \_\_) + 4(y^2 + 4y + \_\_) = -16$$
$$(x^2 - 8x + 16) + 4(y^2 + 4y + 4) =$$
$$-16 + 16 + 4(4)$$

Remember: one half the middle term squared.

$$(x - 4)^2 + 4(y + 2)^2 = 16$$

4. Finish by dividing both sides by 16.

$$\frac{(x - 4)^2}{16} + \frac{(y + 2)^2}{4} = 1$$

Ellipse, horizontal major axis, centered at $(4, -2)$, and vertices at $(0, -2)$ and $(8, -2)$.

(b) $x + 3y^2 + 4y + 7 = 0$

Since this equation contains no x^2 term, this equation is a parabola. The following steps are used in completing the square.

1. Group the y terms and move the x and constant term to the other side.

$$3y^2 + 4y = -x - 7$$

2. Factor out the coefficient of the y^2 term.

$$3\left(y^2 + \tfrac{4}{3}y\right) = -x - 7$$

3. Complete the square in y.

$$3\left(y^2 + \tfrac{4}{3}y + \_\_\right) = -x - 7$$
$$3\left(y^2 + \tfrac{4}{3}y + \tfrac{4}{9}\right) = -x - 7 + 3\left(\tfrac{4}{9}\right)$$
$$3\left(y + \tfrac{2}{3}\right)^2 = -x - \tfrac{17}{3}$$

4. Finish by dividing both sides by 3.

$$\left(y + \tfrac{2}{3}\right)^2 = -\tfrac{1}{3}\left(x + \tfrac{17}{3}\right)$$

Vertex is at $\left(-\tfrac{17}{3}, -\tfrac{2}{3}\right)$.

Section 9.5   Rotation of Axes

Key Ideas

A. General equation and rotation of axes formulas.

B. Equations of shifted conics, general equation of a conic.

A. The **general equation of a conic section** is $Ax^2 + Bxy + Cy^2 + Dx + Ey + F = 0$. To eliminate the xy-term in this equation rotate the axes through the acute angle ϕ that satisfies $\cot 2\phi = \dfrac{A - C}{B}$. This produces a new plane called the XY-plane.

> The coordinates (x, y) and (X, Y) of a point in the xy- and the XY-planes are related by the following Rotation of Axes Formulas:
>
> $$x = X \cos \phi - Y \sin \phi \qquad\qquad X = x \cos \phi + y \sin \phi$$
> $$y = X \sin \phi + Y \cos \phi \qquad\qquad Y = -x \sin \phi + y \cos \phi$$

1. If the coordinates axes are rotated through $\phi = \cos^{-1}\left(\frac{4}{5}\right)$, find the XY-coordinates of the point with xy-coordinates $(-2, 3)$.

   Using the Rotation of Axes Formulas with $x = -2$, $y = 3$, and $\phi = \cos^{-1}\left(\frac{4}{5}\right)$. So $\cos \phi = \frac{4}{5}$ and
   $$\sin \phi = \sqrt{1 - \cos^2\phi} = \sqrt{1 - \left(\frac{4}{5}\right)^2} = \frac{3}{5}.$$
   $$X = x \cos \phi + y \sin \phi = (-2)\left(\tfrac{4}{5}\right) + (3)\left(\tfrac{3}{5}\right)$$
   $$= \tfrac{-8}{5} + \tfrac{9}{5} = \tfrac{1}{5}$$
   $$Y = -x \sin \phi + y \cos \phi = -(-2)\left(\tfrac{3}{5}\right) + (3)\left(\tfrac{4}{5}\right)$$
   $$= \tfrac{6}{5} + \tfrac{12}{5} = \tfrac{18}{5}$$
   The XY-coordinates are $\left(\frac{1}{5}, \frac{18}{5}\right)$.

2. Determine the equation of the conic $2x^2 + y^2 - 4y = 4$ in XY-coordinates when the coordinate axis are rotated through $\phi = 60°$.

   Using the Rotation of Axes Formulas with $\phi = 60°$ we obtain
   $$x = X \cos 60° - Y \sin 60° = \tfrac{1}{2}X - \tfrac{\sqrt{3}}{2}Y \text{ and}$$
   $$y = X \sin 60° + Y \cos 60° = \tfrac{\sqrt{3}}{2}X + \tfrac{1}{2}Y.$$
   Substituting we get
   $$2\left(\tfrac{1}{2}X - \tfrac{\sqrt{3}}{2}Y\right)^2 + \left(\tfrac{\sqrt{3}}{2}X + \tfrac{1}{2}Y\right)^2$$
   $$- 4\left(\tfrac{\sqrt{3}}{2}X + \tfrac{1}{2}Y\right) = 4$$
   $$2\left(\tfrac{1}{4}X^2 - \tfrac{\sqrt{3}}{2}XY + \tfrac{3}{4}Y^2\right) + \tfrac{3}{4}X^2 + \tfrac{\sqrt{3}}{2}XY$$
   $$+ \tfrac{1}{4}Y^2 - 2\sqrt{3}X - 2Y = 4$$
   $$\tfrac{5}{4}X^2 - \tfrac{\sqrt{3}}{2}XY + \tfrac{7}{4}Y^2 - 2\sqrt{3}X - 2Y = 4.$$

B. In the equation $Ax^2 + Bxy + Cy^2 + Dx + Ey + F = 0$, the quantity $B^2 - 4AC$ is called the **discriminant** of the equation.

> The graph of the equation is either a conic or a degenerate conic. In the nondegenerate cases the graph is
>
> | a parabola if | an ellipse if | a hyperbola if |
> |---|---|---|
> | $B^2 - 4AC = 0$ | $B^2 - 4AC < 0$ | $B^2 - 4A > 0$ |

In problem 3 and 4 use the discriminant to determine whether the graph of the equation is a parabola, an ellipse, or a hyperbola. Use rotation of axes to eliminate the xy-term. Sketch the graph of the equation.

3. $x^2 - 2xy + y^2 - 2x + 4y = 0$

$B^2 - 4AC = (-2)^2 - 4(1)(1) = 0$, so this is the equation of a parabola. Since $\cot 2\phi = \frac{1-1}{-2} = 0$, we have $\cos 2\phi = 0 \;\Rightarrow\; 2\phi = \frac{\pi}{2} \;\Leftrightarrow\; \phi = \frac{\pi}{4}$.

Thus $\cos \phi = \frac{\sqrt{2}}{2}$ and $\sin \phi = \frac{\sqrt{2}}{2}$.

Making the substitutions we get

$$x = X\cos\phi - Y\sin\phi = \frac{\sqrt{2}}{2}X - \frac{\sqrt{2}}{2}Y$$
$$= \frac{\sqrt{2}}{2}(X - Y)$$

$$y = X\sin\phi + Y\cos\phi = \frac{\sqrt{2}}{2}X + \frac{\sqrt{2}}{2}Y$$
$$= \frac{\sqrt{2}}{2}(X + Y)$$

$$\left[\frac{\sqrt{2}}{2}(X - Y)\right]^2 - 2\left[\frac{\sqrt{2}}{2}(X - Y)\right]\left[\frac{\sqrt{2}}{2}(X + Y)\right]$$
$$+ \left[\frac{\sqrt{2}}{2}(X + Y)\right]^2 - 2\left[\frac{\sqrt{2}}{2}(X - Y)\right]$$
$$+ 4\left[\frac{\sqrt{2}}{2}(X + Y)\right] = 0$$

$$\tfrac{1}{2}(X^2 - 2XY + Y^2 - 2X^2 + 2Y^2 + X^2 + 2XY$$
$$+ Y^2) - \sqrt{2}X + \sqrt{2}Y + 2\sqrt{2}X + 2\sqrt{2}Y = 0$$

$$\tfrac{1}{2}(4Y^2) + \sqrt{2}X + 3\sqrt{2}Y = 0 \quad\Leftrightarrow$$
$$Y^2 + \frac{3\sqrt{2}}{2}Y = -\frac{\sqrt{2}}{2}Y$$

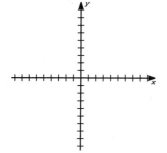

Completing the square, we get

$$\left(Y + \frac{3\sqrt{2}}{4}\right)^2 = -\frac{\sqrt{2}}{2}\left(X - \frac{9\sqrt{2}}{8}\right)$$

4. $14x^2 + 8xy - y^2 + 6 = 0$

$B^2 - 4AC = (8)^2 - 4(14)(-1) = 120.$ The graph of this equation is a hyperbola.

$\cot 2\phi = \frac{14-(-1)}{8} = \frac{15}{8}$

Sketch a right triangle with $\cot 2\phi = \frac{15}{8}.$ We see that

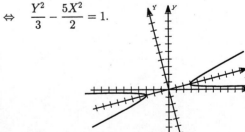

$\cos 2\phi = \frac{15}{17}.$ Using the half-angle formulas we get

$$\cos \phi = \sqrt{\frac{1 + \frac{15}{17}}{2}} = \frac{4}{\sqrt{17}} \text{ and}$$

$$\sin \phi = \sqrt{\frac{1 - \frac{15}{17}}{2}} = \frac{1}{\sqrt{17}}.$$

Making the substitutions, we get

$$x = X \cos \phi - Y \sin \phi = \frac{4}{\sqrt{17}}X - \frac{1}{\sqrt{17}}Y$$

$$= \frac{4X - Y}{\sqrt{17}} \text{ and}$$

$$y = X \sin \phi + Y \cos \phi = \frac{1}{\sqrt{17}}X + \frac{4}{\sqrt{17}}Y$$

$$= \frac{X + 4Y}{\sqrt{17}}.$$

Substituting these values we have:

$$14\left[\frac{4X - Y}{\sqrt{17}}\right]^2 + 8\left[\frac{4X - Y}{\sqrt{17}}\right]\left[\frac{X + 4Y}{\sqrt{17}}\right]$$

$$- \left[\frac{X + 4Y}{\sqrt{17}}\right]^2 + 6 = 0$$

$$\tfrac{14}{17}(16X^2 - 8XY + Y^2) + \tfrac{8}{17}(4X^2 + 15XY - 4Y^2)$$
$$- \tfrac{1}{17}(X^2 + 8XY + 16Y^2) = -6$$

$$224X^2 - 112XY + 14Y^2 + 32X^2 + 120XY$$
$$- 32Y^2 - X^2 - 8XY - 16Y^2 = -102$$

$$255X^2 - 34Y^2 = -102 \quad \Leftrightarrow \quad 15X^2 - 2Y^2 = -6$$

$$\Leftrightarrow \quad \frac{Y^2}{3} - \frac{5X^2}{2} = 1.$$

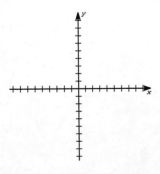

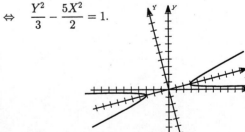

Section 9.6   Polar Coordinates

Key Ideas

A. Relationship between polar and rectangular coordinates.
B. Graphs of polar equations.
C. Symmetry and graphing polar equations with graphing devices.

A. The **polar coordinate system** uses distances and directions to specify the location of points in the plane. Choose a fixed O called the **origin** (or **pole**), then draw a ray starting at O, called the **polar axis**. This axis coincides with the x-axis in rectangular coordinates. Now let P be any point in the plane. Let r be the distance from P to the origin, and let θ be the angle from the polar axis to the segment OP. The ordered pair (r, θ) uniquely specifies the location of P and r and θ are referred to as the **polar coordinates of P**. Although the ordered pair (r, θ) specifies a point, there are infinitely many other ordered pairs that give the same point, $(r, \theta) = (r, \theta + 2n\pi)$, for n an integer and $(r, \theta) = (-r, \theta + \pi)$. When we go from rectangular to polar coordinates we *lose* unique representation of a point but we gain simplification in expressing the equations for some graphs. Polar and rectangular coordinates are related by the following formulas.

Polar to Rectangular:	$x = r\cos\theta$	$y = r\sin\theta$
Rectangular to Polar:	$\tan\theta = \dfrac{y}{x}\ (x \neq 0)$	$r^2 = x^2 + y^2$

1. Find the rectangular coordinates for the point whose polar coordinates are given.

   (a) $\left(3, \frac{\pi}{4}\right)$

   Apply the formulas:
   $$x = r\cos\theta = 3\cos\left(\tfrac{\pi}{4}\right) = 3\left(\tfrac{\sqrt{2}}{2}\right) = \tfrac{3\sqrt{2}}{2}$$
   $$y = r\sin\theta = 3\sin\left(\tfrac{\pi}{4}\right) = 3\left(\tfrac{\sqrt{2}}{2}\right) = \tfrac{3\sqrt{2}}{2}.$$
   Thus the rectangular coordinates are $\left(\tfrac{3\sqrt{2}}{2}, \tfrac{3\sqrt{2}}{2}\right)$.

   (b) $\left(-4, \frac{7\pi}{6}\right)$

   Apply the formulas:
   $$x = r\cos\theta = -4\cos\left(\tfrac{7\pi}{6}\right) = -4\left(-\tfrac{\sqrt{3}}{2}\right) = 2\sqrt{3}$$
   $$y = r\sin\theta = -4\sin\left(\tfrac{7\pi}{6}\right) = -4\left(-\tfrac{1}{2}\right) = 2.$$
   Thus the rectangular coordinates are $\left(2\sqrt{3}, 2\right)$.

2. Convert the given rectangular coordinates to polar coordinates with $r > 0$ and $0 \leq \theta < 2\pi$.

   (a) $\left(4, 4\sqrt{3}\right)$

   Apply the formulas:
   $$r = \sqrt{x^2 + y^2} = \sqrt{(4)^2 + \left(4\sqrt{3}\right)^2}$$
   $$= \sqrt{16 + 48} = \sqrt{64} = 8.$$
   Since $\left(4, 4\sqrt{3}\right)$ is in Quadrant I, we can find θ using $\tan^{-1}$ function. So
   $$\theta = \tan^{-1}\left(\tfrac{y}{x}\right) = \tan^{-1}\left(\tfrac{4\sqrt{3}}{4}\right) = \tan^{-1}\left(\sqrt{3}\right)$$
   $$= \tfrac{\pi}{3}.$$
   Thus the polar coordinates are $\left(8, \tfrac{\pi}{3}\right)$.

(b) $(-5, 12)$

Apply the formulas:
$$r = \sqrt{x^2 + y^2} = \sqrt{(-5)^2 + (12)^2}$$
$$= \sqrt{25 + 144} = \sqrt{169} = 13$$

Since $(-5, 12)$ is in Quadrant II
$$\theta = 180° + \tan^{-1}\left(\frac{y}{x}\right) = 180° + \tan^{-1}\left(\frac{12}{-5}\right)$$
$$= 180° - 67.4° = 112.6°$$

Thus the polar coordinates are $(13, 112.6°)$.
Note we used degrees in this problem because 112.6° is easier to locate than 1.97 radians.

3. Convert the equation to polar form.
$$4x^2 + y^2 = 16$$

(Recall that in rectangular coordinates this is an ellipse.)
Use the substitution $x = r \cos \theta$ and $y = r \sin \theta$
$$4(r \cos \theta)^2 + (r \sin \theta)^2 = 16$$
$$4r^2\cos^2\theta + r^2\sin^2\theta = 16.$$

Solve for r:
$$3r^2\cos^2\theta + r^2(\cos^2\theta + \sin^2\theta) = 16$$
$$3r^2\cos^2\theta + r^2 = 16$$
$$r^2(3\cos^2\theta + 1) = 16$$
$$r^2 = \frac{16}{3\cos^2\theta + 1}$$
$$r = \frac{4}{\sqrt{3\cos^2\theta + 1}}.$$

4. Convert the polar equation to rectangular form.
$$r = 2 \sin \theta - \cos \theta$$

The goal in these problems is to try to have the terms r^2, $r \cos \theta$, and $r \sin \theta$ in the equation since these terms are easily to convert to rectangular variables. In this problem, we can accomplish this by multiplying both sides by r.
$$r = 2 \sin \theta - \cos \theta$$
$$r^2 = 2r \sin \theta - r\cos \theta$$

Making the substitutions, we get
$$x^2 + y^2 = 2y - x.$$

This is the equation of a circle. Completing the square yields the center and radius.
$$x^2 + x + \tfrac{1}{4} + y^2 - 2y + 1 = \tfrac{1}{4} + 1 = \tfrac{5}{4}$$
$$\left(x + \tfrac{1}{2}\right)^2 + (y - 1)^2 = \left(\tfrac{\sqrt{5}}{2}\right)^2$$

So the center is at $\left(-\tfrac{1}{2}, 1\right)$ and it radius is $\tfrac{\sqrt{5}}{2}$.

B. The **graph of a polar equation** $r = f(\theta)$ consists of all points P that have at least one polar representation (r, θ) whose coordinates satisfy the equation. When graphing a polar equation it is important to find the intervals where r is positive and the intervals where r is negative. When r is negative, the point will be located in the direction opposite side to θ.

Several polar graphs have names. Among these graphs are:

> **limaçon** (lima bean shaped and heart shaped): $r = a \pm b \cos \theta$ or $r = a \pm b \sin \theta$
>
> **cardioid** (heart shaped, special case limaçon): $r = a(1 \pm \cos \theta)$ or $r = a(1 \pm \sin \theta)$
>
> **roses** (named because they look like flowers): $r = a \cos n\theta$
> $2n$ leaves when n is even, n leaves when n is odd

5. Sketch the curve whose polar equation is given.

   (a) $r = -4 \cos \theta$

The table below shows the value of r for some convenient values of θ.

θ	0	$\frac{\pi}{6}$	$\frac{\pi}{4}$	$\frac{\pi}{3}$
r	-4	$-2\sqrt{3}$	$-2\sqrt{2}$	-2

θ	$\frac{\pi}{2}$	$\frac{2\pi}{3}$	$\frac{3\pi}{4}$	$\frac{5\pi}{6}$
r	0	2	$2\sqrt{2}$	$2\sqrt{3}$

θ	π	$\frac{7\pi}{6}$	$\frac{5\pi}{4}$	$\frac{4\pi}{3}$
r	4	$2\sqrt{3}$	$2\sqrt{2}$	2

θ	$\frac{3\pi}{2}$	$\frac{5\pi}{3}$	$\frac{7\pi}{4}$	$\frac{11\pi}{6}$
r	0	-2	$-2\sqrt{2}$	$-2\sqrt{3}$

We can see how r changes as θ *sweeps* through the intervals created by these points.

As θ sweeps from 0 to $\frac{\pi}{2}$, r ranges from -4 to 0.
So $r < 0$.

As θ sweeps from $\frac{\pi}{2}$ to π, r ranges from 0 to 4.
So $r > 0$.

As θ sweeps from π to $\frac{3\pi}{2}$, r ranges from 4 to 0.
So $r > 0$.

As θ sweeps from $\frac{3\pi}{2}$ to 2π, r ranges from 0 to -4.
So $r < 0$.

Since $r = -4 \cos \theta \Rightarrow r^2 = -4r \cos \theta \Rightarrow x^2 + y^2 = -4x$, the equation is that of the circle $x^2 + 4x + y^2 = 0 \Leftrightarrow (x+2)^2 + y^2 = 4$.

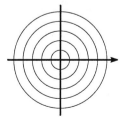

(b) $r = 1 + 3\sin\theta$

$r = 1 + 3\sin\theta = 0 \quad\Rightarrow\quad \sin\theta = -\frac{1}{3}$

Since $\sin^{-1}\left(-\frac{1}{3}\right) \approx -19.5°$, we have $r = 0$ when $\theta \approx -19.5°$ or when $\theta \approx 199.5°$.

The table below shows the value of r for some convenient values of θ.

θ	0	$\frac{\pi}{4}$	$\frac{\pi}{2}$	$\frac{3\pi}{4}$
r	1	3.12	4	3.12

θ	π	$\frac{5\pi}{4}$	$\frac{3\pi}{2}$	$\frac{7\pi}{4}$
r	1	-1.12	-2	-1.12

Thus we have $r > 0$ for $-19.5° < \theta < 199.5°$ and $r < 0$ for $199.5° < \theta < 340.5°$. This is called a limaçon with an inner loop. The graph below also shows the angles at which r changes signs.

C. When sketching the graph of a polar equation, it's often helpful to take advantage of symmetry.

> (1) A graph is **symmetric about the polar axis** if its polar equation is unchanged when we replace θ by $-\theta$.
>
> (2) A graph is **symmetric about the pole** if its polar equation is unchanged when we replace r by $-r$.
>
> (3) A graph is **symmetric about the vertical line $\theta = \dfrac{\pi}{2}$** if its polar equation is unchanged when we replace θ by $\pi - \theta$.

When graphing complicated polar equations of the form $r = f(\theta)$ using a graphing calculator or computer we need to determine the domain for θ. That is, we need to determine n so that $f(\theta + 2n\pi) = f(\theta)$.

6. Determine if the graph of the given polar equation has symmetry.

   (a) $r = -4\cos\theta$

Symmetric about the polar axis?

$-4\cos(-\theta) = -4\cos\theta = r$

Yes, symmetric about the polar axis.

Symmetric about the pole?

$-r = -(-4\cos\theta) = 4\cos\theta \neq r$

Not symmetric about the pole.

Symmetric about the vertical line $\theta = \frac{\pi}{2}$?

$-4\cos(\pi - \theta) = -4(\cos\pi\cos\theta + \sin\pi\sin\theta)$
$= 4\cos\theta \neq r$

Not symmetric about the vertical line.

See the graph in 5a.

(b) $r = 1 + 3\sin\theta$

Symmetric about the polar axis?

$1 + 3\sin(-\theta) = 1 - 3\sin\theta \neq r$

Not symmetric about the polar axis.

Symmetric about the pole?

$-r = -(1 + 3\sin\theta) = -1 - 3\sin\theta \neq r$

Not symmetric about the pole.

Symmetric about the vertical line $\theta = \frac{\pi}{2}$?

$1 + 3\sin(\pi - \theta) = 1 + 3(\sin\pi\cos\theta - \cos\pi\sin\theta)$
$= 1 + 3(0 + \sin\theta) = 1 + 3\sin\theta = r$

Yes, symmetric about the vertical line.
See the graph in 5b.

7. Determine the domain and use a graphing device to sketch.

$r = 3 - 2\sin\frac{3\theta}{4}$

To determine the domain, we must find n so that

$$\sin\left(\frac{3(\theta + 2n\pi)}{4}\right) = \sin\left(\frac{3\theta}{4}\right) \quad \Leftrightarrow$$

$$\sin\left(\frac{3\theta + 6n\pi}{4}\right) = \sin\left(\frac{3\theta}{4}\right).$$

So $\frac{6n\pi}{4} = \frac{3n\pi}{2}$ must be a multiple of 2π. This first happens when $n = 4$. Therefore we use the domain 0 to 8π. Furthermore, since $-1 \leq \sin\theta \leq 1$, we have $1 \leq r \leq 5$.

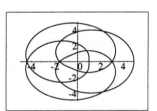

Section 9.7   Polar Equations of Conics

Key Ideas

A. Polar Equations of Conics.

A. Conics can be describe by a fixed positive number e called the **eccentricity**. For any point P on the conic, the eccentricity is the ratio of the distance from P to a fixed point F, called the **focus,** to the distance from P to a fixed line, called the **directrix**. If the focus F is placed at the origin and the directrix is parallel to the y- or x-axis and d units from the origin then we have the following.

> A polar equation of the form
> $$r = \frac{ed}{1 \pm e\cos\theta} \qquad \text{or} \qquad r = \frac{ed}{1 \pm e\sin\theta}$$
> represents a conic with eccentricity e. The conic is
> 1. a parabola if $e = 1$.
> 2. an ellipse if $e < 1$.
> 3. a hyperbola if $e > 1$.

The denominator $1 \pm e\cos\theta$ is used when the directrix is at $x = \pm d$, the negative sign for $x = -d$ and the positive sign when $x = d$. Similarly, the denominator $1 \pm e\sin\theta$ is used when the directrix is at $y = \pm d$, the negative sign for $y = -d$ and the positive sign when $y = d$.

1. Write a polar equation of a conics that has its focus at the origin and satisfies the given conditions.

(a) Ellipse, eccentricity 0.15, directrix $x = -4$.

$e = 0.15$; $d = 4$. Since the directrix is $x = -4$ we use $\cos\theta$ and a negative sign in the denominator.

Thus $r = \dfrac{(0.15)(4)}{1 - (0.15)\cos\theta} = \dfrac{0.6}{1 - 0.15\cos\theta}$.

(b) Hyperbola, eccentricity 3, directrix $y = 4$.

$e = 3$; $d = 4$. Since the directrix is $y = 4$ we use $\sin\theta$ and a positive sign in the denominator.

Thus $r = \dfrac{(3)(4)}{1 + 3\sin\theta} = \dfrac{12}{1 + 3\sin\theta}$.

(c) Parabola, directrix $x = 5$.

$e = 1$ (because it is a parabola); $d = 5$. Since the directrix is $x = 5$ we use $\cos\theta$ and a positive sign in the denominator.

Thus $r = \dfrac{(1)(5)}{1 + (1)\cos\theta} = \dfrac{5}{1 + \cos\theta}$.

In problem 2 and 3, find the eccentricity, identify the conic, and determine the directrix. Sketch the conic.

2. $r = \dfrac{4}{5 - 3\sin\theta}$

We force the leading term of the denominator to be 1 by multiplying the right hand side of the equation by $\dfrac{1/5}{1/5}$.

$$r = \frac{4}{5 - 3\sin\theta} \cdot \frac{1/5}{1/5} = \frac{\frac{4}{5}}{1 - \frac{3}{5}\sin\theta}$$

So $e = \frac{3}{5}$ and the conic is an ellipse.

Then $ed = \frac{4}{5}$ $\Leftrightarrow$ $\left(\frac{3}{5}\right)d = \frac{4}{5}$ $\Leftrightarrow$ $d = \frac{4}{3}$.

Since the sign in the denominator is negative and the trigonometric function is sine, the directrix is

$y = -\frac{4}{3}$.

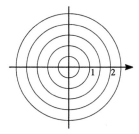

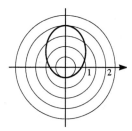

3. $r = \dfrac{2}{2 + 5\cos\theta}$

We force the leading term of the denominator to be 1 by multiplying the right hand side of the equation by $\dfrac{1/2}{1/2}$.

$$r = \frac{2}{2 + 5\cos\theta} \cdot \frac{1/2}{1/2} = \frac{1}{1 + \frac{5}{2}\cos\theta}$$

So $e = \frac{5}{2}$ and the conic is a hyperbola.

Then $ed = 1$ $\Leftrightarrow$ $\left(\frac{5}{2}\right)d = 1$ $\Leftrightarrow$ $d = \frac{2}{5}$.

Since the sign in the denominator is positive and the trigonometric function is cosine, the directrix is

$x = \frac{2}{5}$.

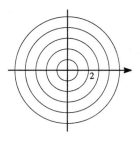

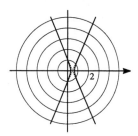

Section 9.8   Parametric Equations

Key Ideas

A. Parametric equations.

B. Using graphing devices to graph parametric curves.

A. In this method, the x- and y-coordinates of points on the curve are given separately as functions of an additional variable t, called the **parameter**:

$$x = f(t) \qquad\qquad y = g(t)$$

These are called **parametric equations** for the curve. Substituting a value of t into each equation determines the coordinates of a point (x, y). As t varies, the point $(x, y) = (f(t), g(t))$ varies and traces out the curve. All of the graphing we have done in the textbook can be done using parametric equations. Often we can *eliminate the parameter* by solving one of the equations for t and substituting this value into the other equation.

1. Find parametric equations for the line passing through $(3, 5)$ and $(5, 4)$.

Since $y = mx + b$, we can let $x = t$ and $y = mt + b$.

We find m: $m = \frac{5-4}{3-5} = -\frac{1}{2}$.

Then using the point-slope equation we have:

$$y - 5 = -\tfrac{1}{2}(x - 3) \quad\Leftrightarrow\quad y = -\tfrac{1}{2}x + \tfrac{3}{2} + 5 \quad\Leftrightarrow$$

$$y = -\tfrac{1}{2}x + \tfrac{13}{2}.$$

Thus this line can be expressed by the parametric equations: $x = t$ and $y = -\tfrac{1}{2}t + \tfrac{13}{2}$.

In problems 2 and 3, (a) sketch the curve represented by the parametric equations, and (b) find a rectangular coordinate equation for the curve by eliminating the parameter.

2. $x = t^3$ and $y = t - 1$

To sketch the graph, we generate points and plot.

t	x	y
-2	-8	-3
-1	-1	-2
0	0	-1
1	1	0
2	8	1
3	27	2

To eliminate the parameter, we solve the 2nd equation for t. So $t = y + 1$. Substituting for t we get the rectangular coordinate equation $x = (y + 1)^3$.

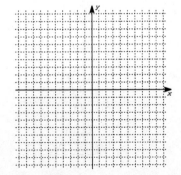

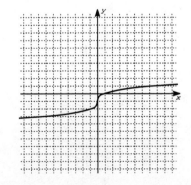

3. $x = \sin t$ and $y = 1 - 2\sin^2 t$

t	x	y
0	0	1
$\frac{\pi}{4}$	$\frac{\sqrt{2}}{2}$	0
$\frac{\pi}{2}$	1	-1
$\frac{3\pi}{4}$	$\frac{\sqrt{2}}{2}$	0
π	0	1
$\frac{5\pi}{4}$	$-\frac{\sqrt{2}}{2}$	0
$\frac{3\pi}{2}$	-1	-1
$\frac{7\pi}{4}$	$-\frac{\sqrt{2}}{2}$	0
2π	0	1

To sketch the graph, we generate points and plot.

To eliminate the parameter, we substitute x for $\sin t$ and obtain the rectangular coordinate equation : $y = 1 - 2x^2$, which is a parabola.

Since $-1 \le \sin t \le 1$ and $0 \le \sin^2 t \le 1$, we have $-1 \le x \le 1$ and $-1 \le y \le 1$. Thus this gives only a portion of the parabola.

B. Most graphing calculators and computer graphing programs can be used to graph parametric equations.

In problems 4 and 5 use a graphing device to sketch the curve represented by the given parametric curves.

4. $x = 2t\cos t, \quad y = \sin^2 t$

Since $0 \le \sin^2 t \le 1$ and x varies wildly, we set the viewing rectangle at $[-10, 10]$ by $[0, 1]$. Shown is the portion of the parametric curve where $-15 \le t \le 15$.

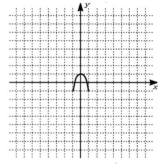

5. $x = 4\cos 3t, \quad y = 3\cos 4t$

This function has a period of 12π. *Remember 12 is the least common multiple of 3 and 4.* Since
$-1 \le \cos t \le 1$,
$-4 \le x \le 4$, and
$-3 \le y \le 3$, we use the viewing rectangle $[-4, 4]$ by $[-3, 3]$.

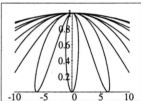

Chapter 10
Sequences and Series

Section 10.1   Sequences and Summation Notation
Key Ideas
A. General sequence.
B. Special sequences-Fibonacci sequence.
C. Partial sums.
D. Sigma notation and finite sums.
E. Finding patterns.

A.   A **sequence** is a set of numbers written in a specific order:

$$a_1, a_2, a_3, a_4, \ldots, a_n, \ldots$$

The number a_1 is called the *first term*, a_2 is the *second term*, and in general, a_n is the *nth term*. A **sequence** can also be defined as a function f whose domain is the set of positive integers. In this form the values $f(1), f(2), f(3), \ldots$ are called the **terms**, where $f(n) = a_n$ is the nth term. Some sequences have formulas that give the nth term directly; these sequences are called **explicit**. Other sequences have formulas that give the nth term based on some previous terms; these sequences are called **recursive**.

1. Find the first five terms, the 10th term, and the 100th term of the sequence defined by the following formulas.

(a) $a_n = \dfrac{n-1}{2n+1}$

$a_n = \dfrac{n-1}{2n+1}$

$n = 1 \qquad a_1 = \dfrac{1-1}{2(1)+1} = 0$

$n = 2 \qquad a_2 = \dfrac{2-1}{2(2)+1} = \frac{1}{5}$

$n = 3 \qquad a_3 = \dfrac{3-1}{2(3)+1} = \frac{2}{7}$

$n = 4 \qquad a_4 = \dfrac{4-1}{2(4)+1} = \frac{3}{9} = \frac{1}{3}$

$n = 5 \qquad a_5 = \dfrac{5-1}{2(5)+1} = \frac{4}{11}$

$n = 10 \qquad a_{10} = \dfrac{10-1}{2(10)+1} = \frac{9}{21} = \frac{3}{7}$

$n = 100 \qquad a_{100} = \dfrac{100-1}{2(100)+1} = \frac{99}{201} = \frac{33}{67}$

(b) $a_n = 2^n + (-1)^n$

$a_n = 2^n + (-1)^n$

$n = 1 \qquad a_1 = 2^{(1)} + (-1)^{(1)} = 2 - 1 = 1$

$n = 2 \qquad a_2 = 2^{(2)} + (-1)^{(2)} = 4 + 1 = 5$

$n = 3 \qquad a_3 = 2^{(3)} + (-1)^{(3)} = 8 - 1 = 7$

$n = 4 \qquad a_4 = 2^{(4)} + (-1)^{(4)} = 16 + 1 = 17$

$n = 5 \qquad a_5 = 2^{(5)} + (-1)^{(5)} = 32 - 1 = 31$

$n = 10 \qquad a_{10} = 2^{(10)} + (-1)^{(10)} = 1024 + 1$
$$= 1025$$

$n = 100 \qquad a_{100} = 2^{(100)} + (-1)^{(100)} = 2^{100} + 1$

(c) $a_n = \dfrac{2n - 3^n}{4^n}$

$a_n = \dfrac{2n - 3^n}{4^n}$

$n = 1$ $a_1 = \dfrac{2(1) - 3^1}{4^1} = \dfrac{2 - 3}{4} = -\dfrac{1}{4}$

$n = 2$ $a_2 = \dfrac{2(2) - 3^2}{4^2} = \dfrac{4 - 9}{16} = -\dfrac{5}{16}$

$n = 3$ $a_3 = \dfrac{2(3) - 3^3}{4^3} = \dfrac{6 - 27}{64} = -\dfrac{21}{64}$

$n = 4$ $a_4 = \dfrac{2(4) - 3^4}{4^4} = \dfrac{8 - 81}{256} = -\dfrac{73}{256}$

$n = 5$ $a_5 = \dfrac{2(5) - 3^5}{4^5} = \dfrac{10 - 243}{1024} = -\dfrac{233}{1024}$

$n = 10$ $a_{10} = \dfrac{2(10) - 3^{10}}{4^{10}} = \dfrac{20 - 59049}{1048576}$

$= -\dfrac{59029}{1048576}$

$n = 100$ $a_{100} = \dfrac{2(100) - 3^{100}}{4^{100}}$

B. The Fibonacci sequence is a classic example of a *recursive* sequence. The Fibonacci sequence is given by: $F_n = F_{n-1} + F_{n-2}$, where $F_1 = F_2 = 1$. That is, the current term is the sum of the two previous terms. (The Fibonacci sequence also has an explicit formula.)

2. Find the first twelve terms of the Fibonacci sequence.

$F_1 = 1$

$F_2 = 1$

$F_3 = F_2 + F_1 = 1 + 1 = 2$

$F_4 = F_3 + F_2 = 2 + 1 = 3$

$F_5 = F_4 + F_3 = 3 + 2 = 5$

$F_6 = F_5 + F_4 = 5 + 3 = 8$

$F_7 = F_6 + F_5 = 8 + 5 = 13$

$F_8 = F_7 + F_6 = 13 + 8 = 21$

$F_9 = F_8 + F_7 = 21 + 13 = 34$

$F_{10} = F_9 + F_8 = 34 + 21 = 55$

$F_{11} = F_{10} + F_9 = 55 + 34 = 89$

$F_{12} = F_{11} + F_{10} = 89 + 55 = 144$

3. Find the first five terms of the recursive sequence defined by $a_n = 2a_{n-1} + 5$, with $a_1 = 3$.

$a_1 = 3$

$a_2 = 2 \cdot a_1 + 5 = 2 \cdot 3 + 5 = 11$

$a_3 = 2 \cdot a_2 + 5 = 2 \cdot 11 + 5 = 27$

$a_4 = 2 \cdot a_3 + 5 = 2 \cdot 27 + 5 = 59$

$a_5 = 2 \cdot a_4 + 5 = 2 \cdot 59 + 5 = 123$

4. Find the first five terms of the recursive sequence defined by $b_n = 3b_{n-1} - b_{n-2}$, with $b_1 = 2$ and $b_2 = 3$.

$b_1 = 2$

$b_2 = 3$

$b_3 = 3 \cdot b_2 - b_1 = 3 \cdot 3 - 2 = 7$

$b_4 = 3 \cdot b_3 - b_2 = 3 \cdot 7 - 3 = 18$

$b_5 = 3 \cdot b_4 - b_3 = 3 \cdot 18 - 7 = 47$

C. For the sequence $a_1, a_2, a_3, \ldots, a_n, \ldots$, the **partial sums** are

$$S_1 = a_1$$
$$S_2 = a_1 + a_2$$
$$S_3 = a_1 + a_2 + a_3$$
$$S_4 = a_1 + a_2 + a_3 + a_4$$
$$\vdots$$
$$S_n = a_1 + a_2 + a_3 + a_4 + \ldots + a_n$$
$$\vdots$$

S_1 is called the **first partial sum**, S_2 the **second partial sum**, and so on. S_n is called the **nth partial sum**. The sequence $S_1, S_2, S_3, S_4, \ldots, S_n, \ldots$ is called the **sequence of partial sums**. The sequence of partial sums is useful in detecting a pattern to find the sum.

5. Find the first five partial sums of the sequence given by $a_n = 2^n + (-1)^n$.

We found the first five terms of this sequence in problem 1(b).

$S_1 = a_1 = 1$

$S_2 = a_1 + a_2 = 1 + 5 = 6$

$S_3 = a_1 + a_2 + a_3 = 1 + 5 + 7 = 13$

$S_4 = a_1 + a_2 + a_3 + a_4 = 1 + 5 + 7 + 17 = 30$

$S_5 = a_1 + a_2 + a_3 + a_4 + a_5$
$\qquad = 1 + 5 + 7 + 17 + 31 = 61$

6. Find the first five partial sums of the sequence given by $a_n = \dfrac{n-1}{2n+1}$.

We found the first five terms of this sequence in problem 1(a).

$S_1 = a_1 = 0$

$S_2 = a_1 + a_2 = 0 + \frac{1}{5} = \frac{1}{5}$

$S_3 = a_1 + a_2 + a_3 = 0 + \frac{1}{5} + \frac{2}{7} = \frac{17}{35}$

$S_4 = a_1 + a_2 + a_3 + a_4 = 0 + \frac{1}{5} + \frac{2}{7} + \frac{1}{3} = \frac{86}{105}$

$S_5 = a_1 + a_2 + a_3 + a_4 + a_5$
$\quad = 0 + \frac{1}{5} + \frac{2}{7} + \frac{1}{3} + \frac{4}{11} = \frac{1366}{1155}$

7. Find the first six partial sums of the sequence given by $a_n = \dfrac{2n}{n+3}$.

$S_1 = a_1 = \frac{2}{4} = \frac{1}{2}$

$S_2 = a_1 + a_2 = \frac{1}{2} + \frac{4}{5} = \frac{13}{10}$

$S_3 = a_1 + a_2 + a_3 = \frac{1}{2} + \frac{4}{5} + \frac{6}{6} = \frac{23}{10}$

$S_4 = a_1 + a_2 + a_3 + a_4 = \frac{1}{2} + \frac{4}{5} + \frac{6}{6} + \frac{8}{7} = \frac{241}{70}$

$S_5 = a_1 + a_2 + a_3 + a_4 + a_5$
$\quad = \frac{1}{2} + \frac{4}{5} + \frac{6}{6} + \frac{8}{7} + \frac{10}{8} = \frac{657}{140}$

$S_6 = a_1 + a_2 + a_3 + a_4 + a_5$
$\quad = \frac{1}{2} + \frac{4}{5} + \frac{6}{6} + \frac{8}{7} + \frac{10}{8} + \frac{12}{9} = \frac{2531}{420}$

8. Find the first four partial sums and then find a formula for the nth partial sum of the sequence given by $a_n = \dfrac{1}{n+1} - \dfrac{1}{n+3}$.

Start by finding the first several partial sums until we find a pattern.

$S_1 = \left(\frac{1}{2} - \frac{1}{4}\right)$

$S_2 = \left(\frac{1}{2} - \frac{1}{4}\right) + \left(\frac{1}{3} - \frac{1}{5}\right)$
$\quad = \left(\frac{1}{2} + \frac{1}{3}\right) - \left(\frac{1}{4} + \frac{1}{5}\right)$

$S_3 = \left(\frac{1}{2} - \frac{1}{4}\right) + \left(\frac{1}{3} - \frac{1}{5}\right) + \left(\frac{1}{4} - \frac{1}{6}\right)$
$\quad = \left(\frac{1}{2} + \frac{1}{3}\right) - \left(\frac{1}{5} + \frac{1}{6}\right)$

At this point a pattern is starting to emerge.

$S_4 = \left(\frac{1}{2} - \frac{1}{4}\right) + \left(\frac{1}{3} - \frac{1}{5}\right) + \left(\frac{1}{4} - \frac{1}{6}\right) + \left(\frac{1}{5} - \frac{1}{7}\right)$
$\quad = \left(\frac{1}{2} + \frac{1}{3}\right) - \left(\frac{1}{6} + \frac{1}{7}\right)$

This type of sum is called a telescoping sum because the middle terms collapse.

$$S_n = \left(\frac{1}{2} + \frac{1}{3}\right) - \left(\frac{1}{n+2} + \frac{1}{n+3}\right)$$

D. **Sigma notation** is a way of expressing the sum of the first n terms of the sequence. $\sum_{k=1}^{n} a_k$ is read as "the sum of a_k from $k = 1$ to $k = n$" and $\sum_{k=1}^{n} a_k = a_1 + a_2 + a_3 + a_4 + \cdots + a_n$. The letter k is called the **index of summation** or the **summation variable** and the idea is to find the sum as k takes on the values $1, 2, 3, 4, \ldots, n$. Any letter can be used for the index of summation, and the index need not start at 1. Sums have the following properties.

> Let $a_1, a_2, a_3, a_4, \ldots$ and $b_1, b_2, b_3, b_4, \ldots$ be sequences. Then for every positive integer n and any constant c, the following properties hold.
>
> 1. $\displaystyle\sum_{k=1}^{n}(a_k + b_k) = \left(\sum_{k=1}^{n} a_k\right) + \left(\sum_{k=1}^{n} b_k\right)$
>
> 2. $\displaystyle\sum_{k=1}^{n}(a_k - b_k) = \left(\sum_{k=1}^{n} a_k\right) - \left(\sum_{k=1}^{n} b_k\right)$
>
> 3. $\displaystyle\sum_{k=1}^{n}(ca_k) = c\left(\sum_{k=1}^{n} a_k\right)$

9. Find the following sums.

   (a) $\displaystyle\sum_{k=1}^{4} \frac{1}{k^2}$

   $$\sum_{k=1}^{4} \frac{1}{k^2} = \frac{1}{1^2} + \frac{1}{2^2} + \frac{1}{3^2} + \frac{1}{4^2}$$
   $$= 1 + \frac{1}{4} + \frac{1}{9} + \frac{1}{16} = \frac{205}{144}$$

   (b) $\displaystyle\sum_{j=3}^{7} (-1)^j j^3$

   $$\sum_{j=3}^{7} (-1)^j j^3 = (-1)^3 3^3 + (-1)^4 4^3 + (-1)^5 5^3$$
   $$(-1)^6 6^3 + (-1)^7 7^3$$
   $$= -27 + 64 - 125 + 216 - 343$$
   $$= -215$$

10. Write the following sum in sigma notation.
    $$\frac{2}{3} + \frac{3}{4} + \frac{4}{5} + \frac{5}{6} + \frac{6}{7}$$

    We find the sequence associated with the terms $\frac{2}{3}, \frac{3}{4}, \frac{4}{5}, \frac{5}{6}$, and $\frac{6}{7}$. Both the numerator and denominator increase by 1, and the denominator is 1 greater than the numerator. So we get that the numerator is k and the denominator is $k + 1$. And the index of summation runs from $k = 2$ to $k = 6$.

    Thus we get the sequence $\displaystyle\sum_{k=2}^{6} \frac{k}{k+1}$.

    (Note there are many solutions, such as $\displaystyle\sum_{j=1}^{5} \frac{j+1}{j+2}$.)

E. Finding the pattern in a sequence of numbers is a very important concept in mathematics, as well as in computer science, the physical sciences, and social sciences. A finite number of terms do not *uniquely* determine a sequence, however, in the next problems we are interested in finding an *obvious sequence* whose first few terms agree with the given ones.

11. Find the formula for the nth term of the sequence whose first terms are given below.

(a) $\frac{2}{3}, \frac{8}{3}, \frac{32}{3}, \frac{128}{3}, \frac{512}{3}$

Some facts appear after observing the sequence.

(1) The denominator is 3.

(2) The numerators are all powers of 2, so we determine the exponent and correlate it to the number of the term (n).

Term, n	1	2	3	4	5
numerator	2	8	32	128	512
exponent	1	3	5	7	9

Thus the formula for the exponents of the numerator is $2n - 1$. Putting it all together we get

$$a_n = \frac{2^{2n-1}}{3}.$$

(b) $\frac{2}{3}, \frac{4}{9}, \frac{6}{27}, \frac{8}{81}$

Some facts appear after observing the sequence.

(1) The numerators are all multiples of 2, and after some trial and error, we find the formula for the numerator is $2n$.

(2) The denominator is 3 to some power. Again, after some trial and error, we find the formula denominator is given by 3^n.

Thus the explicit formula for terms of this sequence is $a_n = \frac{2n}{3^n}$.

Section 10.2   Arithmetic Sequences

Key Ideas

A. Arithmetic Sequences.

B. Partial Sums of an Arithmetic Sequence.

A. An arithmetic sequence is a sequence of the form $a, a+d, a+2d, a+3d, \ldots$. The number a is the first term and the number d is called the common difference. Any two consecutive terms of an arithmetic sequence differ by d. The nth term of the arithmetic sequence is given by the formula: $a_n = a + (n-1)d$.

1. Write the first five terms of an arithmetic sequence where 9 is the first number and 5 is the common difference.

$a_n = a + (n-1)d$

$a_1 = 9$

$a_2 = 9 + (2-1)5 = 9 + 5 = 14$

$a_3 = 9 + (3-1)5 = 9 + 10 = 19$

$a_4 = 9 + (4-1)5 = 9 + 15 = 24$

$a_5 = 9 + (5-1)5 = 9 + 20 = 29$

2. Given the first 4 terms of a sequence are 6, 13, 20, 27. Find a and d, and then the formula for the nth term.

a is first term, so $a = 6$.

d is the common difference between any two terms, so using the first and second term, we have $d = 13 - 6 = 7$.

So $a_n = 6 + (n-1)7$.

3. The 5th term of an arithmetic sequence is 34 and the 8th term is 43. Find a and d and the formula for the nth term.

Use the formula for nth term, $a_n = a + (n-1)d$.

When $n = 5$, we have $a_5 = a + (5-1)d$, so $34 = a + 4d$.

When $n = 8$, we have $a_8 = a + (8-1)d$, so $43 = a + 7d$.

Thus we get the system $\begin{cases} 34 = a + 4d \\ 43 = a + 7d \end{cases}$.

Subtracting the first equation from the second equation, we get

$$\begin{array}{rl} 43 = & a + 7d \\ -34 = & -a - 4d \\ \hline 9 = & 3d \end{array} \quad \Leftrightarrow \quad d = 3.$$

Substitute for a and solve for a.

$34 = a + 4(3) \quad \Leftrightarrow \quad 34 = a + 12 \quad \Leftrightarrow \quad 22 = a$

So $a = 22$, $d = 3$, and $a_n = 22 + (n-1)3$.

B. For the arithmetic sequence $a_k = a + (k-1)d$, the **nth partial sum**

$$S_n = \sum_{k=1}^{n} [a + (k-1)d] = a + [a+d] + [a+2d] + \cdots + [a+(n-1)d].$$

This sum is given by $S_n = \dfrac{n}{2}[2a + (n-1)d] = n\left(\dfrac{a + a_n}{2}\right)$.

4. Find the sum of the first 40 positive multiples of 3.

This is the sequence 3, 6, 9, 12, ..., where $a = 3$ and $d = 3$.

So $a_{40} = 3 + (40-1)3 = 120$.

Then $S_{40} = \sum_{k=1}^{k} a_k = 40\left(\dfrac{3+120}{2}\right) = 2460$.

5. Find the sum of the first five terms of an arithmetic sequence where $a = -2$ and $d = 8$.

Using $S_n = \dfrac{n}{2}[2a + (n-1)d]$, with $n = 5$ and $a = -2$, we get

$S_5 = \frac{5}{2}[2(-2) + (5-1)8] = \frac{5}{2}(-4 + 32) = \frac{5}{2}(28)$

$= 70$.

6. Find the partial sum, S_{10}, of the arithmetic sequence where $a_1 = 0.31$ and $a_{10} = 8.35$.

This time use the formula $S_n = n\left(\dfrac{a + a_n}{2}\right)$

with $n = 10$ to find the partial sum. So
$S_{10} = 10\left(\dfrac{0.31+8.35}{2}\right) = 10\left(\dfrac{8.66}{2}\right) = 43.3$.

Section 10.3   Geometric Sequences

Key Ideas

A. Geometric Sequences.

B. Partial Sums of a Geometric Sequence.

C. The sum of an infinite Geometric Series.

A. A **geometric sequence** is a sequence of the form a, ar, ar^2, ar^3, ar^4, ... The number a is the **first term** and r is the **common ratio**. The ratio of any two consecutive terms is r. The nth term of the sequence is given by the formula: $a_n = ar^{n-1}$.

1. Write the first five terms of a sequence where 9 is the first number and 4 is the common ratio.

$a_n = ar^{n-1}$

$a_1 = 9 \cdot 4^{1-1} = 9$

$a_2 = 9 \cdot 4^{2-1} = 9 \cdot 4 = 36$

$a_3 = 9 \cdot 4^{3-1} = 9 \cdot 16 = 144$

$a_4 = 9 \cdot 4^{4-1} = 9 \cdot 64 = 576$

$a_5 = 9 \cdot 4^{5-1} = 9 \cdot 256 = 2304$

2. The 3rd term of a geometric sequence is 6 and the 6th term is $\frac{2}{9}$. Find the first term and the common ratio.

Substitute the two values for a_n and n into $a_n = ar^{n-1}$.

Since $a_3 = 6$, we get $6 = ar^2$.

And since $a_6 = \frac{2}{9}$, we get $\frac{2}{9} = ar^5$.

Divide a_6 by a_3 to get $\dfrac{\frac{2}{9}}{6} = \dfrac{ar^5}{ar^2}$ $\Leftrightarrow$ $\frac{1}{27} = r^3$.

Solve $r^3 = \frac{1}{27}$ for r, we get $r = \frac{1}{3}$.

Substitute for r and solve for a.

$6 = ar^2$ $\Leftrightarrow$ $6 = a\left(\frac{1}{3}\right)^2$ $\Leftrightarrow$ $6 = \frac{1}{9}a$ $\Leftrightarrow$

$a = 54$.

So the first term is $a = 54$ and the common ratio is $\frac{1}{3}$.

B. For the geometric sequence $a_k = ar^{k-1}$, the **nth partial sum** is

$$S_n = \sum_{k=1}^{n} ar^{k-1} = a + ar + ar^2 + ar^3 + \cdots + ar^n.$$

For $r \neq 1$, this sum is given by $S_n = a\left(\dfrac{1-r^n}{1-r}\right)$.

3. Find the partial sum of the first 20 terms of the geometric sequence for which $a = 5.6$ and $r = 2$.

Using the formula $S_n = a\left(\dfrac{1-r^n}{1-r}\right)$ and substituting the values $n = 20$, $a = 5.6$, and $r = 2$, we get

$$S_{20} = 5.6\left(\tfrac{1-2^{20}}{1-2}\right) = 5.6 \cdot 1048575 = 5872020.$$

4. Find the sum of $\displaystyle\sum_{k=6}^{15} (0.3)^{k-1}$.

To use the formula we first express the sum with the index starting at 1. Since

$$\sum_{k=1}^{5}(0.3)^{k-1} + \sum_{k=6}^{15}(0.3)^{k-1} = \sum_{k=1}^{15}(0.3)^{k-1}, \text{ we get}$$

$$\sum_{k=6}^{15}(0.3)^{k-1} = \sum_{k=1}^{15}(0.3)^{k-1} - \sum_{k=1}^{5}(0.3)^{k-1}.$$

Now,

$$\sum_{k=1}^{15}(0.3)^{k-1} = 1\left(\tfrac{1-(0.3)^{15}}{1-(0.3)}\right) = \tfrac{0.999999986}{0.7}$$
$$= 1.4286$$

$$\sum_{k=1}^{5}(0.3)^{k-1} = 1\left(\tfrac{1-(0.3)^{5}}{1-(0.3)}\right) = \tfrac{0.99757}{0.7}$$
$$= 1.4251$$

$$\sum_{k=6}^{15}(0.3)^{k-1} = 1.4286 - 1.4251 = 0.0035.$$

C. A sum of the form $a + ar + ar^2 + ar^3 + \cdots + ar^{n-1} + \cdots$ is an **infinite series**. From above, the nth partial sum of such a series is $S_n = a\left(\dfrac{1-r^n}{1-r}\right)$, for $r \neq 1$. If $|r| < 1$, then r^n gets close to 0 as n gets large, so S_n gets close to $\dfrac{a}{1-r}$. That is, if $|r| < 1$, then the sum of the infinite series

$$a + ar + ar^2 + ar^3 + \cdots + ar^{n-1} + \cdots = \frac{a}{1-r}.$$

5. Find the sum of the infinite geometric series $\frac{4}{3} + \frac{4}{9} + \frac{4}{27} + \frac{4}{81} + \cdots$.

We need to find the values of a and r, then substitute these values into the formula. a is the first term, so $a = \frac{4}{3}$. r is the common ratio, found by dividing two consecutive terms, so $r = \frac{4/9}{4/3} = \frac{1}{3}$.

Since $|r| < 1$, $S = \dfrac{a}{1-r} = \dfrac{4/3}{1-(1/3)} = 2$.

6. Find the sum of the infinite geometric series $27 + 18 + 12 + 8 + \cdots$.

We need to find the values of a and r, then substitute these values into the formula. a is the first term, so $a = 27$. r is the common ratio, found by dividing two consecutive terms, so $r = \frac{18}{27} = \frac{2}{3}$.

Since $|r| < 1$, $S = \dfrac{a}{1-r} = \dfrac{27}{1-(2/3)} = 81$.

7. Find the sum of the infinite geometric series $\frac{3}{\sqrt{5}} - \frac{6}{5} + \frac{12}{5\sqrt{5}} - \frac{24}{25} + \cdots$.

We need to find the values a and r, then substitute these values into the formula. a is the first term, so $a = \frac{3}{\sqrt{5}}$. r is the common ratio, found by dividing

two consecutive terms, so $r = \dfrac{a_2}{a_1} = \dfrac{-\frac{6}{5}}{\frac{3}{\sqrt{5}}} = -\dfrac{2}{\sqrt{5}}$.

Since $|r| < 1$,

$$S = \frac{a}{1-r} = \frac{\frac{3}{\sqrt{5}}}{1+\frac{2}{\sqrt{5}}} = \frac{\frac{3}{\sqrt{5}}}{\frac{\sqrt{5}+2}{\sqrt{5}}}$$

$$= \frac{3}{\sqrt{5}+2} = \frac{3}{\left(\sqrt{5}+2\right)} \cdot \frac{\left(\sqrt{5}-2\right)}{\left(\sqrt{5}-2\right)} = \frac{3\left(\sqrt{5}-2\right)}{5-4}$$

$$= 3\sqrt{5} - 6.$$

8. Express the repeating decimal 0.090909... as a fraction.

First express as an infinite geometric sum.

$0.09090909\ldots = 0.09 + 0.0009 + 0.000009 + \cdots$

$$= \frac{9}{100} + \frac{9}{100} \cdot \frac{1}{100} + \frac{9}{100} \cdot \frac{1}{100^2} + \cdots$$

So $a = \frac{9}{100}$ and $r = \frac{1}{100}$.

Since $|r| < 1$, $S = \dfrac{a}{1-r} = \dfrac{\frac{9}{100}}{1 - \frac{1}{100}} = \dfrac{\frac{9}{100}}{\frac{99}{100}} = \frac{1}{11}$.

9. Express the repeating decimal $0.3\overline{45}$ as a fraction.

First express as an infinite geometric sum.

$0.3\overline{45} = 0.3 + 0.045 + 0.00045 + 0.0000045 + \cdots$

The geometric series involved is

$0.045 + 0.00045 + 0.0000045 + \cdots =$

$$0.045 + 0.045 \cdot (0.01) + 0.045 \cdot (0.01)^2 + \cdots.$$

So $a = 0.045$ and $r = 0.01$.

Since $|r| < 1$,

$$S = \frac{a}{1-r} = \frac{0.045}{1-0.01} = \frac{0.045}{0.99} = \frac{1}{22}.$$

Thus,

$0.3\overline{45} = 0.3 + 0.045 + 0.00045 + 0.0000045 + \cdots$

$$= \frac{3}{10} + \frac{1}{22} = \frac{38}{110} = \frac{19}{55}.$$

Section 10.4   Annuities and Installment Buying

Key Ideas

A.   Annuities - future value.

B.   Present Value of annuity and installment payments.

A.   An **annuity** is a sum of money that is paid into an account (or fund) in regular equal intervals. The amount of an annuity, A_f , is the sum of all the individual payments from the beginning until the last payment is made, including interest. This is a partial sum of a geometric sequence and results in the formula $A_f = R\dfrac{(1+i)^n - 1}{i}$, where R is the amount of each regular payment, i is the interest per compounding period, and n is the number of payments. Since the value of the money is paid in the future, this is called **future value**.

1. New parents put away \$25 a month into a saving paying 8% compounded monthly. How much money will accumulate in this account in 16 years?	Apply the formula $A_f = R\dfrac{(1+i)^n - 1}{i}$, with $R = \$25$, $n = 16 \cdot 12 = 192$ payments, and $i = \frac{0.08}{12} = 0.00667$. $A_f = (25)\frac{(1+0.00667)^{192}-1}{0.00667} = \$9680.23.$
2. A small company decides to save money to purchase a copying machine that costs \$9,500. How much do they need to deposit in an account that pays 6% compounded monthly if they need to save this money over a 9 month period?	Substitute the values $A_f = 9500$, $n = 9$, and $i = \frac{0.06}{12} = 0.005$ into $A_f = R\dfrac{(1+i)^n - 1}{i}$ and solve for R. $9500 = R\frac{(1.005)^9-1}{0.005} = 9.182115828R$ $R = \$1034.62$ rounding up to the next cent.

B. The **present value**, PV, is today's value of an amount to be paid in n compounded periods in the future. $PV = A(1 + i)^{-n}$, where A is the amount to be paid in the future, i is the interest per compounding period, and n is the number of compounding periods. The **present value of an annuity**, A_p, consisting of n regular equal payments of size R with interest rate i per compounding period, is given by $A_p = R\dfrac{1 - (1 + i)^{-n}}{i}$. This formula can be solved for R to obtain the amount of a regular payment on an installment loan: $R = \dfrac{iA_p}{1 - (1 + i)^{-n}}$. Although it is not possible to algebraically solve the formula for i, it is possible to use a graphing device to find i.

3. Suppose you plan to purchase a DVD player from an electronics super store for $300. If you put the $300 on a charge card at 19.8% interest compounded monthly and pay it off in 10 equal payments, how much will each payment be?

Substitute the values $A_p = 300$, $n = 10$, and $i = \frac{0.198}{12} = 0.0165$ into $R = \dfrac{iA_p}{1 - (1 + i)^{-n}}$ and calculate R.

$R = \frac{0.0165 \cdot 300}{1 - (1.0165)^{-10}} = \frac{4.95}{0.150963872} = \$32.79$

again rounded up to the next cent.

4. Suppose you wish to purchase a car and you have $125.00 a month that you can spend on payments for the next 4 years. If the current interest rate on used cars is 11.4%, how much can the car cost?

Substitute the values $R = 125$, $n = 4 \cdot 12 = 48$, and $i = \frac{0.114}{12} = 0.0095$ into $A_p = R\dfrac{1 - (1 + i)^{-n}}{i}$.

$A_p = 125\frac{1 - (1.0095)^{-48}}{0.0095}$

$= \$4800.26$ (round down to the next cent)

5. A used car dealership advertise a car for $9595 (plus tax and license) or $229.59 a month for 48 months (plus tax and license). What interest rate are they charging?

Let x be the annual interest rate. Then $i = \frac{x}{12}$. So we can graph the function

$$R(x) = \frac{\frac{x}{12}(9595)}{1 - \left(1 + \frac{x}{12}\right)^{-48}}$$ and find where it

intersect the line $y = 229.59$. Shown is the view rectangle $[0.0, 0.1]$ by $[229.25, 229.75]$.

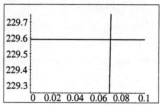

From this graph, we see that $x \approx 0.070$, so the annual interest rate is 7%.

Section 10.5   Mathematical Induction

Key Ideas

A. Principle of mathematical induction.

A. **Mathematical induction** is a type of proof used to show that a statement that depends only on n is true for all natural numbers n. Let $P(n)$ be a statement that depends on n. The principle of mathematical induction is based on showing that the following two conditions are true.

   I. $P(1)$ is true.

   II.   For every natural number k, if $P(k)$ is true, then $P(k+1)$ is true.

Then $P(n)$ is true for all natural numbers n.

Proof by mathematical induction involves the following steps.

> 1.   <u>Basis of induction:</u> *Prove* that the statement, $P(n)$, is true for some initial value; usually this value is 1.
>
> 2.   <u>Induction hypothesis:</u> Assume that the statement $P(k)$ is true.
>
> 3.   <u>The inductive step:</u> Prove that the statement $P(k)$ is true implies that $P(k+1)$ is true.
>
> 4.   <u>Conclusion:</u> Since the two conditions have been shown true, by the principle of mathematical induction, the statement is true for all natural numbers.

---

1. Prove that $n^3 - n + 3$ is divisible by 3 for all natural numbers n.

Here $P(n)$ is the statement "$n^3 - n + 3$ is divisible by 3. "

<u>Basis of induction:</u> $n = 1$
$(1)^3 - (1) + 3 = 3$ which is divisible by 3.
So $P(1)$ is true.

<u>Induction hypothesis:</u> Assume that the statement $P(k)$ is true for some $k \geq 1$, that is, assume that the statement "$k^3 - k + 3$ is divisible by 3" is true.

<u>Inductive step:</u> Show that $P(k+1)$ is true, that is, "$(k+1)^3 - (k+1) + 3$ is divisible by 3".

Proof
$(k+1)^3 - (k+1) + 3$
$$= (k^3 + 3k^2 + 3k + 1) - k - 1 + 3$$
$$= (k^3 - k + 3) + (3k^2 + 3k)$$
$$= (k^3 - k + 3) + 3(k^2 + k)$$
$k^3 - k + 3$ is divisible by 3 (by the induction hypotheses). And $3(k^2 + k)$ is divisible by 3, since it is the product of 3 and an integer. Hence the sum, $(k^3 - k + 3) + 3(k^2 + k)$ is divisible by 3.

<u>Conclusion:</u> $P(k)$ implies $P(k+1)$.
Since the statement "$k^3 - k + 3$ is divisible by 3" implies "$(k+1)^3 - (k+1) + 3$ is divisible by 3", the statement is true for all natural numbers n.

2. Let $\{a_n\}$ be the sequence recursively defined by $a_n = a_{n-1} + 2a_{n-2} - 6$, where $a_1 = 4$ and $a_2 = 5$. Find the first 5 terms of the sequence. Prove that $a_n = 2^{n-1} + 3$.

$a_1 = 4;\ a_2 = 5;$
$a_3 = a_2 + 2a_1 - 6 = 5 + 2(4) - 6 = 7$
$a_4 = a_3 + 2a_2 - 6 = 7 + 2(5) - 6 = 11$
$a_5 = a_4 + 2a_3 - 6 = 11 + 2(7) - 6 = 19$

Basis of induction: $n = 1, n = 2$, and $n = 3$
Note all cases are needed since a_n is based on the terms a_{n-1} and a_{n-2}, and a_3 the first term defined by the recursive definition.
$a_1 = 2^{1-1} + 3 = 2^0 + 3 = 1 + 3 = 4$
$a_2 = 2^{2-1} + 3 = 2^1 + 3 = 2 + 3 = 5$
$a_3 = 2^{3-1} + 3 = 2^2 + 3 = 4 + 3 = 7$
So $P(1)$, $P(2)$, and $P(3)$ are true.

Induction hypothesis:
Assume that $a_j = 2^{j-1} + 3$ is true for $j \leq k$, where $k \geq 3$.
Note: We had to start at $k \geq 2$ because a_2 is not given by the formula $a_j = a_{j-1} + 2a_{j-2} - 6$.
Also note: since the terms a_{n-1} and a_{n-2} are both used in the definition of a_j we need an induction hypotheses that includes both terms.
Assume $a_k = a_{k-1} + 2a_{k-2} - 6 = 2^{k-1} + 3$.

Inductive step: Show that $a_{k+1} = 2^{(k+1)-1} + 3$
$= 2^k + 3$.

Proof
By the recursive definition, $a_{k+1} = a_k + 2a_{k-1} - 6$, and by the inductive hypotheses,
$a_k = 2^{k-1} + 3$ and $a_{k-1} = 2^{k-2} + 3$. Substituting
$a_k = 1 = a_k + 2a_{k-1} - 6$
$\qquad = (2^{k-1} + 3) + 2(2^{k-2} + 3) - 6$
$\qquad = 2^{k-1} + 3 + 2 \cdot 2^{k-2} + 6 - 6$
$\qquad = 2^{k-1} + 2^{k-1} + 3 = 2 \cdot 2^{k-1} + 3$
$\qquad = 2^k + 3.$

Conclusion:
By the principle of mathematical induction,
$a_n = a_{n-1} + 2a_{n-2} - 6 = 2^{n-1} + 3$, where $a_1 = 4$ and $a_2 = 5$, is true for all natural numbers n.

Section 10.6   The Binomial Theorem

Key Ideas
A. Pascal's triangle.

B. Binomial coefficients.

A. An expression of the form $a + b$ is called a **binomial**. When $(a + b)^n$ is expanded, the following pattern emerges:

1. There are $n + 1$ terms, the first being $a^n b^0$ and the last $a^0 b^n$.

2. The *exponents of a decrease by* 1 from term to term while the *exponents of b increase by* 1.

3. The sum of the exponents of a and b in each term is n.

Pascal's triangle is the following triangular array. It gives the coefficients of the terms of a binomial expansion.

$$
\begin{array}{lc}
(a + b)^0 & 1 \\
(a + b)^1 & 1 \quad 1 \\
(a + b)^2 & 1 \quad 2 \quad 1 \\
(a + b)^3 & 1 \quad 3 \quad 3 \quad 1 \\
(a + b)^4 & 1 \quad 4 \quad 6 \quad 4 \quad 1 \\
(a + b)^5 & 1 \quad 5 \quad 10 \quad 10 \quad 5 \quad 1 \\
(a + b)^6 & 1 \quad 6 \quad 15 \quad 20 \quad 15 \quad 6 \quad 1
\end{array}
$$

Every entry (other than a 1) is the sum of the two entries diagonally above it.

1. Use Pascal's triangle above to expand $(3 + 2x)^5$.

$$(a + b)^5 = a^5 + 5a^4 b^1 + 10a^3 b^2 + 10a^2 b^3 + 5a^1 b^4 + b^5$$

So
$$(3 + 2x)^5 = (3)^5 + 5(3)^4(2x) + 10(3)^3(2x)^2 + 10(3)^2(2x)^3 + 5(3)(2x)^4 + (2x)^5$$

$$= 243 + 810x + 1080x^2 + 720x^3 + 240x^4 + 32x^5.$$

2. Use Pascal's triangle above to expand $(3x - 5y)^3$.

$$(a + b)^3 = a^3 + 3a^2 b^1 + 3a^1 b^2 + b^3$$

So
$$(3x - 5y)^3 = (3x)^3 + 3(3x)^2(-5y) + 3(3x)(-5y)^2 + (-5y)^3$$

$$= 27x^3 - 135x^2 y + 225xy^2 - 125y^3.$$

B. The product of the first n natural numbers is denoted by $n!$ and is called $\boldsymbol{n}$ **factorial**. So $n! = 1 \cdot 2 \cdot 3 \cdot \cdots \cdot (n-1) \cdot n$ and we define $0! = 1$. For $0 \le r \le n$, the **binomial coefficient** is denoted by $\binom{n}{r} = \dfrac{n!}{r!(n-r)!}$. Two important relations between binomial coefficients state that $\binom{n}{r} = \binom{n}{n-r}$, and for any nonnegative integers r and k with $r \le k$,

$\binom{k}{r-1} + \binom{k}{r} = \binom{k+1}{r}$. Now, **Pascal's triangle** can also be written as

$$\binom{0}{0}$$

$$\binom{1}{0} \quad \binom{1}{1}$$

$$\binom{2}{0} \quad \binom{2}{1} \quad \binom{2}{2}$$

$$\binom{3}{0} \quad \binom{3}{1} \quad \binom{3}{2} \quad \binom{3}{3}$$

$$\binom{4}{0} \quad \binom{4}{1} \quad \binom{4}{2} \quad \binom{4}{3} \quad \binom{4}{4}$$

$$\binom{5}{0} \quad \binom{5}{1} \quad \binom{5}{2} \quad \binom{5}{3} \quad \binom{5}{4} \quad \binom{5}{5}$$

$$\binom{6}{0} \quad \binom{6}{1} \quad \binom{6}{2} \quad \binom{6}{3} \quad \binom{6}{4} \quad \binom{6}{5} \quad \binom{6}{6}$$

The Binomial theorem states that

$(a+b)^n = \binom{n}{0}a^n b^0 + \binom{n}{1}a^{n-1}b^1 + \binom{n}{2}a^{n-2}b^2 + \cdots + \binom{n}{n-1}a^1 b^{n-1} + \binom{n}{n}a^0 b^n$. The term that contains a^r in the expansion of $(a+b)^n$ is $\binom{n}{r}a^r b^{n-r}$.

3. Evaluate

$$\binom{4}{0} + \binom{4}{1} + \binom{4}{2} + \binom{4}{3} + \binom{4}{4}.$$

The terms

$$\binom{4}{0} + \binom{4}{1} + \binom{4}{2} + \binom{4}{3} + \binom{4}{4}$$

are the expansion of $(1+1)^4$. So

$$\binom{4}{0} + \binom{4}{1} + \binom{4}{2} + \binom{4}{3} + \binom{4}{4}$$

$$= (1+1)^4 = 2^4 = 16.$$

4. Evaluate

$$\binom{4}{0} - \binom{4}{1} + \binom{4}{2} - \binom{4}{3} + \binom{4}{4}.$$

The terms

$$\binom{4}{0} - \binom{4}{1} + \binom{4}{2} - \binom{4}{3} + \binom{4}{4}$$

are the expansion of $(1-1)^4$. So

$$\binom{4}{0} - \binom{4}{1} + \binom{4}{2} - \binom{4}{3} + \binom{4}{4}$$
$$= (1-1)^4 = 0^4 = 0.$$

5. Find the first three terms in the expansion of $(4 - x)^9$.

The first three terms are

$$\binom{9}{0}4^9(-x)^0 + \binom{9}{1}4^8(-x)^1 + \binom{9}{2}4^7(-x)^2$$
$$= 262144 - 589824x + 589824x^2.$$

6. Find the last three terms in the expansion of $(7 + y)^{11}$.

The last three terms

$$\binom{11}{9}7^2(y)^9 + \binom{11}{10}7^1(y)^{10} + \binom{11}{11}7^0(y)^{11}$$
$$= 2695y^9 + 77y^{10} + y^{11}.$$

7. Find the term containing x^8 in the expansion of $(2x - 3)^{15}$.

The term containing a^r is $\binom{n}{r}a^r b^{n-r}$.

So the term containing x^8 is

$$\binom{15}{8}(2x)^8(-3)b^{15-8} = 6435 \cdot 2^8 \cdot x^8 \cdot (-3^7)$$
$$= -3602776320x^8.$$

Chapter 11
Counting and Probability

Section 11.1   Counting Principles

Key Ideas

A. Tree Diagrams.
B. Fundamental counting principle.
C. Counting with cards and dice.

A. A **tree diagram** is a systematic method of determining and listing all the possibilities of a counting or probability problem. At each vertex, all possible choices are drawn as edges and labeled. To create a list of possible outcomes to a counting problem, start at an end vertex and trace your way back to the initial vertex.

1. A family is planning on having three children, list all possible orders of birth.

Here we use a tree diagram to help us count. Let B represent a boy and G represent a girl. This leads to the tree diagram below.

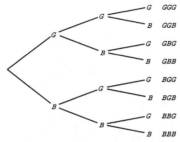

So the 8 possible child births are GGG, GGB, GBG, GBB, BGG, BGB, BBG, BBB.

2. A pet food manufacturer test two types of dog food to see which food is preferred by a dog. A dog is observed at most five times or until it chooses the same food three times. How many different outcomes are possible? Draw a tree diagram to count this question.

Although this count can be made with other methods still to come, using a tree diagram is the easiest. Let A and B represent the two types of dog food. This leads to the tree diagram below.

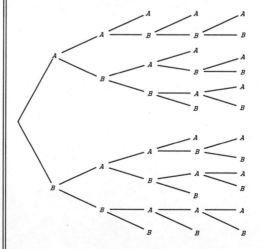

There are 20 possible outcomes.

B. The **fundamental counting principle** states that if two events occur in order and if the first event can occur in m ways and the second in n ways (after the first event has occurred), then the two events can occur in order in $m \times n$ ways. When the events $E_1, E_2, \ldots, E_k$ occur in order and E_1 can occur in n_1 ways, E_2 in n_2 ways, and so on, then the events can occur in order in $n_1 \times n_2 \times \cdots \times n_k$ ways.

3. For security purposes, a company classifies each employee according to five hair colors, four eye colors, six weight categories, five height categories, and two sex categories. How many classifications are possible?

Using the fundamental counting principle there are $5 \cdot 4 \cdot 6 \cdot 5 \cdot 2 = 1200$ classifications.

4. The Uniform Pricing Code (the bar code on a package used to price the item) consists of a 10 digit number. How many different bar codes are possible?

Using the fundamental counting principle there are ten choices for each of the ten digits. As a result there are
$$10 \cdot 10 \cdot 10 \cdot 10 \cdot 10 \cdot 10 \cdot 10 \cdot 10 \cdot 10 \cdot 10 = 10^{10}$$
Uniform Pricing Codes.

5. A car manufacturer makes its lowest priced car in five exterior colors, four interior colors, a choice of manual or automatic transmission, and three levels of options. How many basic models are available?

Using the fundamental counting principle there are $5 \cdot 4 \cdot 2 \cdot 3 = 120$ basic models available.

C. Many counting questions involve a **standard deck of cards**. A standard deck of cards consists of 52 cards broken up into 4 **suits** of 13 cards each. The suits **diamonds** and **hearts** are red, while the suits **clubs** and **spades** are black. There are 13 **kinds** in a deck, Ace, 2, 3, . . . 10, Jack, Queen, King. For example, the cards 3 *of diamonds*, 3 *of hearts*, 3 *of clubs*, and 3 *of spades* form a **kind**. Other common counting questions involve a **die** (or **dice**-plural) which is a six-sided cube with the numbers 1 through 6 painted on the faces. When a pair of dice is rolled there are 36 possible outcomes that can occur. Also, a common "experiment" is to look at the sum of the faces. The table below shows the 36 possible outcomes along with the corresponding sum of the faces.

		Die 1					
		1	2	3	4	5	6
Die 2	1	(1, 1) 2	(1, 2) 3	(1, 3) 4	(1, 4) 5	(1, 5) 6	(1, 6) 7
	2	(2, 1) 3	(2, 2) 4	(2, 3) 5	(2, 4) 6	(2, 5) 7	(2, 6) 8
	3	(3, 1) 4	(3, 2) 5	(3, 3) 6	(3, 4) 7	(3, 5) 8	(3, 6) 9
	4	(4, 1) 5	(4, 2) 6	(4, 3) 7	(4, 4) 8	(4, 5) 9	(4, 6) 10
	5	(5, 1) 6	(5, 2 7	(5, 3) 8	(5, 4) 9	(5, 5) 10	(5, 6) 11
	6	(6, 1) 7	(6, 2) 8	(6, 3) 9	(6, 4) 10	(6, 5) 11	(6, 6) 12

Section 11.2 Permutations and Combinations

Key Ideas

A. Factorial notation.
B. Permutations.
C. Distinguishable Permutations.
D. Combinations.
E. Difference between permutations and combinations.

A. The product of the first n natural numbers is denoted by $n!$ and is called **n factorial**.
$n! = 1 \cdot 2 \cdot 3 \cdot 4 \cdots (n-1) \cdot n$, and $0! = 1$. The term n factorial can also be expressed in a recursive form as $n! = n \cdot (n-1)!$ for $n \geq 1$. For example $6! = 6 \cdot (5!)$.

1. A woman sits down to make four sales calls, in how many ways can she make her calls?

She has 4 choices for the first call, 3 choices for the call, 2 choices for the third call, and 1 choices for the last call.

So there are $4 \cdot 3 \cdot 2 \cdot 1 = 4!$ ways for her to make her sales calls.

B. A **permutation** of a set of distinct objects is an ordering or arrangement of these objects. The number of ways n objects can be arranged is $n!$. When only r of the n objects of the set are arranged, then **the number of permutations of n objects taken r at a time** is denoted by $P(n, r)$ and is given by the

formula $P(n, r) = n \cdot (n-1) \cdot (n-2) \cdots (n-r+1) = \dfrac{n!}{(n-r)!}$.

Permutations are used when (1) order is important, and (2) repetitions are not allowed.

2. A club of 20 people gather for a group photo standing in a row. In how many ways can the photo be taken?

Here we are after an arrangement of all 20 members of the group. So there are $P(20, 20) = 20!$ ways to arrange the group.

$20! \approx 2.432902008 \times 10^{18}$

3. A volleyball team has nine members and six members are on the court at any one time. In how many ways can the six be selected and arranged on the court?

$P(9, 6) = \frac{9!}{(9-6)!} = \frac{9!}{3!}$

$= 9 \cdot 8 \cdot 7 \cdot 6 \cdot 5 \cdot 4 \cdot \frac{3!}{3!} = 60480$

4. A new car dealership has ten different new cars and five pedestals to display cars in the front showroom. In how many ways can the dealership display the cars?

$$P(10,5) = \frac{10!}{(10-5)!} = \frac{10!}{5!}$$
$$= 10 \cdot 9 \cdot 8 \cdot 7 \cdot 6 \cdot \frac{5!}{5!} = 30240$$

C. When considering a set of objects, some of which are of the same kind, then two permutations are **distinguishable**, if one cannot be obtained from the other by interchanging the positions of elements of the same kind. If a set of n objects consists of k different kinds of objects with n_1 objects of the first kind, n_2 objects of the second kind, n_3 objects of the third kind, and so on, where $n_1 + n_2 + n_3 + \cdots + n_k = n$, then the number of distinguishable permutations of these objects is $\frac{n!}{n_1! n_2! n_3! \cdot \cdots \cdot n_k!}$. The numbers $n_1, n_2, n_3, \ldots, n_k$ are called a **partition** of n when $n_1 + n_2 + n_3 + \cdots + n_k = n$.

5. An instructor makes three versions of an exam for her chemistry class with 30 students. If she makes 10 copies of each version, in how many ways can she distribute the exams in her class?

Since all 30 exams are distributed to the class, this is a distinguishable permutation. The number of ways is then $\dfrac{30!}{10! \, 10! \, 10!} = 5.550996791 \times 10^{12}$

6. How many ways can the letters of the word "MOTORSPORTS" be arranged?

Since MOTORSPORTS has 11 letters, 1 "M", 3 "O"'s, 2 "T"'s, 2 "R"'s, 2 "S"'s, and 1 "P", there are $\dfrac{11!}{1! \, 3! \, 2! \, 2! \, 2! \, 1!} = 831,600$ ways to arrange these letters.

7. An ice chest contains six cokes, four orange sodas, eight fruit flavored waters. In how many ways can these be distributed to 18 people at a picnic?

Since we distribute all 18 items in the ice chest, this is a distinguishable permutation. So there are $\dfrac{18!}{6! \, 4! \, 8!} = 9,189,180$ ways to distribute the soft drinks.

314

D. A combination of r elements of a set is any subset of r elements from the set without regard to order. If the set has n elements, then the **number of combinations of n elements taken r at a time** is denoted by $C(n, r)$ and is given by the formula $C(n, r) = \dfrac{n!}{r!\,(n - r)!}$. Combinations are used when (1) order is not important and (2) there are no repetitions. A set with n elements has 2^n subsets.

8. A drawer contains 28 pairs of socks. In how many ways can a person select ten pairs of socks to pack for a vacation?

Order is not important and there are no repeats.

So there are $C(28, 10) = \dfrac{28!}{10!\,(28 - 10)!} = \dfrac{28!}{10!\,18!}$

$= 13123110$ ways to selected the pairs of socks.

9. Twelve cars are stopped at a fruit inspection checkpoint operated by the state of California. In how many ways can the inspectors choose three of the twelve cars to inspect?

Order is not important because the inspectors want a group of cars to inspect. There are no repeats since you will not inspect a car twice.

So there are $C(12, 3) = \dfrac{12!}{3!\,(12 - 3)!} = \dfrac{12!}{3!\,9!} = 220$

ways to select the cars to inspect.

10. A group of 12 friends are at a party.

(a) In how many ways can 3 people be selected to go out and get more soft drinks?

Since a group is selected, order is not important, and obviously there are no repeats. As a result there are $C(12, 3) = \dfrac{12!}{3!\,(12 - 3)!} = \dfrac{12!}{3!\,9!} = 220$ ways to select the three people.

(b) If there are 6 females and 6 males at the party, in how many ways can 2 males and one female be selected to get the soft drinks?

This is still a combination. Since the males can only be selected from the six males and the female can only be selected from the six females, there are $C(6, 2) \cdot C(6, 1) = \dfrac{6!}{2!\,4!} \cdot \dfrac{6!}{1!\,5!} = 15 \cdot 6 = 90$ ways to select the three people.

11. Pauline's Pizza Palace has 14 different kinds of toppings. Pauline's Special Combination is a large pizza with any three different toppings. How many Pauline's Special Combinations are possible?

Since the toppings must be different and order is not not important, this is a combination problem.
The number of Pauline's Special Combinations is

$$C(14, 3) = \frac{14!}{3!\,11!} = 364.$$

12. A hamburger chain advertises that they can make a hamburger in 256 ways. How many different toppings do they need to meet this claim?

Since changing a topping will change the type of hamburger, a subset of toppings will represent a different hamburger. Let n be the number different toppings they have. So we need to solve the equation $2^n = 256$. Thus $n = \log_2 256 = 8$.

13. How many full houses are possible?

A full house is 3 cards of one kind and 2 cards from another kind.
To count this, we first pick 1 of the 13 kinds and 3 of the 4 cards from that kind, then pick another kind from the 12 remaining kinds, and 2 of the 4 cards in that kind. So we get

$$\begin{pmatrix} \text{pick the} \\ \text{kind} \end{pmatrix} \begin{pmatrix} \text{pick 3 of} \\ \text{4 cards} \end{pmatrix} \begin{pmatrix} \text{pick another} \\ \text{kind} \end{pmatrix} \begin{pmatrix} \text{pick 2 of} \\ \text{4 cards} \end{pmatrix}$$

$$= C(13, 1) \cdot C(4, 3) \cdot C(12, 1) \cdot C(4, 2)$$

$$= \frac{13!}{1!\,12!} \cdot \frac{4!}{3!\,1!} \cdot \frac{12!}{1!\,11!} \cdot \frac{4!}{2!\,2!}$$

$$= 13 \cdot 4 \cdot 12 \cdot 6$$

$$= 3744.$$

E. Both permutations and combinations use a set of n distinct objects where r objects are selected at random *without repetitions* (or replacements). The key difference between permutations and combinations is *order*. When order is important, permutations are used and when order is not important, combinations are used. Other key words that help to distinguish combinations are: *group*, *sets*, *subsets*, etc. Key words that distinguish permutations are: *arrange*, *order*, *arrangements*, etc.

14. Fifteen people are selected for a study of diet pills. Eight will be given a new diet pill and seven will be given a placebo. In how many ways can the eight be selected?

This is a combination problem because (1) there are no repeats (a person can only be selected once) and (2) you are interested in a group only.

So there are $C(15, 8) = \dfrac{15!}{8!\,7!} = 6435$ ways to select the test group.

15. Five couples line up outside a movie theater. In how many ways can they line up?

First treat each couple as one item and then arrange the people in the couple.

There are $P(5, 5) = 5!$ ways to arrange the couple.

Then arrange the two people in each couple, there $P(2, 2) = 2!$ ways to arrange each couple.

So there are $5!\,(2!)^5$ ways to arrange the couples in line at the movie theater.

$5!(2!)^5 = 5!\,2^5 = 3840$

16. A newly formed Ad Hoc Committee on Student Retention plans to study ten new students during the Fall semester in order to identify ways to improve student retention. Suppose next fall's incoming class contains 112 new female and 108 new male students.

(a) In how many different ways can 10 new students be chosen at random (with no regard to sex) for this study?

Since there are 220 new students expected, we pick the 10 students from this group.

So there are $C(220, 10) = \dfrac{220!}{10!\,210!}$ ways to pick the study group.

(b) How many possible studies in part (a) have 5 females and 5 males?

The five female students can only be picked from the 112 female students and the five male students can only be picked from the 108 students.

So there

$C(112, 5) \cdot C(108, 5) = \dfrac{112!}{5!\,107!} \cdot \dfrac{108!}{5!\,103!}$ ways to form the study group.

17. A college theater group plans on doing 5 different shows during the year- *Cats*, *The Tempest*, *Annie*, *Streetcar Named Desire*, and *Guys and Dolls*. If *Cats* can only be scheduled last, and *Annie* must be either first or second, how many different schedules are possible?

There are five slots to fill; *Cats* must be placed in the last slot, while *Annie* can be placed in either first or second slot. So there are 2 ways to place *Annie*, 1 way to place *Cats* (in the last slot), and there are remaining slots to fill. Since there are 3! ways to fill fill these remaining slots, there are a total of $2 \cdot 3! \cdot 1 = 12$ ways to schedule the shows.

18. A flush is five cards from the same suit. How many different flushes are possible?

Here we first pick the suit and then 5 cards from that suit.
There are $C(4, 1)$ ways to pick the suit, and there are $C(13, 5)$ ways to pick the 5 cards from the suit. So there are $C(4, 1) \cdot C(13, 5) = 5148$ flushes.

Section 11.3   Probability

Key Ideas

A. Sample spaces and probability.
B. Complements.
C. Mutually exclusive events and unions of events.
D. Probability of the union of two sets.
E. Independent events.

A. An **experiment** is a process that has a finite set of definite outcomes. The set of all possible outcomes of an experiment is called the **sample space**, and we use the letter S to denote it. An **event** is any subset of the sample space. When each individual outcome has an equal chance of occurring, then the probability that event E occurs, written as $P(E)$, is

$$P(E) = \frac{n(E)}{n(S)} = \frac{\text{number of elements in event } E}{\text{number of elements in sample space } S}.$$ As a result, $0 \le P(E) \le 1$. The closer

$P(E)$ is to one the more likely the event E is to happen, whereas the closer to zero the less likely. If $P(E) = 1$, then E is called the **certain event**, that means E is *the only event that can happen*. When $P(E) = 0$ the event is called an **impossible event**.

1. Let S be the experiment where a pair of dice are rolled and the sum of the faces recorded.

   (a) Find the probability that the sum of the numbers showing is 3.

   Using the table on page 229 of this study guide, we see that the sum is 3 in only two of the 36 squares. Thus, $P(\text{sum is } 3) = \frac{2}{36} = \frac{1}{18}$.

   (b) Find the probability that the sum of the faces is odd.

   Again looking at the table on page 229 of this study guide, there are 18 squares where the sum is odd. Thus, $P(\text{sum is odd}) = \frac{18}{36} = .5$.

   (c) Find the probability that the sum of the faces is greater than 9.

   The sum is greater than 9 is the same as the sum is 10, the sum is 11, or the sum is 12. Looking at the table, there are three squares where the sum is 10, two squares where the sum is 11, and one square where the sum is 12, for a total of 6 squares. Thus, $P(\text{sum is greater than } 9) = \frac{6}{36} \approx .167$.

2. A box of 100 ping pong balls contains 50 yellow balls and 50 white balls. If 5 balls are selected at random, what is the probability that

(a) all 5 balls are yellow?

Since there are 100 balls, the sample space, S, consists of all ways to select 5 balls without regard to color. So $n(S) = C(100, 5) = 75287520$.

Since the yellow balls can only be selected from the 50 yellow balls, there are $C(50, 5) = 2118760$ ways to select this group. Thus,

$$P(5 \text{ yellow balls}) = \frac{C(50, 5)}{C(100, 5)} = \frac{2118760}{75287520}$$

$$\approx .02814.$$

(b) 3 balls are yellow and 2 balls are white?

Pick the 3 yellow balls from the 50 yellow balls and pick the 2 white balls from the 50 white balls. There are $C(50, 3) \cdot C(50, 2) = 19600 \cdot 1225 = 24010000$ ways to make these selections. So the probability is

$$P(3 \text{ yellow and 2 white}) = \frac{C(50, 3) \cdot C(50, 2)}{C(100, 5)}$$

$$= \frac{24010000}{75287520} \approx .3189.$$

B. The **complement** of an event E is the set of outcomes of the sample space S that are not in E, and is denoted as E'. The probability that E does not occur is $P(E') = 1 - P(E)$. Sometimes it is easier to find $P(E')$ than to find $P(E)$.

3. A box of 100 calculators contains 3 defective calculators. Suppose five calculators are selected at random. What is the probability that at least one calculator is defective?

The complement of *at least one calculator is defective* is *no calculator is defective*.
There are 97 non-defectives, choose 5. The number of ways of choosing 5 non-defective is $C(97,5)$. So,

$$P(\text{all non-defective}) = \frac{C(97, 5)}{C(100, 5)} \approx .8560. \text{ Then}$$

$$P(\text{at least 1 defective}) = 1 - P(\text{all non-defective})$$

$$\approx 1 - .8560 \approx .1440.$$

4. Five friends compare birth months. What is the probability that at least two will be born in the same month?

Here we assume that each month is equally likely to occur. Since we seek *at least two* this means that two could have the same birth month, three could have the same birth month, two pairs could have the same birth month, etc. The complement, *no two have the same birth month*, is much easier to count. Thus, P(different birth months)

$$= \frac{\left(\begin{array}{c}\text{number of ways to assign}\\ \text{5 different birth months}\end{array}\right)}{\left(\begin{array}{c}\text{number of ways to assign}\\ \text{birth months}\end{array}\right)}$$

$$= \frac{12 \cdot 11 \cdot 10 \cdot 9 \cdot 8}{12 \cdot 12 \cdot 12 \cdot 12 \cdot 12} = \frac{55}{144}$$

So the probability that at least two will have the same birth month is $1 - \frac{55}{144} = \frac{89}{144} \approx .618$.

C. When two events have no outcomes in common they are said to be **mutually exclusive**. The probability of the union of two mutually exclusive events E and F in a sample space S is given by $P(E \cup F) = P(E) + P(F)$.

5. A box contains 6 red building blocks and 4 blue building blocks. An infant selects 6 blocks from the box. Consider the events
   E: the infant selects 3 red & 3 blue blocks,
   F: the infant selects all red blocks, and
   G: the infant select at least four red blocks.
   Which of these events are mutually exclusive?

E and F Since blue building blocks are not red, an outcome cannot satisfy both events, so $E \cap F = \emptyset$, and the events are mutually exclusive.

E and G Since the infant cannot select four red block and three blue block, these events are mutually exclusive.

F and G Since it is possible for the infant to select *at least 4 red blocks* and *six red blocks*, these events are <u>not</u> mutually exclusive.

6. A five card hand is drawn. Consider the following events:

    E: all 5 cards are red,

    F: full house,

    G: two pairs,

    H: all 5 cards are faces cards.

Which of these events are mutually exclusive?

Note: In this problem, events are mutually exclusive if there are no hands that belong to both events. To show that two events are not mutually exclusive, we must show that there are hands that belong to both events.

E and F These events are mutually exclusive because a full house means that there are 3 cards of one kind, at least 1 of these 3 cards cannot be red. So a full house has to have at least one black card.

E and G These events are not mutually exclusive. There are many hands where you have 2 pairs and a fifth card that are all red.

E and H These events are not mutually exclusive. There are three face cards per suit, so there are six red face cards. Thus there are hands that have 5 red face cards.

F and G By definition, two pairs means that the fifth card is from a different kind than the pairs. As a result these events are mutually exclusive.

F and H These events are not mutually exclusive. There are many full houses which involve only face cards.

G and H These events are not mutually exclusive. Since there are three kinds of face cards, so each pair and the fifth card can come from a different kind of face card.

D. When two events are *not* mutually exclusive, the probability of E or F is $P(E \cup F) = P(E) + P(F) - P(E \cap F)$. Remember, when the word "or" is used, you are looking for the union of the events.

7. A five card hand is drawn at random. Find the probability that all red cards are drawn or a full house is drawn.

The sample space is all 5 card hands. The number of elements in the sample space is $C(52, 5)$.

Let E be the event *all 5 cards are red*.

Let F be the event *full house*.

In Problem 6 above, we saw that these events are mutually exclusive.

Since there are 26 red cards from which 5 red cards is drawn, $P(E) = \dfrac{C(26, 5)}{C(52, 5)}$.

$P(F) = \dfrac{3744}{C(52, 5)}$ (see Problem 13, page 233)

$P(E \cup F) = P(E) + P(F)$

$$= \dfrac{C(26, 5)}{C(52, 5)} + \dfrac{3744}{C(52, 5)}$$

8. A five card hand is drawn at random. Find the probability that a full house is drawn or all five cards are face cards.

The sample space is all 5 card hands. The number of elements in the sample space is $C(52, 5)$.

Let F be the event *full house*.

Let H be the event *all five cards are face cards*.

From Problem 6 above, we saw that these events are not mutually exclusive. Here we have

$$P(F) = \frac{3744}{C(52, 5)} \quad \text{(see Problem 13, page 233)}$$

Each suit has 3 face cards and there are 4 suits, so there are 12 face cards. We choose 5; so $n(H) = C(12, 5)$ and

$$P(H) = \frac{C(12, 5)}{C(52, 5)} = \frac{792}{C(52, 5)}.$$

$F \cap H$ is the event that the full house is made from face cards. Since there are 3 kinds of face cards, there are $C(3, 1)$ ways to pick the first kind, $C(4, 3)$ ways to pick 3 of the 4 cards from that kind; there are $C(2, 1)$ ways to pick the second kind, and $C(4, 2)$ ways to pick 2 of the 4 cards from that kind.

$$n(F \cap H) = C(3, 1) \cdot C(4, 3) \cdot C(2, 1) \cdot C(4, 2)$$
$$= 3 \cdot 4 \cdot 2 \cdot 6 = 144$$

$$P(F \cap H) = \frac{144}{C(52, 5)}$$

Thus,

$$P(F \cup H) = P(F) + P(H) - P(F \cap H)$$
$$= \frac{3744}{C(52, 5)} + \frac{792}{C(52, 5)} - \frac{144}{C(52, 5)}$$
$$= \frac{4392}{C(52, 5)}.$$

E. Two events are **independent** when the occurrence of one event does not affect the probability of another event occurring. When events E and F are independent, then $P(E \cap F) = P(E) \cdot P(F)$. Many students get *independent events* and *mutually exclusive events* confused. The difference is that when two events are mutually exclusive they have nothing in common, but when the events are independent then the events do not influence each other. If two events are independent they are *not* mutually exclusive.

9. A pair of dice is rolled 5 times. What is the probability of rolling five 7's in a row?

Since rolling a 7 on any one roll does not influence the next roll, each roll is independent, this probability is given by

$\frac{1}{6} \cdot \frac{1}{6} \cdot \frac{1}{6} \cdot \frac{1}{6} = \frac{1}{6^5}$.

10. A driver on a certain stretch of interstate has a .0001 probability of getting a radar ticket. Suppose a driver drives this stretch of road 6 times, what is the probability that the driver gets at least one radar ticket?

These events are independent. Since the complement is easier to find, we find it first. The complement of *getting at least one ticket* is *getting NO tickets.*

$P(\text{no ticket}) = 1 - .001 = .999$

So $(.999)^6 = .994$ is the probability of no tickets in 6 trips. Thus, $P(\text{getting a ticket}) = 1 - .994 = .006$.

Section 11.4 Expected Value
Key Ideas
A. What Is Expected Value?

A. The **expected value** of an experiment is a weighted average of *payoffs* based on the outcomes of the experiment. The weight applied to each payoff is the probability of the event (that corresponds to the payoff) occurring. Expected value E is $E = a_1 p_1 + a_2 p_2 + \cdots + a_n p_n$. Here a_i is the payoff and p_i is the probability.

1. A game consists of rolling a die. You receive the face value in dollars if an odd number is rolled. If the game costs $1.50 to play, what is the expected value to you?

The payoffs and their respective probabilities are given in the following table.

Payoff	Event Description	Probability
$-$1.50$	an even number is rolled	$\frac{1}{2}$
-0.50	a one is rolled $(-1.50 + 1)$	$\frac{1}{6}$
1.50	a 3 is rolled $(-1.50 + 3)$	$\frac{1}{6}$
3.50	a 5 is rolled $(-1.50 + 5)$	$\frac{1}{6}$

$$E = -\$1.50 \cdot \tfrac{1}{2} - 0.50 \cdot \tfrac{1}{6} + 1.50 \cdot \tfrac{1}{6} + 3.50 \cdot \tfrac{1}{6}$$
$$= 0$$

2. A sales clerk at a car stereo store works for commission only. If 20% of the people purchase an item where the clerk gets a $3 commission, if 10% purchase an item where the clerk gets a $5 commission, and if 5% purchase an item where the clerk gets a $8 commission, what is the expected value per customer who enters the store?

The payoffs (commissions)and their respective probabilities are given in the following table.

Payoff	Probability
$3	0.20
$5	0.10
$8	0.05
$0	the rest

$$E = 3(0.20) + 5(0.10) + 8(0.05) + 0$$
$$= 0.6 + 0.5 + 0.4$$
$$= 1.50$$

3. A bowl contains eight chips which cannot be distinguished by touch alone. Five chips are marked $1 each and the remaining three chips are marked $4 each. A player is blindfolded and draws two chips at random without replacement from the bowl. The player is then paid the sum of the two chips. Find the expected payoff of this game.

The possible payoff are $2 (two $1 chips are selected), $5 ($1 chip and a $4 chip) and $8 (two $4 chips are selected).

The probability of getting $2 is $\dfrac{C(5,2)}{C(8,2)} = \dfrac{10}{28}$.

The probability of getting $5 is
$$\dfrac{C(5,1) \cdot C(3,1)}{C(8,2)} = \dfrac{15}{28}.$$

The probability of getting $8 is $\dfrac{C(3,2)}{C(8,2)} = \dfrac{3}{28}$.

$E = 2 \cdot \frac{10}{28} + 5 \cdot \frac{15}{28} + 8 \cdot \frac{3}{28}$

$\quad = \frac{119}{28} = \$4.25$

4. A box of 10 light bulbs contains 3 defective light bulbs. If a random sample of 2 bulbs is drawn from the box, what is the expected number of defective bulbs?

We use the number of defective bulbs as the payoff, so a sample of two bulbs can have 0, 1, or 2 defective bulbs.

Payoff	Probability
0	$\dfrac{C(7,2)}{C(10,2)} = \frac{7}{15}$
1	$\dfrac{C(7,1) \cdot C(3,1)}{C(10,2)} = \frac{7}{15}$
2	$\dfrac{C(3,2)}{C(10,2)} = \frac{1}{15}$

$E = 0 \cdot \frac{7}{15} + 1 \cdot \frac{7}{15} + 2 \cdot \frac{1}{15}$

$\quad = \frac{9}{15} = \frac{3}{5}$

Chapter 12
Limits: A Preview of Calculus

Section 12.1   Finding Limits Numerically and Graphically

Key Ideas
A. Definition of limits.
B. When a limit fails to exist.
C. One-sided limits.

A. We write

$$\lim_{x \to a} f(x) = L$$

and say "the limit of $f(x)$, as x approaches a, equals L" if we can make the values of $f(x)$ arbitrarily close to L by taking x to be sufficiently close to a, but not equal to a. Roughly speaking, this says that the values of $f(x)$ get closer and closer to the number L as x gets closer and closer to the number a (from either side of a) but $x \neq a$. In this section we will use tables and graphs to estimate the limit as x gets close to a. Care must be taken that we do not exceed the capabilities of the calculator, for example see Exercise 28 in the main text.

For problems 1-5 complete the table of values (to five decimal places) and use the table to estimate the limit. Then graph in an appropriate viewing rectangle to verify.

1. $\lim\limits_{x \to 4} \dfrac{x^2 - 3x - 2}{x - 5}$

x	3.9	3.99	3.999
$f(x)$			

x	4.1	4.01	4.001
$f(x)$			

Completing the table we get:

x	3.9	3.99	3.999
$f(x)$	−1.37273	−1.93079	−1.99301

x	4.1	4.01	4.001
$f(x)$	−2.78889	−2.07081	−2.00701

From the table we get $\lim\limits_{x \to 4} \dfrac{x^2 - 3x - 2}{x - 5} = -2.$

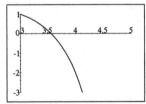

2. $\lim\limits_{x \to 3} \dfrac{3 - x}{x^2 - 2x - 3}$

x	2.9	2.99	2.999
$f(x)$			

x	3.1	3.01	3.001
$f(x)$			

Completing the table we get:

x	2.9	2.99	2.999
$f(x)$	−0.25641	−0.25063	−0.25006

x	3.1	3.01	3.001
$f(x)$	−0.24390	−0.24938	−0.24994

From the table we get $\lim\limits_{x \to 3} \dfrac{3 - x}{x^2 - 2x - 3} = -0.25.$

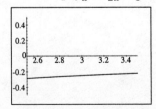

3. $\lim\limits_{x \to -1} \dfrac{x^2 + 3x + 2}{x + 1}$

x	-1.1	-1.01	-1.001
$f(x)$			

x	-0.9	-0.99	-0.999
$f(x)$			

Completing the table we get:

x	-1.1	-1.01	-1.001
$f(x)$	0.90000	0.99000	0.99900

x	-0.9	-0.99	-0.999
$f(x)$	1.10000	1.01000	1.00100

From the table we get $\lim\limits_{x \to -1} \dfrac{x^2 + 3x + 2}{x + 1} = 1$.

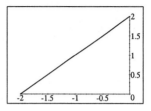

4. $\lim\limits_{x \to 0} x \sin \dfrac{1}{x}$

x	-0.1	-0.01	-0.001
$f(x)$			

x	0.1	0.01	0.001
$f(x)$			

Completing the table we get:

x	-0.1	-0.01	-0.001
$f(x)$	-0.05440	-0.00506	0.00083

x	0.1	0.01	0.001
$f(x)$	-0.05440	-0.00506	0.00083

From the table we get $\lim\limits_{x \to 0} x \sin \dfrac{1}{x} = 0$.

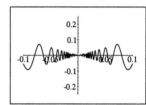

5. $\lim\limits_{x \to 1} \dfrac{\ln x^6}{x - 1}$

x	0.9	0.99	0.999
$f(x)$			

x	1.1	1.01	1.001
$f(x)$			

Completing the table we get:

x	0.9	0.99	0.999
$f(x)$	6.32163	6.03020	6.00300

x	1.1	1.01	1.001
$f(x)$	5.71861	5.97020	5.99700

From the table we get $\lim\limits_{x \to 1} \dfrac{\ln x^6}{x - 1} = 6$.

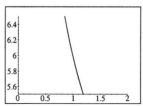

329

B. Functions do not necessarily approach a finite value at every point. There *are* many different reasons for limits to fail to exist. The three most common reasons are listed below.

1. The limit of $f(x)$ as x approaches a from the right is different from the limit of $f(x)$ as x approaches a from the left.

2. Values of $f(x)$ do not approach a fixed number as x approaches a.

3. Values of $f(x)$ do not approach a finite number.

6. Graph each function and explain why the limit does not exist.

(a) $\lim\limits_{x \to 2} \dfrac{x^2 - 3x + 2}{x^3 - 2x^2 - 4x + 8}$

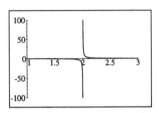

Here the function fails to go to any finite number. Thus $\lim\limits_{x \to 2} \dfrac{x^2 - 3x + 2}{x^3 - 2x^2 - 4x + 8}$ does not exist.

(b) $\lim\limits_{x \to 0} e^{\cos(1/x)}$

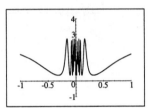

Here the function oscillates between $\dfrac{1}{e}$ and e as x approaches 0. Thus $\lim\limits_{x \to 0} e^{\cos(1/x)}$ does not exist.

C. We write

$$\lim_{x \to a^-} f(x) = L$$

and say "the limit of $f(x)$, as x approaches a from the left, equals L" if we can make the values of $f(x)$ arbitrarily close to L by taking x to be sufficiently close to a and $a < x$. Similarly, we write

$$\lim_{x \to a^+} f(x) = L$$

and say "the limit of $f(x)$, as x approaches a from the right, equals L" if we can make the values of $f(x)$ arbitrarily close to L by taking x to be sufficiently close to a and $a > x$. Thus, we get the following statement concerning two-sided and one-sided limits.

$$\lim_{x \to a} f(x) = L \text{ if, and only if, } \lim_{x \to a^-} f(x) = L \text{ and } \lim_{x \to a^+} f(x) = L$$

As a consequence, if $\lim\limits_{x \to a^-} f(x) \neq \lim\limits_{x \to a^+} f(x)$, then $\lim\limits_{x \to a} f(x)$ does not exist.

7. For the function f whose graph is given below, state the value of the given quantity, if it exists. If it does exist, explain why.

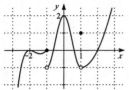

(a) $\lim\limits_{x \to -1^-} f(x)$

$\lim\limits_{x \to -1^-} f(x) = 0$

(b) $\lim\limits_{x \to -1^+} f(x)$

$\lim\limits_{x \to -1^+} f(x) = -1$

(c) $\lim\limits_{x \to -1} f(x)$

$\lim\limits_{x \to -1} f(x)$ does not exist since

$\lim\limits_{x \to -1^-} f(x) \neq \lim\limits_{x \to -1^+} f(x).$

(d) $\lim\limits_{x \to 1^-} f(x)$

$\lim\limits_{x \to 1^-} f(x) = -1$

(e) $\lim\limits_{x \to 1^+} f(x)$

$\lim\limits_{x \to 1^+} f(x) = -1$

(e) $\lim\limits_{x \to 1} f(x)$

$\lim\limits_{x \to 1} f(x) = -1$

Section 12.2   Finding Limits Algebraically

Key Ideas

A. Limit Laws.

B. Special limits and limits by direct substitution.

C. Limits by direct substitution.

A. We start by listing the following property of limits.

> Suppose that c is a constant and that the following limits exist:
> $$\lim_{x \to a} f(x) \quad \text{and} \quad \lim_{x \to a} g(x)$$
> Then
>
> 1. $\lim_{x \to a} [f(x) + g(x)] = \lim_{x \to a} f(x) + \lim_{x \to a} g(x)$ Sum Law
>
> 2. $\lim_{x \to a} [f(x) - (g(x)] = \lim_{x \to a} f(x) - \lim_{x \to a} g(x)$ Difference Law
>
> 3. $\lim_{x \to a} [cf(x)] = c \lim_{x \to a} f(x)$ Constant Multiple Law
>
> 4. $\lim_{x \to a} [f(x)g(x)] = \lim_{x \to a} f(x) \cdot \lim_{x \to a} g(x)$ Product Law
>
> 5. $\lim_{x \to a} \dfrac{f(x)}{g(x)} = \dfrac{\lim_{x \to a} f(x)}{\lim_{x \to a} g(x)}$ if $\lim_{x \to a} g(x) \neq 0$ Quotient Law
>
> 6. $\lim_{x \to a} [f(x)]^n = \left[\lim_{x \to a} f(x) \right]^n$ where n is a positive integer   Power Law
>
> 7. $\lim_{x \to a} \sqrt[n]{f(x)} = \sqrt[n]{\lim_{x \to a} f(x)}$ where n is a positive integer   Root Law
>
>   If n is even, we assume that $\lim_{x \to a} f(x) > 0$.

These laws will be proved in an advanced mathematics class.

1. Use the Limit Laws and the graphs of f and g on the right to evaluate the following limits, if they exist.

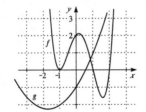

(a) $\lim_{x \to -1} [f(x) + g(x)]$

$$\lim_{x \to -1} [f(x) + g(x)] = \lim_{x \to -1} f(x) + \lim_{x \to -1} g(x)$$
$$= 0 + (-2)$$
$$= -2$$

(b) $\lim_{x \to 0} [f(x)g(x)]$

$$\lim_{x \to 0} [f(x)g(x)] = \lim_{x \to 0} f(x) \cdot \lim_{x \to 0} g(x)$$
$$= (2)(-1)$$
$$= -2$$

(c) $\lim_{x \to 1} \dfrac{g(x)}{f(x)}$

$$\lim_{x \to 1} \dfrac{g(x)}{f(x)} = \dfrac{\lim_{x \to 1} g(x)}{\lim_{x \to 1} f(x)}$$

Since $\lim_{x \to 1} f(x) = 0$, $\lim_{x \to 1} \dfrac{g(x)}{f(x)}$ does not exist.

B. The following four special limits are need to apply the Limit Laws.

1. $\lim\limits_{x \to a} c = c$

2. $\lim\limits_{x \to a} x = a$

3. $\lim\limits_{x \to a} x^n = a^n$ where n is a positive integer

4. $\lim\limits_{x \to a} \sqrt[n]{x} = \sqrt[n]{a}$ where n is a positive integer and $a > 0$

When we apply the Limits Laws and these special limits to a polynomial or a rational function we find that we get the same value as direct substitution gives.

If f is a polynomial or a rational function and a is in the domain of f, then $\lim\limits_{x \to a} f(x) = f(a)$.

2. Find the following limits by direct substitution.

(a) $\lim\limits_{x \to 4} \dfrac{x^2 - 3x - 2}{x - 5}$

Since the function $f(x) = \dfrac{x^2 - 3x - 2}{x - 5}$ is a rational function and 4 is in its domain,

$$\lim_{x \to 4} \frac{x^2 - 3x - 2}{x - 5} = \frac{(4)^2 - 3(4) - 2}{(4) - 5}$$

$$= \frac{16 - 12 - 2}{-1} = -2.$$

Compare to your result in Problem 1 in Section 12.1.

(b) $\lim\limits_{x \to 3} \sqrt{25 - x^2}$

We apply the Roots Law first, so

$$\lim_{x \to 3} \sqrt{25 - x^2} = \sqrt{\lim_{x \to 3} (25 - x^2)}.$$

Since the function $g(x) = 25 - x^2$ is a polynomial, by direct substitution we get:

$$\lim_{x \to 3} (25 - x^2) = 25 - (3)^2 = 25 - 9 = 16. \text{ Thus}$$

$$\lim_{x \to 3} \sqrt{25 - x^2} = \sqrt{\lim_{x \to 3} (25 - x^2)} = \sqrt{16} = 4.$$

C. Often we can not apply the Quotient Law immediately, since the limit of the denominator is 0. In these cases we need to do some preliminary algebra. Usually the preliminary algebra is to factor or rationalize either the numerator or the denominator. After common factors are canceled, we can then find the limits by direct substitution.

3. $\lim\limits_{x \to -1} \dfrac{x^2 + 3x + 2}{x + 1}$

Since both the numerator and denominator are 0, we find the common factor and cancel.

$$\lim_{x \to -1} \frac{x^2 + 3x + 2}{x + 1} = \lim_{x \to -1} \frac{(x + 1)(x + 2)}{x + 1}$$

$$= \lim_{x \to -1} (x + 2) = (-1) + 2$$

$$= 1$$

Compare this result to the one in Problem 3, Section 12.1.

4. $\lim\limits_{x \to 3} \dfrac{3 - x}{x^2 - 2x - 3}$

Since both the numerator and denominator are 0, we find the common factor and cancel.

$$\lim_{x \to 3} \frac{3 - x}{x^2 - 2x - 3} = \lim_{x \to 3} \frac{-(x - 3)}{(x + 1)(x - 3)}$$

$$= \lim_{x \to 3} \frac{-1}{x + 1} = -\frac{1}{4}$$

Compare this result to the one in Problem 2, Section 12.1.

C. Some limits are best calculated by first finding the left- and right-handed limits. Remember *a two-sided limit exists if and only if both of the one-sided limits exist and are equal.*

5. Suppose $f(x) = \begin{cases} x^2 + 4x - 3 & x \le 1 \\ x + 1 & x > 1 \end{cases}$.

Find the following limit.

$\lim\limits_{x \to 1} f(x)$

Since $f(x) = x^2 + 4x - 3$ for $x \le 1$, we have

$\lim\limits_{x \to 1^-} f(x) = \lim\limits_{x \to 1^-} (x^2 + 4x - 3)$

$\qquad = (1)^2 + 4(1) - 3 = 2$.

Since $f(x) = x + 1$ for $x > 1$, we have

$\lim\limits_{x \to 1^+} f(x) = \lim\limits_{x \to 1^+} (x + 1)$

$\qquad = (1) + 1 = 2$.

So $\lim\limits_{x \to 1^-} f(x) = \lim\limits_{x \to 1^+} f(x) = 2$.

Therefore, $\lim\limits_{x \to 1} f(x) = 2$

Section 12.3   Tangent Lines and Derivatives
Key Ideas
A.   Tangent lines.
B.   Derivative of a function.
C.   Instantaneous rate of change.

A.   We use limits to find the tangent line to a curve or the instantaneous rate of change of a function. A **tangent line** is a line that *just* touches a curve. The slope of the tangent line is the limit of the slopes of the secant line.

> The **tangent line** to the curve $y = f(x)$ at the point $P(a, f(a))$ is the line through P with slope
> $$m = \lim_{x \to a} \frac{f(x) - f(a)}{x - a}$$
> provided that this limit exists.

1.   Find an equation of the tangent line to the function $y = x^2 - 2$ at the point $(2, 2)$.

Let $f(x) = x^2 - 2$. Then the slope of the tangent line at $(2, 2)$ is

$$m = \lim_{x \to 2} \frac{f(x) - f(2)}{x - 2}$$

$$= \lim_{x \to 2} \frac{(x^2 - 2) - (2)}{x - 2} = \lim_{x \to 2} \frac{x^2 - 4}{x - 2}$$

$$= \lim_{x \to 2} \frac{(x - 2)(x + 2)}{x - 2} = \lim_{x \to 2} (x + 2) = 4.$$

Then an equation of the tangent line at $(2, 2)$ is

$y - 2 = 4(x - 2)$ which simplifies to

$-4x + y + 6 = 0$.

2.   Find an equation of the tangent line to the function $y = x^3$ at the point $(1, 1)$.

Let $f(x) = x^3$. Then the slope of the tangent line at $(1, 1)$ is

$$m = \lim_{x \to 1} \frac{f(x) - f(1)}{x - 1} = \lim_{x \to 1} \frac{x^3 - 1}{x - 1}$$

$$= \lim_{x \to 1} \frac{(x - 1)(x^2 + x + 1)}{x - 1}$$

$$= \lim_{x \to 1} (x^2 + x + 1) = 1 + 1 + 1 = 3.$$

Then an equation of the tangent line at $(1, 1)$ is

$y - 1 = 3(x - 1)$ which simplifies to

$-3x + y + 2 = 0$.

B.

The **derivative of a function** f **at a number** a, denoted by $f'(a)$, is

$$f'(a) = \lim_{h \to 0} \frac{f(a+h) - f(a)}{h}$$

if this limit exists.

3. Find the derivative of the function $f(x) = 3x^2 - 4x$ at the number -1.

According to the definition of a derivative, with $a = -1$, we have

$$f'(-1) = \lim_{h \to 0} \frac{f(-1+h) - f(-1)}{h}$$

$$= \lim_{h \to 0} \frac{[3(-1+h)^2 - 4(-1+h)] - [3+4]}{h}$$

$$= \lim_{h \to 0} \frac{3 - 6h + 3h^2 + 4 - 4h - 7}{h}$$

$$= \lim_{h \to 0} \frac{-10h + 3h^2}{h} = \lim_{h \to 0} \frac{h(-10 + 3h)}{h}$$

$$= \lim_{h \to 0} (-10 + 3h) = -10$$

Thus $f'(-1) = -10$.

4. Find the derivative of the function $f(x) = \dfrac{x+3}{x}$ at $x = a$, where a is in the domain of f.

According to the definition of a derivative, we have

$$f'(a) = \lim_{h \to 0} \frac{f(a+h) - f(a)}{h}$$

$$= \lim_{h \to 0} \frac{\left[\frac{a+h+3}{a+h}\right] - \left[\frac{a+3}{a}\right]}{h}$$

$$= \lim_{h \to 0} \frac{\left[\frac{a+h+3}{a+h} - \frac{a+3}{a}\right](a+h)(a)}{h(a+h)(a)}$$

$$= \lim_{h \to 0} \frac{a^2 + ah + 3a - a^2 - ah - 3a - 3h}{ah(a+h)}$$

$$= \lim_{h \to 0} \frac{-3h}{ah(a+h)} = \lim_{h \to 0} \frac{-3}{a(a+h)} = \frac{-3}{a^2}$$

Thus $f'(a) = \dfrac{-3}{a^2}$.

C. In Section 2.4 the average rate of change of a function f between the number a and b was defined as

$$\text{average rate of change} = \frac{\text{change in } y}{\text{change in } x} = \frac{f(x) - f(a)}{x - a}$$

The limit of these average rate of change is called the instantaneous rate of change.

If $y = f(x)$, the **instantaneous rate of change of f with respect to x** at $x = a$ is the limit of the average rate of change as x approaches a:

$$\text{instantaneous rate of change} = \frac{\text{change in } y}{\text{change in } x} = \lim_{x \to a} \frac{f(x) - f(a)}{x - a} = f'(a)$$

---

5. A ball is rolled down an incline; its distance (in meters) from the starting point after t seconds is given by $s(t) = t^2 + 6t$.

(a) Find the velocity of the ball when $t = a$.

Using the definition of instantaneous rate of change we get:

$$f'(a) = \lim_{x \to a} \frac{f(t) - f(a)}{t - a}$$

$$= \lim_{x \to a} \frac{t^2 + 6t - a^2 - 6a}{t - a}$$

$$= \lim_{x \to a} \frac{(t^2 - a^2) + (6t - 6a)}{t - a}$$

$$= \lim_{x \to a} \frac{(t - a)(t + a) + 6(t - a)}{t - a}$$

$$= \lim_{x \to a} \frac{(t - a)(t + a + 6)}{t - a} = \lim_{x \to a} (t + a + 6)$$

$$= 2a + 6$$

(b) Find the velocity of the ball after 2 and 5 seconds.

Substituting for a we have
$f'(2) = 2(2) + 6 = 10$ meters/sec.

Substituting for a we have
$f'(5) = 2(5) + 6 = 16$ meters/sec.

(c) When will the velocity of the ball reach 30 meters/sec?

Solve $f'(a) = 30$ for a. So

$30 = 2a + 6 \quad \Leftrightarrow \quad 2a = 24 \quad \Leftrightarrow \quad a = 12.$

So the ball reaches 30 meters/sec in 12 seconds.

Section 12.4   Limits at Infinity; Limits of Sequences

Key Ideas

A. Limits at infinity.
B. Numerical and graphical methods.
C. Limits of sequences.

A. Here we want to know what happens to a function as x becomes arbitrarily large.

Let f be a function defined on some interval (a, ∞). Then
$$\lim_{x \to \infty} f(x) = L$$
means that the values $f(x)$ can be made arbitrarily close to L by taking x sufficiently large.

Let f be a function defined on some interval $(-\infty, a)$. Then
$$\lim_{x \to -\infty} f(x) = L$$
means that the values $f(x)$ can be made arbitrarily close to L by taking x sufficiently large negative.

The line $y = L$ is called a **horizontal asymptote** of the curve $y = f(x)$ if either $\lim_{x \to \infty} f(x) = L$ or $\lim_{x \to -\infty} f(x) = L$. The Limit Laws of Section 12.2 also apply to limits at infinity. As a result we have the following important rule.

If k is a positive integer, then $\lim_{x \to \infty} \dfrac{1}{x^k} = 0$ and $\lim_{x \to -\infty} \dfrac{1}{x^k} = 0$.

We utilize this rule when finding the limit at infinity of a rational function. In this case we divide by the highest power of x.

Find the limit in Problems 1 and 2.

1. $\lim\limits_{x \to \infty} \dfrac{x^3 - 5x^2}{x^3 + 6x - 8}$

$$\lim_{x \to \infty} \frac{x^3 - 5x^2}{x^3 + 6x - 8} = \lim_{x \to \infty} \frac{(x^3 - 5x^2)}{(x^3 + 6x - 8)} \cdot \frac{\frac{1}{x^3}}{\frac{1}{x^3}}$$

$$= \lim_{x \to \infty} \frac{1 - \dfrac{5}{x}}{1 + \dfrac{6}{x^2} - \dfrac{8}{x^3}}$$

$$= \frac{\lim\limits_{x \to \infty} \left(1 - \dfrac{5}{x}\right)}{\lim\limits_{x \to \infty} \left(1 + \dfrac{6}{x^2} - \dfrac{8}{x^3}\right)}$$

$$= \frac{\lim\limits_{x \to \infty} 1 - \lim\limits_{x \to \infty} \dfrac{5}{x}}{\lim\limits_{x \to \infty} 1 + \lim\limits_{x \to \infty} \dfrac{6}{x^2} - \lim\limits_{x \to \infty} \dfrac{8}{x^3}}$$

$$= \frac{1 - 0}{1 + 0 + 0} = \frac{1}{1} = 1$$

2. $\displaystyle\lim_{x\to\infty}\frac{3x^3+7x^4}{x^4-8x^5+x^6}$

$$\lim_{x\to\infty}\frac{3x^3+7x^4}{x^4-8x^5+x^6}=\lim_{x\to\infty}\frac{\left(3x^3+7x^4\right)}{\left(x^4-8x^5+x^6\right)}\cdot\frac{\dfrac{1}{x^6}}{\dfrac{1}{x^6}}$$

$$=\lim_{x\to\infty}\frac{\dfrac{3}{x^3}+\dfrac{7}{x^2}}{\dfrac{1}{x^2}-\dfrac{8}{x}-1}$$

$$=\frac{\displaystyle\lim_{x\to\infty}\left(\dfrac{3}{x^3}+\dfrac{7}{x^2}\right)}{\displaystyle\lim_{x\to\infty}\left(\dfrac{1}{x^2}-\dfrac{8}{x}-1\right)}$$

$$=\frac{\displaystyle\lim_{x\to\infty}\dfrac{3}{x^3}-\lim_{x\to\infty}\dfrac{7}{x^2}}{\displaystyle\lim_{x\to\infty}\dfrac{1}{x^2}-\lim_{x\to\infty}\dfrac{8}{x}-\lim_{x\to\infty}1}$$

$$=\frac{0+0}{0-0-1}=\frac{0}{-1}=0$$

B. Making tables and graphing functions are important tools in determining limits. Care must be taken to ensure a sufficiently large enough window is chosen so that limits that are large are not missed. For example of a limit that is large see Exercise 18 in the text.

3. Make a table of values to estimate the limit. Then use a graphing device to confirm your result graphically.

$$\lim_{x\to\infty}\frac{\ln x}{\sqrt{x}}$$

Constructing a table we get:

x	100	10,000	1,000,000	10^{10}
$f(x)$	0.46052	0.09210	0.01382	0.00023

From the table $\displaystyle\lim_{x\to\infty}\frac{\ln x}{\sqrt{x}}=0$. The graph below verifies this limit.

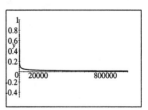

4. Make a table of values to estimate the limit. Then use a graphing device to confirm your result graphically.

$$\lim_{x\to\infty} \frac{\sqrt{3x^5 - 10x + 9}}{x^2 + 30x + 100}$$

Constructing a table we get:

x	100	10^4	10^8
$f(x)$	13.22	172.7	17320.5

From the table $\lim_{x\to\infty} \dfrac{\sqrt{3x^5 - 10x + 9}}{x^2 + 30x + 100}$ increases without bound, therefore the limit does not exist. The graph below verifies that the limit does not exist.

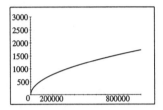

C. Sequences were introduced in Section 10.1. In this section we are interested in what happens to a_n as n becomes large.

> A sequence $a_1, a_2, a_3, \ldots$ has **limit L** and we write
> $$\lim_{n\to\infty} a_n = L \quad \text{or} \quad a_n \to L \text{ as } n \to \infty$$
> if the nth term a_n of the sequence can be made arbitrarily close to L by taking n sufficiently large.

If $\lim_{n\to\infty} a_n$ exists, we say the sequence **converges** (or is **convergent**). Otherwise, we say the sequence **diverges** (or is **divergent**). The limit of a sequence at infinity and the limit of a function as x goes to infinity are related by the following statement.

$$\text{If } \lim_{x\to\infty} f(x) = L \text{ and } f(n) = a_n \text{ when } n \text{ is an integer, then } \lim_{n\to\infty} a_n = L.$$

When using a graphing device to graph sequences, be sure to turn the "connect" mode off.

5. If the sequence is convergent, find it limit. If it is divergent, explain why.

$$a_n = \frac{5}{n^3}\left[\frac{n(n+1)(2n+1)}{6}\right]$$

We start by simplifying a_n.

$$a_n = \frac{5}{n^3}\left[\frac{n(n+1)(2n+1)}{6}\right]$$

$$= \frac{5}{6} \cdot \frac{n}{n} \cdot \frac{n+1}{n} \cdot \frac{2n+1}{n}$$

$$= \frac{5}{6}\left(1 + \frac{1}{n}\right)\left(2 + \frac{1}{n}\right)$$

$$\lim_{n\to\infty} a_n = \lim_{n\to\infty} \frac{5}{6}\left(1 + \frac{1}{n}\right)\left(2 + \frac{1}{n}\right)$$

$$= \frac{5}{6} \cdot \lim_{n\to\infty}\left(1 + \frac{1}{n}\right) \cdot \lim_{n\to\infty}\left(2 + \frac{1}{n}\right)$$

$$= \frac{5}{6}(1)(2) = \frac{5}{3}$$

6. If the sequence is convergent, find it limit. If it is divergent, explain why.
$$a_n = \left(3 + \frac{1}{n^4}\right)\left(\frac{n(n+1)}{2}\right)$$

We start by simplifying a_n.
$$a_n = \left(3 + \frac{1}{n^4}\right)\left(\frac{n(n+1)}{2}\right)$$
$$= \frac{3}{2} \cdot n(n+1) + \frac{1}{2n^2} \cdot \frac{n}{n} \cdot \frac{n+1}{n}$$
$$= \frac{3}{2}n(n+1) + \frac{1}{2n^2}\left(1 + \frac{1}{n}\right)$$
$$\lim_{n \to \infty} a_n = \lim_{n \to \infty}\left[\frac{3}{2}n(n+1) + \frac{1}{2n^2}\left(1 + \frac{1}{n}\right)\right]$$
$$= \frac{3}{2}\lim_{n \to \infty} n(n+1) + \lim_{n \to \infty}\frac{1}{2n^2}\left(1 + \frac{1}{n}\right)$$

which diverges since $\lim_{n \to \infty} n(n+1)$ grows

without bound.

7. Write out the first eight terms of the sequence $a_n = \dfrac{6^n + 5^n}{8^n}$. Estimate $\lim_{n \to \infty} a_n$ and use a graphing device to confirm your estimate.

We start by computing the first eight terms

$a_1 = 1.375 \qquad a_2 = 0.95313 \qquad a_3 = 0.66602$

$a_4 = 0.46899 \qquad a_5 = 0.33267 \qquad a_6 = 0.23758$

$a_3 = 0.17074 \qquad a_8 = 0.12340$

Since $a_n > 0$, in appears that $\lim_{n \to \infty} \dfrac{6^n + 5^n}{8^n} = 0$.

We graph a_n for $n = 1$ to $n = 12$ in the window below. Remember to turn the "connect" mode off.

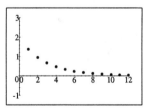

Section 12.5   Areas

Key Ideas

A. Using rectangles to estimate area.

B. Definition of area and summation formula.

A. We consider a central problem in calculus called the *area problem*. We estimate the area under the curve $y = f(x)$ from a to b with constant-width rectangles that approximate the area. We let R_n represent the sum of the n rectangles using n equal subintervals and the right endpoints for the height of the rectangles. We also let L_n represent the sum of the n rectangles using n equal subintervals and the left endpoints for the height of the rectangles.

> Let $\Delta x = \dfrac{b-a}{n}$ and $a_0 = a,\ a_1 = a + \Delta x,\ \ldots,\ a_i = a + i\Delta x,\ \ldots,\ a_n = a + n\Delta x = b$. Then
>
> $$R_n = \sum_{i=1}^{n}[f(a_i) \cdot \Delta x] = \frac{b-a}{n}[f(a_1) + f(a_2) + \cdots + f(a_n)] \text{ and}$$
>
> $$L_n = \sum_{i=0}^{n-1}[f(a_i) \cdot \Delta x] = \frac{b-a}{n}[f(a_0) + f(a_1) + \cdots + f(a_{n-1})].$$

1. (a) Estimate the area under the curve
$f(x) = x^2 - 4x + 5$ from $x = 1$ to $x = 3$ using
four approximating rectangles and left endpoints.
Sketch the graph and the rectangles.

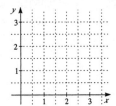

We are asked for L_4. We start by graphing $y = f(x)$
and the four left endpoint rectangles.

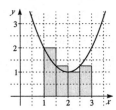

$L_4 = [f(a_0) + f(a_1) + f(a_2) + f(a_3)] \cdot \Delta x,$

where $\Delta x = \frac{3-1}{4} = 0.5$. Constructing a table of
values for $f(x)$ we get:

x	$f(x)$
$a_0 = 1$	2
$a_1 = 1.5$	1.25
$a_2 = 2$	1
$a_3 = 2.5$	1.25

Thus

$L_4 = [2 + 1.25 + 1 + 1.25](0.5)$

$\quad = (5.5)(0.5)$

$\quad = 2.75.$

(b) Estimate the area under the curve
$f(x) = x^2 - 4x + 5$ from $x = 1$ to $x = 3$ using four approximating rectangles and right endpoints. Sketch the graph and the rectangles.

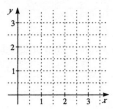

We are asked for R_4. We start by graphing $y = f(x)$ and the four right endpoint rectangles.

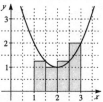

$R_4 = [f(a_1) + f(a_2) + f(a_3) + f(a_4)] \cdot \Delta x$,
where $\Delta x = \frac{3-1}{4} = \frac{1}{2} = 0.5$ and

a_i	$f(a_i)$
$a_1 = 1.5$	1.25
$a_2 = 2$	1
$a_3 = 2.5$	1.25
$a_4 = 3$	2

Thus
$L_4 = [1.25 + 1 + 1.25 + 2](0.5) = (5.5)(0.5)$
$= 2.75.$

(c) Estimate the area under the curve
$f(x) = x^2 - 4x + 5$ from $x = 1$ to $x = 3$ using eight approximating rectangles and left endpoints.

We are asked for L_8.
$L_8 = [f(a_0) + f(a_1) + \cdots + f(a_7)] \cdot \Delta x$,
where $\Delta x = \frac{3-1}{8} = 0.25$ and

a_i	$f(a_i)$	a_i	$f(a_i)$
$a_0 = 1$	2	$a_4 = 2$	1
$a_1 = 1.25$	1.5625	$a_5 = 2.25$	1.0625
$a_2 = 1.5$	1.25	$a_6 = 2.5$	1.25
$a_3 = 1.75$	1.0625	$a_7 = 2.75$	1.5625

Thus $L_8 = (10.75)(0.25) = 2.6875.$

(d) Estimate the area under the curve
$f(x) = x^2 - 4x + 5$ from $x = 1$ to $x = 3$ using eight approximating rectangles and right endpoints.

We are asked for R_8.
$R_8 = [f(a_1) + f(a_2) + \cdots + f(a_8)] \cdot \Delta x$,
where $\Delta x = \frac{3-1}{8} = 0.5$ and

a_i	$f(a_i)$	a_i	$f(a_i)$
$a_1 = 1.25$	1.5625	$a_5 = 2.25$	1.0625
$a_2 = 1.5$	1.25	$a_6 = 2.5$	1.25
$a_3 = 1.75$	1.0625	$a_7 = 2.75$	1.5625
$a_4 = 2$	1	$a_8 = 3$	2

Thus $R_8 = (10.75)(0.25) = 2.6875.$

2 Estimate the area under the curve $f(x) = 1 + e^x$ from $x = -1$ to $x = 1$ using ten rectangles and right endpoints. Repeat using left endpoints

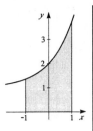

We are asked to determine L_{10} and R_{10}.

$L_{10} = [f(a_0) + f(a_1) + \cdots + f(a_9)] \cdot \Delta x$ and
$R_{10} = [f(a_1) + f(a_2) + \cdots + f(a_{10})] \cdot \Delta x$,

where $\Delta x = \dfrac{1 - (-1)}{10} = 0.2$.

We next construct a table of functional values.

x	$f(x)$	x	$f(x)$
$a_0 = -1$	1.36788	$a_6 = 0.2$	2.22140
$a_1 = -0.8$	1.44933	$a_7 = 0.4$	2.49182
$a_2 = -0.6$	1.54881	$a_8 = 0.6$	2.82212
$a_3 = -0.4$	1.67032	$a_9 = 0.8$	3.22554
$a_4 = -0.2$	1.81873	$a_{10} = 1$	3.71828
$a_5 = 0$	2		

Thus $L_{10} = (20.61595)(0.2) = 4.12319$ and
$R_{10} = (22.96635)(0.2) = 4.59327$.

B. We obtain better estimates by increasing the number of rectangles or strips. Using limits we can define the area of a region in the following way.

> The **area** A of the region S that lies under the graph of a continuous function f from a to b is the limit of the sum of the areas of approximating rectangles:
>
> $$A = \lim_{n \to \infty} R_n = \lim_{n \to \infty} [f(x_1)\,\Delta x + f(x_2)\,\Delta x + \cdots + f(x_n)\,\Delta x]$$
>
> Using sigma notation, we write this as
>
> $$A = \lim_{n \to \infty} \sum_{k=1}^{n} f(x_k) \cdot \Delta x.$$

In using this formula for area, remember that Δx is the width of an approximating rectangle, x_k is the right endpoint of the kth rectangle, and $f(x_k)$ is its height. So

Width: $\qquad \Delta x = \dfrac{b-a}{n}$

Right endpoint: $\qquad x_k = a + k\,\Delta x$

Height: $\qquad f(x_k) = f(a + k\,\Delta x)$

We list two of the needed properties of sums from Section 10.1

$$\sum_{k=1}^{n}(a_k \pm b_k) = \left(\sum_{k=1}^{n} a_k\right) \pm \left(\sum_{k=1}^{n} b_k\right) \quad \text{and} \quad \sum_{k=1}^{n}(ca_k) = c\left(\sum_{k=1}^{n} a_k\right)$$